METHODS IN MOLECULAR BIOLOGY™

For other titles published in this series, go to
www.springer.com/series/7651

cDNA Libraries

Methods and Applications

Edited by

Chaofu Lu

Department of Plant Sciences and Plant Pathology, Montana State University, Bozeman, MT, USA

John Browse and James G. Wallis

Institute of Biological Chemistry, Washington State University, Pullman, WA, USA

Editors
Chaofu Lu
Department of Plant Sciences
and Plant Pathology
Montana State University
Bozeman, MT
USA
clu@montana.edu

James G. Wallis
Institute of Biological Chemistry
Washington State University
Pullman, WA
USA
wallis@wsu.edu

John Browse
Institute of Biological Chemistry
Washington State University
Pullman, WA
USA
jab@wsu.edu

ISSN 1064-3745 e-ISSN 1940-6029
ISBN 978-1-61779-064-5 e-ISBN 978-1-61779-065-2
DOI 10.1007/978-1-61779-065-2
Springer New York Dordrecht Heidelberg London

Library of Congress Control Number: 2011922530

Printed on acid-free paper

Humana Press is part of Springer Science+Business Media (www.springer.com)

Preface

The discovery of reverse transcriptase by Howard Temin and David Baltimore in 1970 launched a revolution in molecular biology that was unmatched until the advent of DNA amplification by polymerase chain reaction (PCR). Not only did the discovery overturn the "central dogma" that information coded in DNA flowed through RNA to protein, but also the activities of RNA-dependent DNA polymerase were, and are, important in medicine and other fields. The utility of these enzymes as instruments to identify coding sequences of genes and to analyze their expression was quickly realized.

The applications of cDNA technology have changed dramatically as the technology has advanced. In the early days of cDNA analysis, it revealed which genes were expressed and often the tissue specificity of gene expression. Analysis of cDNA molecules showed the chromosomal patterns of introns and exons and revealed the predicted protein sequences of countless genes. While the techniques for isolating RNA, generating cDNA, and analyzing the cloned cDNA were crude by today's standards, they provided many insights into the workings of plants, animals, and other organisms. Not unlike the effects of PCR in biology, the cDNA revolution continues as the basic techniques are revised and new uses for the technology are developed.

Previous volumes in this series have supplied many techniques that continue to be important; in this volume, we provide current techniques that reflect the most recent advances in the construction and application of cDNA libraries. Broadly, the techniques we describe can be divided into two classes. The first class includes improved approaches to some of the most basic elements of creating cDNA libraries, while the second class is much wider and includes visionary applications of cDNA technology which were either unforeseen or technically impractical until recently.

Some of the most important advances in cDNA technology are new approaches to challenges that have been inherent in the production and analysis of cDNA libraries from the earliest days of the technology. These limits have been rolled back by dramatic technical advances in several previously limiting processes. Advances in separation of complex tissues into their several components for analysis and the ability to create suitable cDNA libraries from minuscule tissue samples, even from a single cell, have greatly expanded the range of practical experiments. A suite of technical improvements have made full-length, normalized libraries with reduced bias available. These libraries are suitable for expression analysis, and the ease of library construction has been enhanced by adopting *in vitro* recombination methods, greatly expanding the numbers of clones available for expression and increasing confidence that a library adequately represents the genes expressed in the source tissue. Analysis of the information gleaned from cDNA libraries has been continuously refined, and new bioinformatic approaches provide a more complete description of the genes transcribed in each tissue used in library construction.

The second class of additions to the array of cDNA library technology arises from visionary application of these refined techniques to new areas of research. Small RNA molecules recently and strikingly emerged as important participants in cell regulation and development; extension of cDNA libraries to this realm of small RNA has become critical in understanding these regulators. Novelty that is artificially induced, either through intentional error-prone synthesis or through shuffling of domains, is another source of new material for library construction. The ability to easily generate full-length expression libraries has made new kinds of experiments possible, whether they rely on gain-of-function analysis in transgenic organisms or on high-throughput functional screening of libraries in test organisms. Expression libraries are also critical in the development of YSD (yeast surface display) analysis of protein interactions with ligands, a technique whose emergence may have far-reaching applications.

The ability of cDNA technology to analyze the transcriptome of an organism or tissue also points to a new era of cDNA applications. Discovery of SNP markers for genetic traits through pyrosequencing, transcriptome analysis for gene discovery or splicing analysis, and whole-genome expression analysis of organisms whose genome has not been sequenced are all techniques described in this volume.

The developments of cDNA technology described herein demonstrate that the technology continues to advance and to provide answers to fundamental questions of biology. Even more sophisticated technical improvements coupled to the scientific vision to apply the techniques to an expanded range of problems have kept the construction, analysis, and use of cDNA libraries on the cutting edge of biology, and we expect improvements will continue in the future.

We thank the publishers for inspiring the collection of this material. We heartily thank the contributors for taking time from their continuing scientific endeavors to share their acquired skills with a wider audience, and we hope that the contents of this volume will speed its readers to even more successful scientific discovery.

Bozeman, MT *Chaofu Lu*
Pullman, WA *John Browse*
Pullman, WA *James G. Wallis*

Contents

Contributors

VERONIKA ANISIMOVA • *Shemiakin and Ovchinnikov Institute of Bioorganic Chemistry RAS, Moscow, Russia; Evrogen JSC, Moscow, Russia*
W. BRAD BARBAZUK • *Department of Biology and the Genetics Institute, University of Florida, Gainesville, FL, USA*
EKATERINA V. BARSOVA • *Shemiakin and Ovchinnikov Institute of Bioorganic Chemistry RAS, Moscow, Russia*
DIANA BELLIN • *Department of Biotechnology, University of Verona, Verona, Italy*
ROY BICKNELL • *Cancer Research UK Angiogenesis Group, Institute for Biomedical Research, Schools of Immunity and Infection and Cancer studies, College of Medicine and Dentistry, University of Birmingham, Birmingham, UK*
SCOTT BIDLINGMAIER • *UCSF Comprehensive Cancer Center, University of California at San Francisco, San Francisco, CA, USA*
EKATERINA A. BOGDANOVA • *Shemiakin and Ovchinnikov Institute of Bioorganic Chemistry RAS, Moscow, Russia*
JOHN BROWSE • *Institute of Biological Chemistry, Washington State University, Pullman, WA, USA*
MASSIMO DELLEDONNE • *Department of Biotechnology, University of Verona, Verona, Italy*
ELMAR ENDL • *Institutes of Molecular Medicine and Experimental Immunology, University of Bonn, Bonn, Germany*
CAROLA ENGLER • *Icon Genetics GmbH, Biozentrum Halle, Halle, Germany*
ALBERTO FERRARINI • *Department of Biotechnology, University of Verona, Verona, Italy*
XIANG-DONG FU • *Department of Cellular and Molecular Medicine, Stem Cell Program, University of California, San Diego, La Jolla, CA, USA*
ANDREAS GRATZ • *Bioanalytics, Institute of Pharmaceutical and Medicinal Chemistry, Heinrich-Heine-University, Düsseldorf, Germany*
JOHN M.J. HERBERT • *Cancer Research UK Angiogenesis Group, Institute for Biomedical Research, Schools of Immunity and Infection and Cancer studies, College of Medicine and Dentistry, University of Birmingham, Birmingham, UK*
MIEKO HIGUCHI • *RIKEN Plant Science Center, Yokohama, Kanagawa, Japan*
KENZO HIROSE • *Department of Neurobiology, Graduate School of Medicine, The University of Tokyo, Tokyo, Japan*
TAKANARI ICHIKAWA • *RIKEN Plant Science Center, Yokohama, Kanagawa, Japan*
JOACHIM JOSE • *Bioanalytics, Institute of Pharmaceutical and Medicinal Chemistry, Heinrich-Heine-University, Düsseldorf, Germany*
SEISHI KATO • *Department of Rehabilitation Engineering, Research Institute, National Rehabilitation Center for Persons with Disabilities, Tokorozawa, Japan*
YOUICHI KONDOU • *RIKEN Plant Science Center, Yokohama, Kanagawa, Japan*

HAI-RI LI • *Department of Cellular and Molecular Medicine, Stem Cell Program, University of California, San Diego, La Jolla, CA, USA*
BIN LIU • *Department of Anesthesia, UCSF Comprehensive Cancer Center, University of California at San Francisco, San Francisco, CA, USA*
MICHAEL T. LOVCI • *Department of Cellular and Molecular Medicine, Stem Cell Program, University of California, San Diego, La Jolla, CA, USA*
CHAOFU LU • *Department of Plant Sciences and Plant Pathology, Montana State University, Bozeman, MT, USA*
CHENG LU • *Delaware Biotechnology Institute, University of Delaware, Newark, DE, USA*
SERGEY A. LUKYANOV • *Shemiakin and Ovchinnikov Institute of Bioorganic Chemistry RAS, Moscow, Russia*
CHRISTINA J. MAIER • *Division of Molecular Dermatology, Department of Dermatology, Paracelsus Private Medical University Salzburg, Salzburg, Austria*
RICHARD H. MAIER • *Division of Molecular Dermatology, Department of Dermatology, Paracelsus Private Medical University Salzburg, Salzburg, Austria*
SYLVESTRE MARILLONNET • *Icon Genetics GmbH, Biozentrum Halle, Halle, Germany*
MINAMI MATSUI • *RIKEN Plant Science Center, Yokohama, Kanagawa, Japan*
MANUELA MURA • *Cancer Research UK Angiogenesis Group, Institute for Biomedical Research, Schools of Immunity and Infection and Cancer studies, College of Medicine and Dentistry, University of Birmingham, Birmingham, UK*
HIROSHI NOJIMA • *Department of Molecular Genetics and DNA-chip Development Center for Infectious Diseases, Research Institute for Microbial Diseases, Osaka University, Osaka, Japan*
RICHARD O'CONNELL • *Department of Plant–Microbe Interactions, Max-Planck-Institute for Plant Breeding Research, Köln, Germany*
KUNIYO OHTOKO • *Hitachi High-Technologies Co., Hitachinaka, Japan*
KAMIL ÖNDER • *Division of Molecular Dermatology, Department of Dermatology, Paracelsus Private Medical University Salzburg, Salzburg, Austria*
MIO OSHIKAWA • *Department of Rehabilitation Engineering, Research Institute, National Rehabilitation Center for Persons with Disabilities, Tokorozawa, Japan*
ALEXANDER SCHEGLOV • *Shemiakin and Ovchinnikov Institute of Bioorganic Chemistry RAS, Moscow, Russia*
PATRICK S. SCHNABLE • *Department of Agronomy, Center for Plant Genomics, Iowa State University, Ames, IA, USA*
DMITRY A. SHAGIN • *Shemiakin and Ovchinnikov Institute of Bioorganic Chemistry RAS, Moscow, Russia*
IRINA A. SHAGINA • *Evrogen JSC, Moscow, Russia*
VIKAS SHEDGE • *Center for Plant Science Innovation, University of Nebraska, Lincoln, NE, USA; Dupont Experiment Station, Wilmington, DE, USA*
DOV J. STEKEL • *School of Biosciences, University of Nottingham, Nottingham, UK*
KOHTAROH SUGAO • *Department of Neurobiology, Graduate School of Medicine, The University of Tokyo, Tokyo, Japan*
MICHAIL SYCHEV • *Moscow State University of Railway Engineering, Moscow, Russia*

HIROYUKI TAKAHARA • *Department of Bioproduction Science, Ishikawa Prefectural University, Ishikawa, Japan*
TAKAHIRO TOUGAN • *Department of Molecular Genetics and DNA-chip Development Center for Infectious Diseases, Research Institute for Microbial Diseases, Osaka University, Osaka, Japan*
LAURA L. VAGNER • *Shemiakin and Ovchinnikov Institute of Bioorganic Chemistry RAS, Moscow, Russia; Evrogen JSC, Moscow, Russia*
JAMES G. WALLIS • *Institute of Biological Chemistry, Washington State University, Pullman, WA, USA*
GENE W. YEO • *Department of Cellular and Molecular Medicine, Stem Cell Program, University of California, San Diego, La Jolla, CA, USA*

Part I

Enhanced Approaches to Inherent Problems

Chapter 1

Isolation of Fungal Infection Structures from Plant Tissue by Flow Cytometry for Cell-Specific Transcriptome Analysis

Hiroyuki Takahara, Elmar Endl, and Richard O'Connell

Abstract

Many plant pathogenic fungi differentiate a series of highly specialized infection structures to invade and colonize host tissues. Especially at early stages of infection, the ratio of fungal to plant biomass is very low. To investigate cell-specific patterns of gene expression, it is necessary to purify the fungal structures of interest from infected plants. We describe here a method to isolate the biotrophic hyphae of *Colletotrichum higginsianum* from *Arabidopsis* leaves, based on a combination of pre-enrichment by isopycnic centrifugation followed by further purification by fluorescence-activated cell sorting. This protocol efficiently eliminates contamination by plant components and nontarget fungal cell-types. Moreover, the isolated cells remain alive, providing high-quality RNA for library construction. The method can be readily adapted for cell-specific transcriptome analysis in other plant–microbe interactions.

Key words: Fluorescence-activated cell sorting, Flow cytometry, *Colletotrichum higginsianum*, *Arabidopsis thaliana*, Biotrophy, Transcriptome

1. Introduction

To successfully penetrate and colonize host tissues, many plant pathogenic fungi sequentially produce a whole series of highly specialized cell-types or "infection structures." In the case of some biotrophic parasites, feeding structures called haustoria or intracellular hyphae develop inside living plant cells, after penetration through the plant cell wall (1). The construction of cDNA libraries is a powerful approach for identifying genes that are differentially expressed at specific stages of fungal morphogenesis and plant infection (2–5). However, sampling the transcriptome of infection structures formed in planta is hindered by the low ratio of fungal to plant biomass, especially at early stages of infection. In order to

Chaofu Lu et al. (eds.), *cDNA Libraries: Methods and Applications*, Methods in Molecular Biology, vol. 729,
DOI 10.1007/978-1-61779-065-2_1, © Springer Science+Business Media, LLC 2011

investigate cell-specific patterns of gene expression, it is necessary to isolate the fungal structures of interest from infected plants.

Previously, fungal structures have been isolated from plants using techniques such as density gradient centrifugation, lectin affinity chromatography, immunomagnetic separation, and laser capture microdissection (6–9). Fluorescence-activated cell sorting (FACS) is a form of flow cytometry that allows a heterogeneous mixture of cells to be sorted, one cell at a time, based on their specific light-scattering and fluorescence characteristics (10). However, although FACS is widely used in animal and plant cell biology for cell-specific expression profiling (11–14), it has rarely been applied to plant pathogens. In this chapter, we present a protocol for isolating the biotrophic hyphae of *Colletotrichum higginsianum* from infected *Arabidopsis* leaves, based on a combination of isopycnic centrifugation and FACS. The work-flow is summarized schematically in Fig. 1a. We expect that the method can be modified for isolating the infection structures of many other plant pathogens.

Fungal hyphae are first released from host epidermal cells by mechanical homogenization and then partially enriched by isopycnic centrifugation. This pre-enrichment step is crucial to minimize the sorting time required for the subsequent FACS purification (see Note 1). The density gradient medium used for centrifugation is Percoll, a colloidal suspension of silica particles coated with polyvinylpyrrolidone, which has low osmolarity and is nontoxic toward cells (8). Cell sorting is based on the specific labeling of intact, viable hyphae by a green fluorescent vital dye, and fluorescein diacetate, while contaminating plant chloroplasts are removed on the basis of their red autofluorescence (15) (Fig. 1b). Other fluorescent markers could be used in place of FDA, for example, fluorescent reporter proteins (in the case of fungi that can be genetically transformed) and fluorochrome-tagged lectins or antibodies that label cell-specific surface epitopes (7, 16, 17).

Nearly all the cells isolated by our method remain alive, so that high-quality RNA can be extracted for cDNA library construction or expression profiling experiments (Fig. 2). Moreover, cell sorting efficiently eliminates contamination by plant components and nontarget fungal cell-types, yielding hyphae with 94% purity, on average (15) (Fig. 1d). This provides an enormous enrichment of mRNA from a single fungal cell type, and deep-sequencing of a cDNA library prepared from FACS-purified hyphae with Roche 454 GS FLX technology revealed that out of 404,000 ESTs only 0.03% had homology to plant sequences (unpublished data). The isolation method is invasive and takes approximately 1.5 h to complete. In order to minimize transcriptional changes, we perform all steps of the procedure at or below 4°C and limit cell sorting runs to 30-min duration. Libraries prepared from the FACS-isolated hyphae appear to accurately represent the fungal transcriptome in planta, because out of 78 *C. higginsianum* genes selected from a cDNA library, all were expressed at the equivalent stage of plant

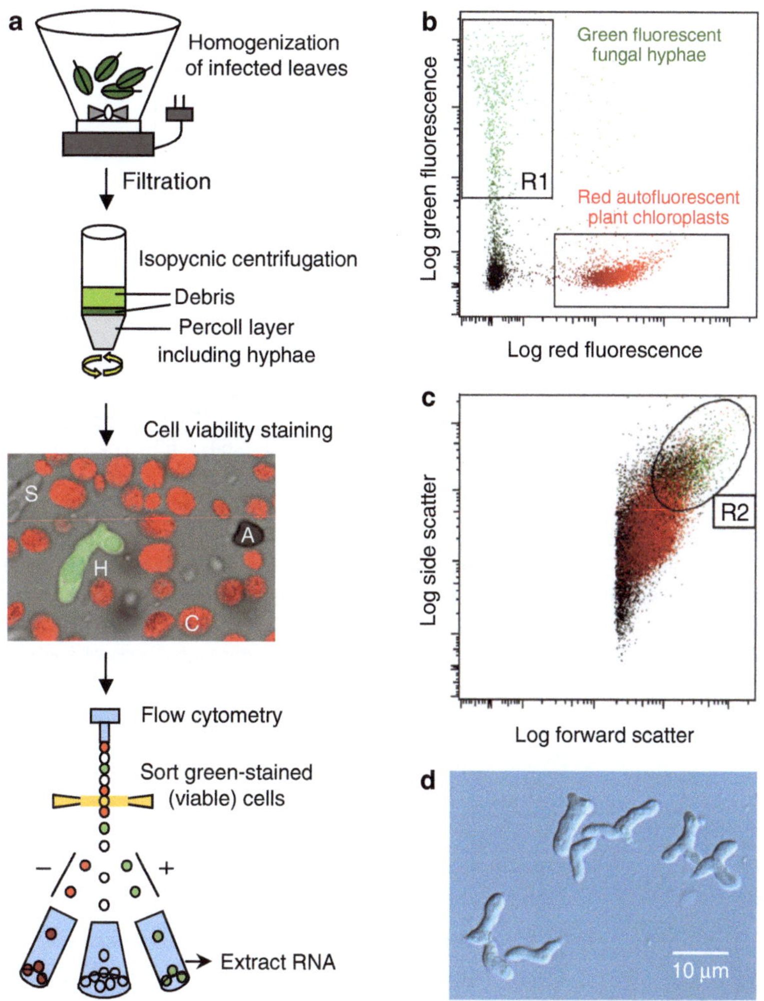

Fig. 1. Isolation of *Colletotrichum* infection hyphae from *Arabidopsis* plants. (**a**) Work-flow of the isolation procedure. Fungal hyphae are first released from infected leaves by homogenization, then pre-enriched by isopycnic centrifugation, stained with a *green* fluorescent vital dye, and purified by FACS. *A* appressorium, *S* spore, *H* hypha, *C* chloroplast. (**b, c**) *Dot plot* cytograms showing the gating strategy used for sorting. Cells combining strong *green fluorescence* (R1) and high forward- and side-scatter (R2) were selected. (**d**) Light micrograph showing hyphae purified by FACS. Reproduced from ref. 15 with permission from Blackwell Publishing Ltd.

infection (15) (unpublished data). Nevertheless, we cannot exclude the possibility that transcript levels are modified during isolation, and while the method is a powerful tool for gene discovery, it may be less suitable for expression profiling.

2. Materials

2.1. Culture Medium for Colletotrichum

1. Mathur's agar medium: dissolve 2.8 g glucose, 1.22 g $MgSO_4 \cdot 7H_2O$, 2.72 g KH_2PO_4, 2.18 g mycological peptone (Oxoid Ltd, Basingstoke, UK), and 30 g agar in 1 L deionized

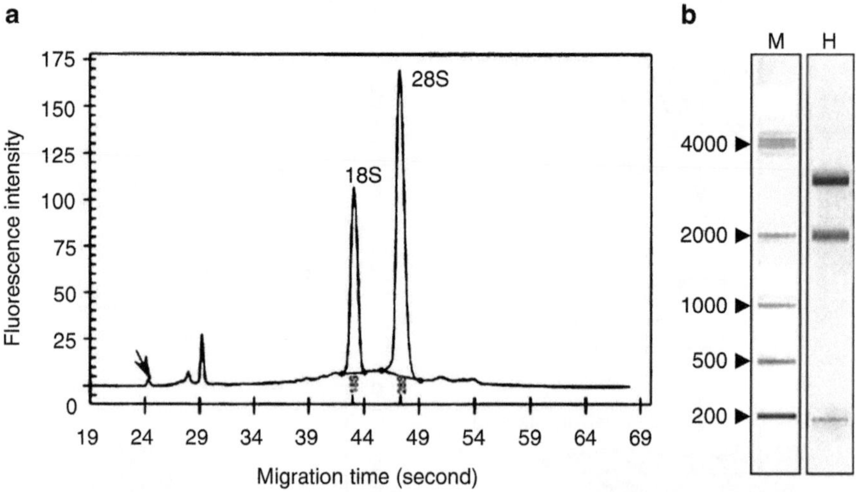

Fig. 2. Assessment of quality and quantity of total RNA extracted from FACS-purified hyphae of *Colletotrichum higginsianum*. (**a**) Bioanalyzer electropherogram showing a low baseline and high, narrow peaks corresponding to the 28S and 18S ribosomal RNAs. The 28S:18S ratio = 1.8, indicating good RNA integrity. The first peak represents a 50-bp marker (*arrow*) added to the sample. (**b**) Corresponding capillary gel electrophoresis images, showing the DNA sizing ladder (M) and total RNA extracted from FACS-purified hyphae (H). Reproduced from ref. 15 with permission from Blackwell Publishing Ltd.

water. Dispense 100 mL aliquots into 250-mL Erlenmeyer flasks, seal with a cotton wool plug, cover with aluminum foil, and autoclave. Allow the flasks to cool on a flat surface.

2.2. Fungal and Plant Materials

1. Fungal strain: *C. higginsianum* IMI 349063 (CABI-Europe, Egham, UK).
2. Susceptible *Arabidopsis thaliana* accession: Columbia-0 glabrous mutant Col-*gl1*-1 (WT-1, Lehle Seeds Round Rock, TX) (see Note 2).

2.3. Stock Solutions

1. 10× Isolation buffer: 2 M sucrose in 0.2 M 3-(*N*-morpholino) propane sulfonic acid (MOPS) buffer, pH 7.2. Store at –20°C.
2. 1× Isolation buffer: one part 10× isolation buffer diluted with nine parts of sterile deionized water. Store at 4°C.
3. Percoll stock solution: one part 10× isolation buffer mixed with nine parts of Percoll (Sigma, St. Louis, MO). Store at 4°C.
4. Percoll working solution: 5 mL Percoll stock solution mixed with 5.93 mL of 1× isolation buffer to produce a specific gravity of 1.085 g/mL. Store at 4°C.
5. Fluorescein diacetate vital dye: 5 mg/mL stock solution of fluorescein diacetate (FDA; Sigma) in acetone. Store at –20°C. Prepare a working solution (0.01% w/v) in 1× isolation buffer immediately before use.
6. FACSFlow phosphate-buffered saline sheath solution (BD Bioscience).
7. 70% (v/v) Ethanol in deionized water.

2.4. Molecular Biology Kits and Reagents

1. PicoPure RNA isolation kit (Arcturus Bioscience).
2. RNA 6000 Pico assay kit (Agilent Technologies).
3. SMART PCR cDNA synthesis kit (Takara Bio-Clontech).
4. RNase-free DNase I solution (Qiagen).
5. QIAquick PCR purification Kit (Qiagen).

2.5. Other Supplies

1. Nylon mesh, 40- and 50-μm pore size (Bückmann GmbH, Germany).
2. Polystyrene Petri dishes, 145-mm diameter (Greiner Bio-One).
3. Falcon polypropylene centrifuge tubes, 50-mL (BD Bioscience).
4. Glass round-bottomed FACS tubes, 6-mL (BD Bioscience).

3. Methods

3.1. Plant Growth and Inoculation

1. Culture *C. higginsianum* for 10–12 days at 25°C on Mathur's medium and harvest conidia (spores) by adding 10 mL of sterile deionized water to the culture flask and shaking vigorously. Adjust the concentration of the spore suspension to 5×10^6 spores/mL, using a hemocytometer slide to count the cells.
2. Grow *Arabidopsis* plants for 5–6 weeks in a peat-based compost using a controlled-environment chamber (10-h light period, 180 $\mu E/m^2/s$, 23°C, 65% humidity).
3. Excise the fully expanded rosette leaves and arrange 15–20 leaves in the base of a large Petri dish (145-mm diameter). Inoculate the lower (abaxial) leaf surface with approximately 100–200 μL of conidial suspension, using small pieces (1 × 2 cm) of 50-μm nylon mesh and an artist's brush to evenly distribute the liquid across the hydrophobic leaf surface (see Note 3).
4. Line the lid of the Petri dish with wet tissue paper, seal the dish with Parafilm to maintain 100% humidity, and incubate in the dark at 25°C for 40 h (see Note 4).

3.2. Isopycnic Centrifugation

Buffer solutions should be ice-cold, and all homogenization and filtration steps should be performed in a cold room.

1. Remove ungerminated conidia from the leaf surface by rinsing in 1 L of deionized sterile water in a beaker.
2. Blot off surplus water using tissue paper and place approximately 500 leaves (60 g fresh weight) into the precooled jar of a Waring blender (or similar) together with 200 mL of 1× isolation buffer. Homogenize at high speed for 1 min (see Note 5).

3. Filter the homogenate through 50-μm nylon mesh to remove plant cell wall debris and collect the filtrate in a precooled beaker.
4. Re-homogenize material retained on the filter in 100 mL of 1× isolation buffer for 1 min, re-filter, and rinse the residue with a further 100-mL buffer. Pool the filtrates from steps 3 and 4, giving 400 mL in total.
5. Transfer the pooled filtrate into eight 50-mL Falcon tubes and centrifuge at 1,080 × *g* for 15 min at 4°C in a refrigerated centrifuge equipped with a swing-out rotor.
6. Discard the supernatants and resuspend each pellet in 5 mL of 1× isolation buffer.
7. Using a Pasteur pipette, slowly layer the cell suspensions onto 5-mL aliquots of 1.085 g/mL Percoll working solution in eight 50-mL Falcon tubes (see Note 6). The cell suspension should float on top of the Percoll "cushion." Take care to avoid mixing at the interface between the two layers.
8. Centrifuge the tubes at 720 × *g* for 15 min at 4°C with no braking (see Note 7). Using a Pasteur pipette, carefully remove the upper aqueous layer and the dense green layer of chloroplasts floating above the Percoll cushion (see Note 8).
9. Dilute each Percoll cushion to 50 mL with 1× isolation buffer and centrifuge at 1,080 × *g* for 15 min at 4°C.
10. Using a Pasteur pipette connected to a vacuum pump, remove and discard the supernatants. Resuspend the pellets in 1 mL 1× isolation buffer and pool them together.
11. Estimate the concentration of all cell types/particles present in the suspension, including contaminants such as plant chloroplasts, using a hemocytometer slide. Optionally, check cell viability by fluorescence microscopy after staining with 0.01% (w/v) FDA (see Note 9).
12. Maintain the sample on ice until ready to start cell sorting.

3.3. Fluorescence-Activated Cell Sorting

The sorting parameters given below are optimized for the FACS DiVa cell sorter (BD Bioscience), which has a "stream-in-air" flow system. However, the strategy used for detection and sorting should be applicable to instruments from other manufacturers. Cell sorting is performed using excitation from a laser emitting 150 mW at 488 nm. Forward- and side-scatter signals are collected through 488/10 nm band-pass filters. Green fluorescence emission from fluorescein diacetate is collected with a 530/20-nm band-pass filter and the red autofluorescence of chloroplasts with a 630/22-nm band-pass filter. All pulses are displayed on a logarithmic scale to obtain the full dynamic range. Preliminary studies should be performed on stained samples to optimize threshold and gain settings for the forward- and side-scatter detectors.

1. Increase sensitivity of the forward- and side-scatter detectors until electronic noise and impurities in the sheath fluid are detected. Then decrease sensitivity to just above background.
2. Dilute the sample with 1× isolation buffer to obtain a suspension in which the total concentration of all cells/particles (including chloroplasts and other contaminants) is in the range $2.5–5.0 \times 10^6$/mL. The target cells (biotrophic hyphae) should comprise 5–10% of the total particles present in the sample.
3. Filter the sample using a 40-μm pore size nylon mesh to remove any large particles or aggregates, such as plant cell wall fragments, which could block the injection nozzle and disrupt the flow stream.
4. Immediately prior to analysis, label the living fungal cells by adding FDA stock solution to the sample to give a final concentration of 0.01% (w/v).
5. Run the sample on the cell sorter and record a sufficient number of events to see fluorescein-labeled fungal cells and autofluorescent chloroplasts in a display of green versus red fluorescence (Fig. 1b). Adjust the spill-over of fluorescein fluorescence into the red channel using standard procedures to compensate for spectral overlap.
6. Set a region of interest (R1) on the cells exhibiting green fluorescence in a two-parameter display of the fluorescence signals, and then use the software to display these events in a dot plot of forward- versus side-scatter (Fig. 1c).
7. Apply a second region of interest (R2) in the forward- versus side-scatter plot that includes most the hyphae identified in step 6. An appropriate combination of regions R2 and R1 is used to precisely define the population of fluorescein-labeled hyphae for subsequent cell sorting (see Note 10). Verify the purity of sorted cells by collecting a sample of the positive flow stream onto a slide and view with light microscopy (Fig. 1d).
8. Sort the sample using a 90-μm injection nozzle at an event rate of 5,000/s, with the sort mode optimized for purity and a sheath fluid pressure of 25 psi. Make use of the mechanical agitator to prevent settling and aggregation of the cells, and the water-cooling system to maintain the cells at 4°C during sorting.
9. Collect the sorted cells into glass FACS tubes containing 2 mL 1× isolation buffer.
10. Transfer the cells into RNase-free 1.5-mL Eppendorf tubes and centrifuge at $5,000 \times g$ for 10 min. Remove the supernatant, snap-freeze the pellet in liquid nitrogen, and store at −80°C.

3.4. RNA Extraction and Quality Assessment

The cell sorting protocol described above yields a relatively small number of cells (approximately 4×10^5 in a typical experiment). We present here a method for extracting total RNA based on the PicoPure RNA isolation kit (Arcturus Bioscience), which is suitable for recovering total RNA from single cells or laser capture microdissection samples, as well as larger samples containing up to 100 μg RNA.

1. Pipette 100 μl of PicoPure Extraction Buffer into tubes containing a frozen pellet of sorted cells and resuspend by gentle pipetting.
2. Incubate the cell suspension at 42°C for 30 min (see Note 11).
3. Centrifuge at 3,000×*g* for 2 min to remove cell debris and transfer the supernatant into a fresh RNase-free Eppendorf tube.
4. Add an equal volume (approx. 100 μL) of 70% ethanol and mix thoroughly by pipetting.
5. Proceed with the RNA isolation according to the manufacturer's instructions. Remove any contaminating genomic DNA by treatment with RNase-free DNase I (Qiagen).
6. Elute the total RNA from the Picopure Purification Column using the minimum recommended volume of Elution Buffer (11 μL).
7. Use a 2-μL aliquot to estimate the quality and quantity of the extracted RNA using an Agilent 2100 Bioanalyzer with the RNA 6000 Pico assay kit (Fig. 2), according to the manufacturer's instructions (see Note 12).
8. Store the RNA at −80°C until starting cDNA library preparation.

3.5. cDNA Library Construction

Using the RNA extraction procedure described above, it is possible to obtain approximately 0.55 μg of total RNA from 4×10^5 FACS-purified biotrophic hyphae. To generate an amplified, oligo-dT-primed cDNA library, the SMART PCR cDNA synthesis kit (Takara Bio-Clontech) can be used according to the manufacturer's instructions with some minor modifications.

1. As starting material for first-strand cDNA synthesis, use 150 ng of total RNA in a total reaction volume of 10 μL.
2. Use a 2-μL aliquot from the first-strand synthesis for subsequent PCR amplification in a total reaction volume of 100 μL. Determine the optimal number of PCR cycles by agarose gel electrophoresis (see Note 13).
3. Purify the amplified cDNAs using the QIAquick PCR purification Kit (Qiagen).

4. Digest the purified cDNAs with *Sca*I to cleave the 3′ end of the oligo-dT adaptor sequences (see Note 14).
5. After further purification using the QIAquick Kit and A-Tailing by DNA polymerase in the presence of dATP, clone the cDNAs into the pGEM-T Easy vector (Promega) and transform them into *Escherichia coli* DH5α competent cells.

4. Notes

1. The time required to sort a given number of target cells is directly related to their concentration in the sample mixture (10). For example, at a sort rate of 10,000 events per second, it would take 47 min to sort 10^6 target cells from a mixture in which they comprise 1% of the total, but only 17 min if they comprise 10% of the total.
2. Brush-inoculation of excised *Arabidopsis* leaves with fungal spore suspension is facilitated by use of the Col-0 *glabrous* mutant, which lacks trichomes.
3. These conditions result in heavy infection, where most *Arabidopsis* epidermal cells contain ten or more biotrophic hyphae of *Colletotrichum*. For other plant–fungal interactions, optimize the inoculation conditions to obtain the maximum possible number of target infection structures in the plant tissue available for extraction.
4. Apply the inoculum and seal the Petri dish as quickly as possible to avoid dessication of the excised leaves. Incubation of the inoculated leaf tissue in the dark facilitates later removal of plant chloroplasts by isopycnic centrifugation, probably by increasing their buoyant density. At 40 h after inoculation, most infections should consist of biotrophic hyphae inside host epidermal cells. Verify this by light microscopy as follows: clear the leaf tissue for 30 min in a 1:3 mixture of chloroform:ethanol, mount in lactophenol under a coverslip, and view with differential interference contrast microscopy.
5. Mechanical homogenization is used to release the fungal structures from plant tissue. Bulbous determinate structures, such as fungal haustoria and intracellular hyphae, and unicellular structures, such as spores and yeast cells, survive this process and can be isolated intact. However, long filamentous hyphae are likely to be fragmented, resulting in loss of cytoplasm unless retained between septa.
6. The choice of density for the Percoll cushion depends on the buoyant density of the cells of interest. To optimize the method for infection structures of other fungal pathogens,

layer the tissue homogenate onto a stepped density gradient comprising five steps of 1.040, 1.065, 1.090, 1.115, and 1.140 g/mL Percoll. After centrifugation, determine by microscopy in which Percoll layer the target cells are concentrated.

7. Avoid rapid braking of the centrifuge rotor, which results in mixing between the Percoll layer and plant debris floating above it.
8. During aspiration, take care to avoid transferring parts of the compacted chloroplast layer into the Percoll. Try to remove as little of the Percoll layer as possible because this contains the cells of interest.
9. Cells stained by FDA retain cytoplasmic esterase activity and an intact plasma membrane (18).
10. The threshold and sensitivity for forward- and side-scatter detectors should be adjusted so that all events, including chloroplasts and other small particles, are identified. If the threshold is set too high, the instrument will ignore events that might be crucial for the correct sorting decision, resulting in greater contamination. For purifying infection structures of fungi other than *Colletotrichum*, it will be necessary to optimize the forward- and side-scatter settings according to their size and optical properties.
11. Incubation in the PicoPure Extraction Buffer efficiently recovers total RNA from *C. higginsianum* hyphae without any mechanical disruption of the cells. Other RNA extraction methods, such as TRIzol reagent (Invitrogen), could be used if a larger number of cells are available.
12. Assess total RNA integrity by inspecting the Agilent Bioanalyzer electropherogram. Intact RNA should present 18S and 28S ribosomal RNA peaks that are high and narrow, with a 28S:18S ratio between 1.8 and 2.0 (19). Low baseline fluorescence also indicates the absence of RNA degradation products. If available, use the Agilent software to determine the RNA integrity number.
13. Perform PCR amplification with a range of cycle numbers, for example, 15, 18, 21, and 24, and electrophorese 5-μL aliquots of the products.
14. The supplied 3′ BD SMART CDS primer II A contains a *Sca*I site (AGTACT) just after the oligo-$dT_{(30)}$ sequence (AAGCA GTGGTATCAACGCAG<u>AGTACT</u>$_{(30)}$VN-3′). This site is present in the 3′ oligo-dT primer but is absent from the 5′ oligonucleotide primer. Even after bidirectional cloning into the cloning vector, it is therefore possible to perform directional sequencing from the 5′ end using the 5′ PCR Primer II (AAGCAGTGGTATCAACGCAGAGT).

Acknowledgments

The authors thank Andreas Dolf and Peter Wurst for expert technical assistance with flow cytometry. This work was supported by funding from the Max Plank Gesellschaft and Deutsche Forschungsgemeinschaft (Grant OC104/1-1, SPP1212-PlantMicro).

References

1. O'Connell, R. J. and Panstruga, R. (2006) Tête à tête inside a plant cell: establishing compatibility between plants and biotrophic fungi and oomycetes. *New Phytol.* **171,** 699–718.
2. Hahn, M. and Mendgen, K. (1997) Characterization of *in planta*-induced rust genes isolated from a haustorium-specific cDNA library. *Mol. Plant Microbe Interact.* **10,** 427–37.
3. Catanzariti, A. M., Dodds, P. N., Lawrence, G. J., Ayliffe, M. A., and Ellis, J. G. (2006) Haustorially expressed secreted proteins from flax rust are highly enriched for avirulence elicitors. *Plant Cell* **18,** 243–56.
4. Kleemann, J., Takahara, H., Stuber, K., and O'Connell, R. J. (2008) Identification of soluble secreted proteins from appressoria of *Colletotrichum higginsianum* by analysis of expressed sequence tags. *Microbiology* **154,** 1204–17.
5. Soanes, D. M. and Talbot, N. J. (2006) Comparative genomic analysis of phytopathogenic fungi using expressed sequence tag (EST) collections. *Mol. Plant Pathol.* **7,** 61–70.
6. Mackie, A. J., Robert, A. M., Callow, J. A., and Green, J. R. (1991) Molecular differentiation in pea powdery mildew haustoria-identification of 62 kDa *N*-linked glycoprotein unique to the haustorial plasma membrane. *Planta* **183,** 399–408.
7. Hahn, M. and Mendgen, K. (1992) Isolation by ConA binding of haustoria from different rust fungi and comparison of their surface qualities. *Protoplasma* **170,** 95–103.
8. Pain, N. A., Green, J. R., Gammie, F., and O'Connell, R. J. (1994) Immunomagnetic isolation of viable intracellular hyphae of *Colletotrichum lindemuthianum* from infected bean leaves using a monoclonal antibody. *New Phytol.* **127,** 223–32.
9. Tang, W., Coughlan, S., Crane, E., Beatty, M., and Duvick, J. (2006) The application of laser microdissection to *in planta* gene expression profiling of the maize anthracnose stalk rot fungus *Colletotrichum graminicola*. *Mol. Plant Microbe Interact.* **19,** 1240–50.
10. Fisher, D., Francis, G. E., and Rickwood, D. (1998) *Cell Separation: A Practical Approach.* Oxford University Press, Oxford, UK.
11. Birnbaum, K., Jung, J. W., Wang, J. Y., Lambert, G. M., Hirst, J. A., Galbraith, D. W., and Benfey, P. N. (2005) Cell type-specific expression profiling in plants via cell sorting of protoplasts from fluorescent reporter lines. *Nat. Methods* **8,** 615–19.
12. Galbraith, D. W. and Birnbaum, K. (2006) Global studies of cell type-specific gene expression in plants. *Annu. Rev. Plant Biol.* **57,** 451–75.
13. Lobo, M. K., Karsten, S. L., Gray, M., Geschwind, D. H., and Yang, X. W. (2006) FACS-array profiling of striatal projection neuron subtypes in juvenile and adult mouse brains. *Nat. Neurosci.* **9,** 443–52.
14. Shigenobu, S., Arita, K., Kitadate, Y., Noda, C., and Kobayashi, S. (2006) Isolation of germline cells from *Drosophila* embryos by flow cytometry. *Dev. Growth Differ.* **48,** 49–57.
15. Takahara, H., Dolf, A., Endl, E., and O'Connell, R. (2009) Flow cytometric purification of *Colletotrichum higginsianum* biotrophic hyphae from *Arabidopsis* leaves for stage-specific transcriptome analysis. *Plant J.* **59,** 672–83.
16. Czymmek, K. J., Bourett, T. M., and Howard, R. J. (2005) Fluorescent protein probes in fungi. In: Savidge T, Pothoulakis C, eds. *Methods in Microbiology*, Vol. 34. Microbial Imaging Elsevier, Amsterdam, 27–62.
17. Pain, N. A., O'Connell, R. J., Mendgen, K., and Green, J. R. (1994) Identification of glycoproteins specific to biotrophic intracellular hyphae formed in the *Colletotrichum*-bean interaction. *New Phytol.* **127,** 233–42.
18. Rotman, B. and Papermaster, B. W. (1966) Membrane properties of living mammalian cells as studied by enzymatic hydrolysis of fluorogenic esters. *Proc. Natl. Acad. Sci. USA* **55,** 134–41.
19. Sambrook, J. and Russel, D. W. (2001) *Molecular Cloning: A Laboratory Manual.* Cold Spring Harbor Laboratory Press, Cold Spring Harbor, NY.

Chapter 2

Preparation of a High-Quality cDNA Library from a Single-Cell Quantity of mRNA Using Chum-RNA

Hiroshi Nojima and Takahiro Tougan

Abstract

Unlike exponential amplification using polymerase chain reaction (PCR), linear RNA amplification using T7 RNA polymerase is advantageous for genome-wide analysis of gene expression and for cDNA library preparation from single-cell quantities of RNA. However, the use of RNA polymerase requires a large amount of RNA, as the optimum concentration of the substrate (mRNA), or the Michaelis constant (K_m), is one millionfold higher than the single-cell amount of mRNA. To circumvent this K_m problem, we designed a small mRNA-like dummy molecule, termed chum-RNA, which can be easily removed after the completion of the reaction. Chum-RNA allowed the preparation of a high-quality cDNA library from single-cell quantities of RNA after four rounds of T7-based linear amplification, without using PCR amplification. The use of chum-RNA may also facilitate quantitative reverse-transcription (qRT)-PCR from small quantities of substrate.

Key words: Small RNA, Single-cell cDNA library, T7 RNA polymerase, Sense mRNA amplification, RT-PCR

1. Introduction

Amplification of the RNA isolated from a limited amount of specimen is obligatory to compare gene expression patterns among cells and/or tissues for microarray analysis or cDNA library preparation. Although the polymerase chain reaction (PCR), which is based on exponential amplification, is a powerful method for amplifying a single target DNA, it often produces a biased product because of distinct efficiencies of amplification for transcripts (or cDNAs) of differing lengths, abundance, and/or diversity (1).

The linear amplification method, which was first developed by Van Gelder, Eberwine, and coworkers (2, 3), is an alternative

Chaofu Lu et al. (eds.), *cDNA Libraries: Methods and Applications*, Methods in Molecular Biology, vol. 729, DOI 10.1007/978-1-61779-065-2_2, © Springer Science+Business Media, LLC 2011

to PCR that is particularly useful for the amplification of RNA, as it is considered to generate non-biased RNA pools. This technique and subsequent improved protocols, with or without combination with PCR (4–8), have allowed the genome-wide microarray analysis of gene expression using a single-cell amount of RNA as the starting material (9–12). However, most cDNA library preparations from single-cell amounts of mRNA are performed with at least partial assistance of PCR amplification (6, 13, 14), as the use of RNA polymerase alone requires 1 μg of total RNA after two-round amplification of complementary RNA (cRNA) (15).

We noticed that this is mainly because the optimum concentration, or the Michaelis constant (K_m), of most of the enzymes used in cDNA library preparation is more than 1 μM; this value exceeds the single-cell amount of mRNA by one millionfold (15). To circumvent this K_m problem, we recently designed a small mRNA-like molecule, termed chum-RNA, and synthesized sense RNA (sRNA) successfully using RNA amplification without the aid of PCR to prepare a high-quality cDNA library from a single-cell amount of mRNA (15). Chum-RNA can be added to the reaction mixture to increase the effective quantity of substrate, thereby increasing the substrate conversion rate of the enzyme, and can be easily removed after the completion of the reaction. Here, we present the detailed protocol for the application of chum-RNA to the preparation of a cDNA library and for reverse transcriptase (RT)-PCR using a small amount of mRNA.

2. Materials

2.1. Phenol/Chloroform Extraction and Ethanol Precipitation

1. Phenol/chloroform: Add 100 g of crystallized DNA-grade phenol and 0.1 g of 8-hydroxyl quinoline to 100 mL of 1 M Tris–HCl (pH 7.5), mix, and dissolve in a water bath at 50°C (see Note 1). Place the solution on a bench for 10 min and then remove the supernatant. Confirm the neutral pH of the supernatant using a pH test strip. If it is still acidic, repeat this step until the pH of the supernatant becomes neutral. Add an equal volume of chloroform, cover the bottle in aluminum foil, and store at 4°C. As 8-hydroxyl quinoline acts as an antioxidant, a partial inhibitor of RNase, and an indicator of the organic phase of the solution (as indicated by its bright yellow color), it is recommended to prepare a fresh stock of phenol/chloroform solution when its color turns to dark yellow.
2. Glycogen carrier: Use glycogen solution as a carrier for the ethanol precipitation of nucleic acids and CHROMA SPIN centrifugation.
3. Dry-ice ethanol bath: Perform ethanol precipitation by cooling the sample (one volume of sample, one-tenth volume of 3 M

Na acetate, and two volumes of ethanol) in a dry-ice ethanol bath for more than 5 min. Alternatively, immerse the sample into ethanol that is kept cool in a vacuum-insulated stainless steel container in a deep freezer set to −85°C.

2.2. Agarose Gel Electrophoresis

1. TAE buffer (50×): Prepare a 50× stock solution using 726 g Trizma base, 171.3 mL glacial acetic acid, 55.8 g EDTA-2Na, and bring the volume to 3 L with Nanopure water (see Note 2). Store at room temperature. Before use, dilute 20 mL of this solution in 980 mL of Nanopure water.
2. 1% Agarose gel: 1 g Agarose in 100 mL TAE buffer (1×).
3. DNA molecular weight marker: *Sty*I-digested lambda phage DNA, which comprises DNA fragments of the following sizes (kb): 19.33, 7.74, 6.22, 4.26, 3.47, 2.69, 1.88, 1.49, 0.93, and 0.42.
4. DNA visualization: Immerse the agarose gel in water containing ethidium bromide (EtBr) at 1 μg/mL and visualize the fluorescent signal under UV light.

2.3. RT-PCR

1. Enzyme: *ExTaq*™ DNA polymerase (TaKaRa Bio Inc., Ohtsu, Japan) (see Note 3).
2. Primers for glyceraldehyde-3-phosphate dehydrogenase (GAPDH) detection: Forward (HsGAPDH-F) 5′-CGA GAT CCC TCC AAA ATC AA-3′ and reverse (HsGAPDH-R) 5′-AGG GGT CTA CAT GGC AAC TG-3′.
3. It is recommended to perform PCR in a PCR tube (0.2 mL thermo-strip; ABgene Epsom, UK).

2.4. Preparation of a Single-Cell cDNA Library (see Notes 1 and 4)

1. Chum-RNA: 5′-AAU UCG UCU GGA CAC G$(A)_{25}$-3′. Chum-RNA (1 μg/μl) can be purchased from Gene Design Inc., Osaka, Japan (http://www.saito.tv/e/lsp/LSP_GuideList/English/GeneDesign.htm?m3).
2. Enzymes: RNase H, DNA polymerase I, T4 DNA polymerase, T4 DNA ligase, RNase-free DNase I, T7 RNA polymerase, *Xho*I, *Not*I, and RNase inhibitor. Reverse transcriptase (SuperScript III; Invitrogen, San Diego, CA).
3. Nucleotides: NTP and 10 mM rATP (TaKaRa Bio). Prepare a 25 mM NTP mix from ATP, CTP, GTP, and UTP (100 mM each).
4. Linker primer (HPLC grade): 1.6 mg/mL of 5′-$(GA)_{10}$ A CGC GTC GAC TCG AGC GGC CGC GGA CCG $(T)_{18}$-3′.
5. Amplification adaptor: Sense T7: 5′CAC TAG TAC GCG TAA TAC GAC TCA CTA TAG GGA ATT CCC CGG G-3′; antisense T7: 5′-pCCC GGG GAA TTC CCT ATA GTG AGT CGT ATT ACG CGT ACT AGT GAG CT-3′.

6. Library adaptor (TaKaRa Bio): *Bam*HI(*Bgl*II)–*Sma*I d(GATCCCCGGG) and p*Sma*I linker: d(pCCCGGG).
7. 10× First-strand buffer: 500 mM Tris–HCl (pH 8.3), 750 mM KCl, and 30 mM $MgCl_2$.
8. First-strand mixture: 10 mM dATP, dGTP, and dTTP, and 5 mM 5-methyl-dCTP.
9. 10× Second-strand buffer: 188 mM Tris–HCl (pH 8.3), 906 mM KCl, and 46 mM $MgCl_2$.
10. Second-strand nucleotide mixture: 10 mM dATP, dGTP, and dTTP, and 25 mM dCTP.
11. 10× T4 DNA polymerase buffer: 500 mM Tris–HCl (pH 8.3), 100 mM $MgCl_2$, 500 mM NaCl, and 100 mM dithiothreitol (DTT).
12. 10× Ligase buffer: 500 mM Tris–HCl (pH 7.5), 70 mM $MgCl_2$, and 10 mM DTT.
13. 10× *Not*I buffer supplement: 278 mM NaCl, 8 mM $MgCl_2$, 1.8 mM DTT, 0.018% BSA, and 0.018% Triton X-100.
14. 10× *Bgl*II buffer: 100 mM Tris–HCl (pH 7.5), 1.0 M NaCl, 70 mM $MgCl_2$, and 10 mM DTT.
15. 10× T7 Pol buffer: 400 mM Tris–HCl (pH 8.0), 80 mM $MgCl_2$, 20 mM spermidine, and 50 mM DTT.
16. 10× Bacterial alkaline phosphatase (BAP) buffer: 500 mM Tris–HCl (pH 8.0) and 10 mM $MgCl_2$.
17. 10× STE: 1 M NaCl, 100 mM Tris–HCl (pH 8.0), and 10 mM ethylenediaminetetraacetic acid (EDTA).
18. Centrifuge column: CHROMA SPIN-400 (Clontech, Palo Alto, CA).
19. Vector DNA: pAP3*neo* (TaKaRa Bio).
20. TE buffer: 10 mM Tris–HCl (pH 7.5) and 1 mM EDTA (see Note 4).
21. 1/10 TE buffer: 1 mM Tris–HCl (pH 7.5) and 0.1 mM EDTA (see Note 4).
22. Centrifuge filter: Ultrafree-C3 (Cat. #UFCP3TK50; Millipore, Bedford, MA) or MINICENT-30 (Cat. #08627; Tosoh SMD, Grove City, OH) (see Note 5).
23. QuickPrep Micro mRNA purification (QMP) kit: Cat. #27 -9255-01 (GE Healthcare Bio-Sciences Corp., Piscataway, NJ).
24. Plasmid Maxi kit: Cat. #12162 (QIAGEN, Hilden, Germany).
25. Other reagents: DTT, sodium dodecyl sulfate (SDS), EDTA, dimethyl sulfoxide (DMSO), NaCl, sodium acetate (NaAc), and ethanol.

2.5. Electroporation and Propagation of Escherichia coli Cells

1. Electro-MAX DH12S cells: Cat. #18312-017 (Invitrogen).
2. L-Broth (Luria–Bertani medium): Mix 10 g Bacto tryptone (Difco), 5 g Bacto yeast extract (Difco), and 5 g NaCl. Make up the volume of the mixture to ~990 mL using Nanopure water (see Note 2) and stir. Adjust pH to 7.4 with NaOH and then adjust the volume with Nanopure water up to 1 L; stir to mix. Pour into a 1-L bottle and autoclave for 20 min.
3. L-Broth ampicillin plates: Add 15 g Bacto agar (Difco) to 1 L of pH-adjusted L-broth and autoclave for 30 min. Cool to about 65°C, add 0.15 g of ampicillin mix, and pour into plates.
4. L-Broth top agar: Add 0.7 g Bacto agar (Difco) to 100 mL L-broth and autoclave for 20 min.
5. SOB: Mix 20 g Bacto tryptone (Difco), 5 g Bacto yeast extract (Difco), 2 mL 5 M NaCl, and 1.25 mL 2 M KCl. Adjust the volume with Nanopure water up to 1,000 mL and stir. Pour into 1-L bottle and autoclave for 20 min. After autoclaving, add 10 mL of 2 M Mg solution (1 M $MgSO_4 \cdot 7H_2O$ and 1 M $MgCl_2 \cdot 6H_2O$, which were autoclaved separately for 20 min).
6. SOC: Add 1 mL of 2 M glucose to 100 mL of SOB. Filter using a 0.22-μm bottle-top filter. Store at 4°C.

3. Methods

3.1. Preparation of mRNA from a Single Mammalian Cell

Chum-RNA may be useful for protecting mRNA from digestion by RNase or from nonspecific binding to the wall of the tubes. Thus, it is recommended to add chum-RNA into the solution used for mRNA purification. This section is a modified version of the QMP kit manufacturer's protocol for the use of chum-RNA in mRNA purification from a single mammalian cell.

1. Transfer a small amount of specimen to a sterile 1.5-mL microfuge (or microcentrifuge) tube containing 0.4-mL of QMP kit extraction buffer and mix by vortexing for 30 s.
2. Add 0.8-mL QMP kit elution buffer and vortex for 5 s.
3. Add 5 μg of chum-RNA if the amount of specimen is very small, to protect the sample RNA from degradation or from nonspecific binding to the wall of the tubes.
4. From the bottle of oligo(dT)-cellulose solution, which was gently shaken to resuspend the cellulose, transfer 1 mL to a sterile 1.5-mL microfuge tube.
5. Centrifuge the samples (see steps 2 and 4 in Subheading 3.1) at maximum speed (20,000 × *g*; 16,000 × *g*) for 1 min.

6. Remove the supernatant from the cellulose-containing tube (see step 4 in Subheading 3.1) and add the supernatant (~1.2 mL) from the specimen-containing tube (see step 2 in Subheading 3.1).
7. Mix the solution by gently inverting the tube for 3 min to trap the mRNA within the oligo(dT)-cellulose.
8. Centrifuge at 20,000 × *g* for 10 s.
9. Remove the supernatant, add 1-mL QMP kit high-salt buffer and then repeat steps 7 and 8 in Subheading 3.1.
10. Repeat steps 7–9 in Subheading 3.1 five times.
11. Remove the supernatant, add 1-mL low-salt buffer, mix the solution by inverting the tube (to wash the mRNA), and then centrifuge at 20,000 × *g* for 10 s.
12. Repeat step 11 in Subheading 3.1.
13. Remove the supernatant, add 0.3-mL low-salt buffer, and transfer the solution to a microfuge cup inserted into the 2-mL receptacle tube.
14. Centrifuge the tube in a microfuge at 20,000 × *g* for 5 s to remove the wash buffer from the column.
15. Discard the wash buffer (~0.3 mL) in the receptacle tube.
16. Add 0.3-mL low-salt buffer to the microfuge cup and repeat steps 14 and 15 in Subheading 3.1.
17. Repeat step 16 in Subheading 3.1.
18. Transfer the microfuge cup to a fresh receptacle tube.
19. Add 0.2-mL of warmed (65°C) elution buffer and centrifuge at 20,000 × *g* for 5 s to elute the mRNA from the column.
20. Repeat see step 19 in Subheading 3.1.
21. Add 5 μg chum-RNA, 40 μL 5 M NaCl, and 0.8 mL ethanol to the 0.4 mL elution buffer collected in the receptacle tube. Mix by vortexing and chill in a dry ice/ethanol bath for 15 min.
22. Centrifuge the pellet in a microfuge for 10 min at 20,000 × *g*.
23. Add 500 μL ice-cold 70% ethanol and centrifuge again for 1 min at 20,000 × *g*.
24. Remove the 70% ethanol and centrifuge again for 1 min at 20,000 × *g*.
25. Remove the residual 70% ethanol from around the pellet.
26. The precipitated mRNA can be stored as a pellet in a deep freezer.

3.2. Synthesis of cDNA from a Small Amount of mRNA Using Chum-RNA

Chum-RNA is useful to promote cDNA synthesis using a very small amount of mRNA (15). As chum-RNA may also serve to protect mRNA from degradation and as a carrier during ethanol precipitation, it is practical to add chum-RNA to the reaction mixture from the beginning of the procedure if the amount of specimen is very small. The procedure described in the next section is illustrated schematically in Figs. 1a-i–iii and 2a.

1. Mix the following reagents in a sterile 0.5-mL microfuge tube:

5 mM Tris–HCl (pH 7.5)	3.5 μL
RNA or mRNA sample	1.0 μL (or pellet from step 26 in Subheading 3.1)
Chum-RNA (1.0 μg/μL)	5.0 μL (5.0 μg)

2. Warm the tube at 65°C for 5 min and then quickly immerse in ice-cold water.

3. Add the following reagents and mix by vortexing:

10× First-strand buffer	2.5 μL
0.1 M DTT	2.5 μL
First-strand mixture	1.5 μL
Linker primer (1.6 μg/μL)	1.0 μL(1.6 μg) (see Note 6)
RNase inhibitor	0.5 μL
H_2O (see Note 1)	6.5 μL (to 25 μL with RTase)

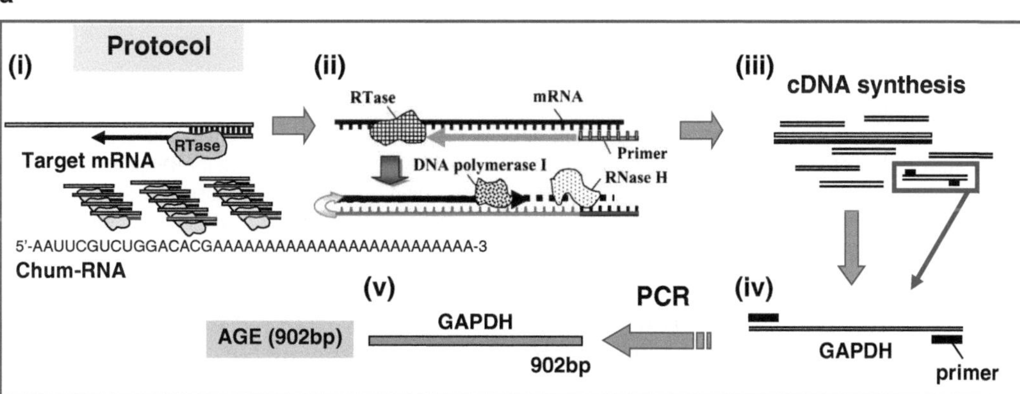

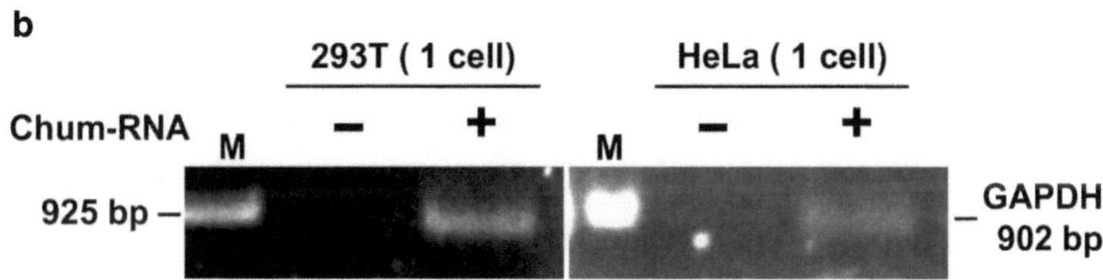

Fig. 1. Chum-RNA facilitates sense-strand mRNA amplification from a single-cell amount of mRNA. The amplified mRNA may be useful for subsequent PCR amplification. (**a**) Schematic illustration of the protocols described in Subheadings 3.2 and 3.3. (**b**) Synthesis of cDNA using a single-cell-derived amount of mRNA from 293T (*left*, 10.5 pg) and HeLa (*right*, 10.1 pg) cells in the presence (+) or absence (–) of chum-RNA (3 μM). Successful cDNA synthesis was confirmed by the detection of the GAPDH cDNA, which was detected by PCR (50 cycles at 50°C) using the reaction product of step 26, Subheading 3.2 and was observed as a band of 902 bp on AGE. M denotes the molecular size marker (925 bp).

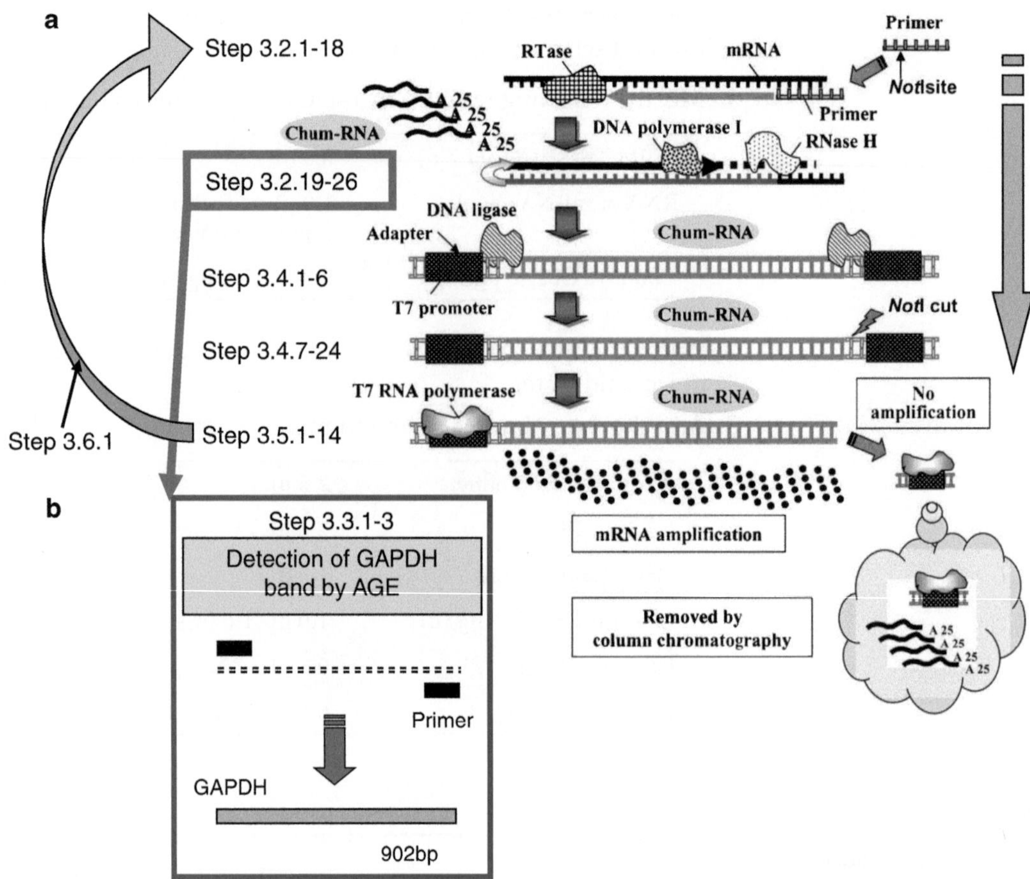

Fig. 2. Schematic depiction of the procedure used for a single round of sense-strand mRNA amplification (from step 1, Subheading 3.2, to step 18, Subheading 3.5). (**a**) Chum-RNA is present in the reaction mixture throughout. A fraction of the chum-RNA may be converted into chum-cDNA during the procedure. Subsequently, the chum-RNA, the chum-cDNA, and the adapter that cuts by *Not*I digestion are removed using column chromatography (described at step 5 to 17, Subheading 3.5). DNA is indicated by a *gray line* and mRNA is indicated by a *black line*. See text for details. (**b**) Successful cDNA synthesis may be confirmed by PCR (steps 1–3, Subheading 3.3).

4. Incubate at room temperature (~18°C) for 10 min to allow the linker primer to anneal to the mRNA.
5. Add 1.0 μL RTase (SuperScript III) and incubate at 42°C for 45 min.
6. Add 1.0 μL RTase and incubate at 55°C for 30 min.
7. Immerse the sample tube in ice-cold water and cool for 5 min.
8. Add the following reagents:

10× Second-strand buffer	20 μL
0.1 M DTT	7.5 μL
Second-strand nucleotide mixture	3.0 μL
H_2O (ice cold) (see Note 1)	133 μL (to 200 μL with RNase H)

9. Incubate in ice-cold water for 5 min.
10. Add 1.5 μL (2 U) RNase H and 10 μL (50 U) *E. coli* DNA Polymerase I.
11. Incubate at 16°C for 150 min.
12. Add an equal volume (200 μL) of phenol/chloroform and vortex for 5 s.
13. Centrifuge the tube in a microfuge for 1 min at 20,000×*g*.
14. Transfer the supernatant to a fresh 1.5-mL microfuge tube. Add 0.1 volumes (20 μL) of 3 M sodium acetate (NaAc) and 2.5 volumes (500 μL) of ice-cold ethanol. Mix and chill in a dry ice/ethanol bath for 15 min.
15. Centrifuge the pellet in a microfuge for 10 min at 20,000×*g*.
16. Add 500 μL ice-cold 70% ethanol and centrifuge again for 1 min at 20,000×g.
17. Remove the 70% ethanol and centrifuge again for 1 min at 20,000×*g*.
18. Remove the residual 70% ethanol from around the pellet.
19. Add the following reagents to the pellet:

10× T4 DNA polymerase buffer	10 μL
2.5 mM dNTP mixture	5 μL
H_2O (see Note 1)	81.5 μL (to 100 μL with enzyme)

20. Incubate in ice-cold water for 5 min.
21. Add 3.5 μL (5 U) of T4 DNA polymerase and incubate at 37°C for 30 min.
22. Add an equal volume (100 μL) of phenol/chloroform and vortex for 5 s.
23. Centrifuge the tube in a microfuge for 1 min at 20,000×*g*.
24. Transfer the supernatant to a fresh 1.5-mL microfuge tube and add 0.1 volumes (10 μL) of 3 M sodium acetate (NaAc) and 2.5 volumes (250 μL) of ice-cold ethanol. Mix and chill in a dry ice/ethanol bath for 15 min.
25. Centrifuge the pellet in a microfuge for 10 min at 20,000×*g*.
26. The reaction can be stopped here: refer to either Subheading 3.3 or 3.4. The precipitated double-stranded DNA (dsDNA) can be stored at –80°C.

3.3. Application of Chum-RNA to RT-PCR Using a Small Amount of mRNA

Chum-RNA may be useful for RT-PCR using a small amount of mRNA (Figs. 1a-iv and -v and 2b). Thus, the sample obtained at the end of step 26 in Subheading 3.2 may be used for PCR amplification, as described below. Two examples of the result of this procedure are shown in Fig. 1b.

1. Dissolve the pellet (see step 26 in Subheading 3.2) in 50 μL TE buffer; this template will be useful for 50× PCR.
2. Add the following reagents to a PCR tube (see Subheading 2.3):

10× *ExTaq*™ buffer	1.0 μL (see Note 3)
dNTP	0.8 μL
Primer Fw (10 pmol/μL)	1.0 μL (see Subheading 2.3)
Primer Rv (10 pmol/μL)	1.0 μL (see Subheading 2.3)
ExTaq™ DNA polymerase	0.1 μL (see Note 3)
Template cDNA	1.0 μL (see step 1 in Subheading 3.3)
H_2O (see Note 1)	5.1 μL (to 10 μL)

3. Perform PCR at an annealing temperature of 50 or 55°C and 30, 40, or 50 amplification cycles.
4. Analyze the reaction product corresponding to the GAPDH cDNA (at 902 bp) using 1% agarose gel electrophoresis, as shown in Fig. 1b.

3.4. Adapter Ligation, NotI Digestion, and Spin Centrifugation

1. Mix the following reagents with the precipitate (see step 26 in Subheading 3.2):

dsDNA (step 26 in Subheading 3.2)	Precipitate
10× Ligase buffer	2.0 μL
10 mM rATP	2.0 μL
Amplification adapter (0.35 μg/μL)	1.0 μL (0.35 μg) (see Note 7)
H_2O (see Note 1)	13.5 μL (to 18.5 μL with DNA ligase)

2. Add 1.5 μL (4 U) T4 DNA ligase.
3. Incubate at 8°C overnight.
4. Heat the tube at 70°C for 30 min to denature the T4 DNA ligase.
5. Centrifuge in a microfuge for 1 min at 20,000 × *g*.
6. Transfer the supernatant to a fresh 0.5-mL microfuge tube.
7. Add the following reagents:

*Not*I buffer supplement	27 μL (see Note 8)
*Not*I	3.0 μL (~50 U)

8. Incubate at 37°C for 90 min.

9. Add 5 μL 10× STE and 1 μL glycogen carrier. Mix by vortexing. Incubate in ice-cold water until use (see step 15 of this Subheading 3.4 below).
10. Mix a CHROMA SPIN-400 column by inverting it a couple of times, cut off the top and bottom lids, and then allow it to stand on a receptacle tube (a 1.5-mL microfuge tube with lid removed) at room temperature for 10 min to drain the extra TE solution from the column.
11. Add 1 mL of 1× STE to the top of the column, set the column and receptacle tube into a 15-mL plastic tube, and centrifuge for 3 min at 700 × *g* using, for example, Beckman's J6-HC centrifuge with a swing-basket rotor 20 cm diameter rotating at 1,800 rpm (700 × *g*).
12. Discard the solution that was spun into the receptacle tube.
13. Centrifuge again for 3 min at 700 × *g* to remove the residual solution from the column completely.
14. Discard the small amount of solution that was spun into the receptacle tube.
15. Transfer the column to a fresh receptacle tube and load the sample (step 9 of this section) onto the top (at the center) of the column, drop by drop (~10 μL per drop). Never let the solution touch the inner wall of the column, to avoid the solution from passing free of the resin packed into the column.
16. Set the column and receptacle tube into a 14-mL round-bottom plastic tube (used as a protector to hold the column tubes in a swing basket) and centrifuge for 3 min at 700 × *g*.
17. Transfer the fractionated sample solution collected at the bottom of the receptacle tube (~50 μL) to a fresh 0.5-mL microfuge tube, add 50 μL of phenol/chloroform, and vortex for 5 s.
18. Centrifuge the tube in a microfuge for 1 min at 20,000 × *g*.
19. Transfer the supernatant to a fresh 0.5-mL microfuge tube. Add 1 μL glycogen carrier, 4 μL 5 M NaCl, and 100 μL ice-cold ethanol. Mix and chill in a dry ice/ethanol bath for 15 min.
20. Centrifuge the pellet in a microfuge for 10 min at 20,000 × *g*.
21. Add 500 μL ice-cold 70% ethanol and centrifuge for 1 min at 20,000 × *g*.
22. Remove the 70% ethanol and centrifuge for 1 min at 20,000 × *g*.
23. Remove the residual 70% ethanol from around the pellet.
24. The wet precipitated DNA can be stored at −20 or −85°C until it is used in the procedure described in Subheading 3.5.

3.5. Amplification of mRNA by T7 RNA Polymerase Using Chum-RNA

1. Mix the following reagents with the precipitate (see step 24 in Subheading 3.4):

dsDNA (from step 24 in Subheading 3.4)	Precipitate
10× T7 Pol buffer	10 μL
Chum-RNA (1.0 μg/μL)	5.0 μL (5.0 μg)
25 mM NTP mix	8.0 μL
H_2O (see Note 1)	72 μL (to 95 μL)

2. Add 5 μL (50 U) of T7 RNA polymerase.
3. Incubate at 37°C for 90 min.
4. Add 1 μL (10 U) T7 RNA polymerase and incubate at 37°C for 30 min (go to step 10 of this section below).
5. During the reaction, mix a CHROMA SPIN-400 column (Clontech) by inverting it a couple of times, cut off the top and bottom lids, and then allow it to stand on a receptacle tube (a 1.5-mL microfuge tube with lid removed) at room temperature for 10 min, to drain out the extra TE solution from the column.
6. Add 1 mL of 1× STE to the top of the column, set the column and receptacle tube into a 15-mL plastic tube, and centrifuge for 3 min at 700×*g* (at 1,800 rpm using Beckman J6-HC centrifuge, in a swing-basket rotor 20 cm in diameter).
7. Discard the solution that was spun into the receptacle tube.
8. Centrifuge for 3 min at 700×*g* to remove the residual solution from the column completely.
9. Discard the small amount of solution that was spun into the receptacle tube.
10. Transfer the column to a fresh receptacle tube and load the sample (step 4 of this section) onto the top (at the center) of the column, drop by drop (~10 μL per drop). Never let the solution touch the inner wall of the column to avoid the solution from passing free of the resin packed into the column.
11. Set the column and receptacle tube into a 14-mL round-bottom plastic tube (used as a protector to hold the column tubes in a swing basket) and centrifuge for 3 min at 700×*g*.
12. Transfer the fractionated sample solution collected at the bottom of the receptacle tube (~50 μL) to a fresh 0.5-mL microfuge tube, add 50 μL of phenol/chloroform, and vortex for 5 s.
13. Centrifuge the tube in a microfuge for 1 min at 20,000×*g* to obtain a pellet.
14. Add 0.5-mL ice-cold 70% ethanol and centrifuge for 1 min at 20,000×*g*.

15. Remove the 70% ethanol and centrifuge for 1 min at 20,000×*g*.
16. Remove the residual 70% ethanol around the pellet.
17. The wet pellet (i.e., precipitated amplified mRNA) can be stored at −20°C until use.

3.6. Synthesis of cDNA Using Amplified mRNA for Construction of Single-Cell cDNA Library

1. Mix the following reagents in a sterile 0.5-mL microfuge tube:

5 mM Tris–HCl (pH 7.5)	3.5 μL
RNA or mRNA sample	1.0 μL (or pellet from step 14 in Subheading 3.5)
Chum-RNA (1.0 μg/μL)	5.0 μL (5.0 μg)

2. Incubate at 65°C for 5 min and then quickly immerse tube in ice-cold water.
3. Mix the following reagents:

10× First-strand buffer	2.5 μL
0.1 M DTT	2.5 μL
First-strand mixture	1.5 μL
Linker primer (1.6 μg/μL)	1.0 μL (1.6 μg)
RNase inhibitor	0.5 μL
H_2O (see Note 1)	6.5 μL (to 24 μL)

4. Repeat steps 4–6 in Subheading 3.2 to synthesize double-stranded cDNA using amplified mRNA.
5. The amplification procedure (steps 1–4 in Subheading 3.5) may be repeated four times to obtain a quantity of amplified mRNA that would be sufficient for the preparation of a cDNA library (Fig. 3).
6. The wet precipitated amplified mRNA can be stored at −20°C until use (see step 1 in Subheading 3.8).

3.7. Preparation of an Insertion-Ready Vector

1. In a sterile 0.5-mL microfuge tube (on ice), mix the following reagents:

Vector DNA (pAP3*neo*)	100 μg
10× *Not*I buffer	20 μL
H_2O (see Note 1)	To 200 μL

2. Add 50 U *Not*I, incubate the reaction for 2 h at 37°C, add 20 U of *Not*I, and incubate for 1 h at 37°C.
3. Add an equal volume (~210 μL) of phenol/chloroform and vortex for 5 s.

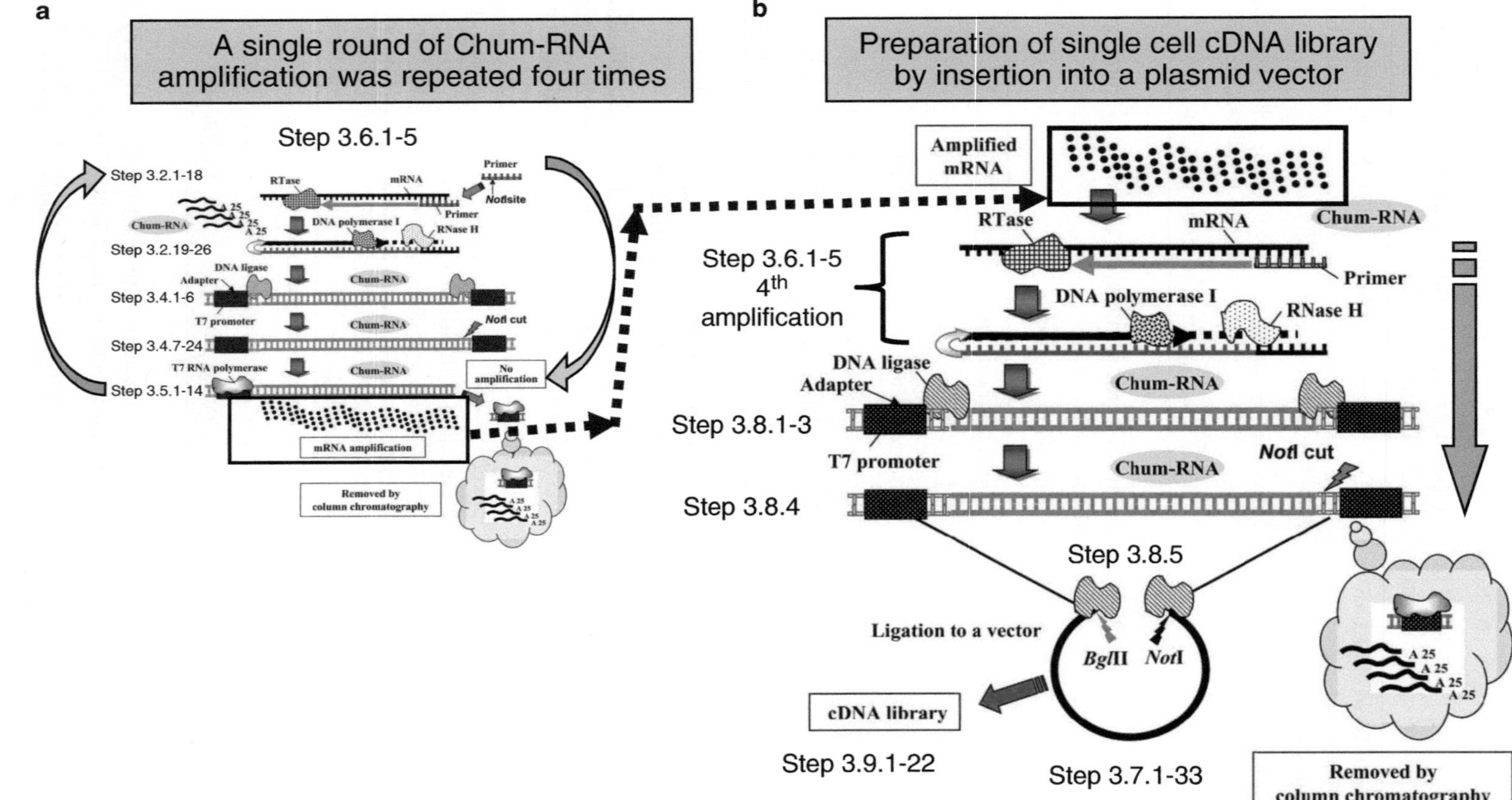

Fig. 3. Schematic representation of the procedure used for the preparation of a cDNA library from a single-cell-derived amount of mRNA using chum-RNA (Subheadings 3.6–3.9). **(a)** Illustration of the procedure used for a single round of chum-RNA amplification. **(b)** Illustration of the procedure used for the preparation of a single-cell cDNA library after four rounds of chum-RNA amplification.

4. Centrifuge the tube in a microfuge for 1 min at 20,000 × *g*.
5. Transfer the supernatant to a fresh 0.5-mL microfuge tube, add an equal volume (210 μL) of phenol/chloroform, and vortex for 5 s.
6. Centrifuge the tube in a microfuge for 1 min at 20,000 × *g*.
7. Transfer the supernatant to a fresh 1.5-mL microfuge tube, add 17 μL of 3 M sodium acetate (NaAc) and 420 μL of ice-cold ethanol, and then mix and chill in a dry ice/ethanol bath for 15 min.
8. Centrifuge the pellet in a microfuge for 10 min at 20,000 × *g*.
9. Add 500 μL of ice-cold 70% ethanol to wash the pellet, and centrifuge for 1 min at 20,000 × *g*.
10. Remove the 70% ethanol and centrifuge for 1 min at 20,000 × *g*.
11. Remove the residual 70% ethanol from around the pellet.
12. Add the following reagents to the wet pellet and mix by vortexing:

10× *Not*I buffer	20 μL
H_2O (see Note 1)	To 200 μL

13. Add 20 U of *Not*I and incubate for 1 h at 37°C.
14. Confirm the complete digestion of DNA by *Not*I using 1% agarose gel electrophoresis. A single band at 4.3 kb is expected.
15. Repeat steps 3–11 in Subheading 3.7.
16. Add the following reagents to the wet pellet and mix by vortexing:

10× BAP buffer	20 μL
H_2O (see Note 1)	To 200 μL

17. Add 1 U of BAP and incubate for 30 min at 65°C.
18. Repeat steps 3–11 in Subheading 3.7.
19. Add the following reagents to the wet pellet and mix by vortexing:

10× *Bgl*II Buffer	20 μL
H_2O (see Note 1)	To 200 μL

20. Add 100 U of *Bgl*II and incubate for 1 h at 37°C.
21. Add 10 μL 10% SDS and 20 μL 0.25 EDTA and mix by vortexing.

22. Add an equal volume (~240 μL) of phenol/chloroform and vortex for 5 s.
23. Centrifuge the tube in a microfuge for 1 min at 20,000 × *g*.
24. Transfer the supernatant to a fresh 0.5-mL microfuge tube, add an equal volume (~240 μL) of chloroform, and vortex for 5 s to remove the phenol from the sample.
25. Centrifuge the tube in a microfuge for 1 min at 20,000 × *g*.
26. Transfer the supernatant to a centrifuge filter (Ultrafree-C3 or Minicent-30) and centrifuge in a microfuge for 20 min (or until all solutions are spun to the receptacle tube) at 13,000 × *g*, which is the maximum speed that allows avoidance of filter fracture.
27. Add TE (100 μL) to the upper chamber of the filter cup and centrifuge for 20 min at 13,000 × *g*.
28. Repeat step 27 in Subheading 3.7.
29. Add 1/10 TE (90 μL) to the upper chamber of the filter cup and mix by pipetting three to five times using a 200-μL pipette tip.
30. Transfer the solution to a fresh 0.5-mL microfuge tube, add 10 μL of 10× *Bgl*II buffer and 10 U of *Bgl*II, and then incubate for 1 h at 37°C.
31. Add 10 μL of 10× STE to the reaction mixture and mix by vortexing.
32. Remove the small DNA fragments produced by the *Not*I–*Bgl*II digestion using a CHROMA SPIN-400 column as described in steps 10–24 in Subheading 3.4.
33. Add TE to yield 0.1 μg/μL of *Not*I–*Bgl*II-digested vector solution by measuring the optical density (OD) at 260 nm. Note that $OD_{260} = 1.0$ equals 50 μg of DNA.

3.8. Adapter Ligation and Insertion to a Cloning Vector

1. Mix the following reagents with the precipitate from step 6 in Subheading 3.6:

dsDNA (step 6 in Subheading 3.7)	Pellet
10× Ligase buffer	2.0 μL
10 mM rATP	2.0 μL
Library adapter (0.35 μg/μL) (see Note 9)	1.0 μL (0.35 μg)
H_2O (see Note 1)	13.5 μL (to 18.5 μL)

2. Add 1.5 μL (4 U) of T4 DNA ligase and incubate at 8°Ct overnight.
3. Repeat steps 4–24 in Subheading 3.4.

4. Add the following reagents to the wet precipitate:

10× Ligase buffer	3 μL
10 mM rATP	3 μL
*Not*I/*Bgl*II digested vector (step 33 in Subheading 3.7)	1 μL (100–300 ng)
H_2O (see Note 1)	22 μL (to 30 μL)

5. Add 1 μL (4 U) T4 DNA ligase and incubate at 12°C overnight.
6. The reaction product can be directly used in the next step (see step 1 in Subheading 3.9) or stored at −20°C until use.

3.9. Electroporation to Generate a Single-Cell cDNA Library

Dispense 2 mL of SOC into each of the five 14-mL polypropylene round-bottom tubes (Falcon #2059; Becton, Dickinson and Company, Franklin Lakes, NJ) and incubate at 37°C.

1. Warm the sample from step 6 in Subheading 3.8 at 70°C for 10 min.
2. Add TE (70 μL) and phenol/chloroform (100 μL), and mix by vortexing for 5 s.
3. Centrifuge in a microfuge at 20,000 × *g* for 1 min.
4. Transfer the supernatant to a fresh 0.5-mL microfuge tube, add 100 μL chloroform, and mix by vortexing for 5 s to remove the phenol.
5. Centrifuge in a microfuge at 20,000 × *g* for 1 min.
6. Transfer the supernatant (~100 μL) to a centrifuge filter (Ultrafree-C3 or Minicent-30) and centrifuge in a microfuge for 20 min at 13,000 × *g*, which is the maximum speed that allows avoidance of filter fracture (see Note 5).
7. Add TE (100 μL) to the upper chamber of the filter cup and centrifuge for 20 min at 13,000 × *g*.
8. Repeat step 7 in Subheading 3.9.
9. Add TE (20 μL) to dissolve the unfiltered DNA that remains in the upper chamber of the filter cup and mix by pipetting three to five times using a 200-μL pipette tip.
10. Invert the 1.5-mL tube over the filter cup and centrifuge at 13,000 × *g* for 10 s to collect the DNA-containing solution (20~25 μL) (see Note 10).
11. Transfer two or three tubes (see Note 10) of ElectroMAX DH12S cells (100 μL; GIBCO-BRL) from the deep freezer into ice-cold water, add 10 μL of sample (step 10 in Subheading 3.9) to each tube, gently mix by vortexing for 1 s, and keep the tubes in ice-cold water.

12. Transfer ~55 μL of sample into an electroporation cuvette (2-mL Gene Pulser cuvette, Bio-Rad, Hercules, CA) and immediately set it in the electroporation apparatus (Gene Pulser, Bio-Rad, or Glectro cell manipulator, BTX, Holliston, MA). Begin pulsing at 2.5 kV and 129 Ω.
13. Transfer the solution into 2 mL of pre-warmed SOC (see Subheading 2.5) using a sterile Pasteur pipette or a 10-μL pipette tip. Rinse the cuvette bottom with SOC to increase the recovery rate.
14. Shake vigorously in a rotary shaker (200–250 rpm) at 37°C for ~60 min.
15. Repeat steps 13–14 in Subheading 3.9 for the remainder of the sample solutions, one at a time (step 12 of this section).
16. Transfer the SOC of all tubes (step 15 of this section) into 100 mL L-broth placed in a 500-mL flask supplemented with 50 mg/L ampicillin, and mix. Take aliquots (4, 20, 100, and 500 μL) and plate onto LB-ampicillin plates after mixing with pre-warmed (50°C) LB top agar, incubate at 37°C overnight, and count the number of *E. coli* colonies. Eight colonies on a 4-μL plate indicate a cDNA library complexity (independent colony) of one million.
17. Shake the 500-mL flask in a rotary shaker (~200 rpm) at 37°C for several hours (or overnight), until OD_{600} reaches a value of 1.0–2.0.
18. Transfer 30 mL the *E. coli* culture into SOC, add 2.1 mL DMSO, mix, and dispense into a 1.5-mL stock tube for storage in liquid nitrogen (or in a deep freezer).
19. *E. coli* cells in DMSO can be stored for years in liquid nitrogen (−196°C), but 90% die within months in a deep freezer (−80°C).
20. Prepare plasmid DNA from the remainder of the 70 mL of the *E. coli* culture using the Qiagen plasmid DNA purification kit.
21. Digest the plasmid DNA with appropriate restriction enzymes (*Bam*HI for pAP3*neo* vector) to assess the size distribution of the cDNA inserts.
22. Add an equal volume of ethanol to the remaining plasmid DNA and store at −20°C. This can be stored for years in a deep freezer.

4. Notes

1. Unless otherwise stated, all solutions used in the biochemical reactions were prepared in water that was purified using the

following method. First, pre-filtered tap water was purified using the reverse osmosis (RO) method, which was further purified using an electric heating Distilled Water Apparatus (A4D, Sigma-Aldrich, Milwaukee, WI). This was further purified using a Barnstead's Nanopure system (Thermo Fisher Scientific Inc., Waltham, MA; this water has a resistivity of 18.2 MΩ cm and its total organic content is very low). This standard is referred to as "H_2O" in this text. For small-scale reactions, it is recommended to store H_2O in aliquots in 1.5-mL microfuge tubes at −20°C, which may be thawed immediately prior to use.

2. For preparation of agarose gel buffer (TAE) and *E. coli* propagation solutions, we skipped the process of "double-distilled water" preparation (see Note 1). This standard is referred to as "Nanopure water" in this text.

3. *ExTaq*™ DNA polymerase and *ExTaq*™ buffer can be replaced by other PCR kits available commercially.

4. The solutions used for biochemical reactions, including TE, should be stored frozen in aliquots (in 1.5-mL microfuge tubes) at −20°C, which may be thawed immediately prior to use.

5. Minicent-30 (molecular weight cut-off: 30,000 Da) can be purchased from Tosoh SMD (Cat. #08627), Grove City, OH.

6. Oligo$(dT)_{18-22}$ primers or random primers may not be useful for chum-RNA-mediated amplification, as the chum-RNA/primer hybrid lacks the protruding portion of oligonucleotide that is used as a scaffold for the enzyme.

7. Mix the following oligonucleotides in equal molar ratios to reach 0.35 μg/μL in the annealing buffer (10 mM Tris–HCl [pH 7.5], 1 mM EDTA, and 10 mM $MgCl_2$); 5′CAC TAG TAC GCG TAA TAC GAC TCA CTA TAG GGA ATT CCC CGG G-3′ (sense T7) and 5′-pCCC GGG GAA TTC CCT ATA GTG AGT CGT ATT ACG CGT ACT AGT GAG CT-3′ (antisense T7). Incubate at 65°C for 2 min, at 37°C for 10 min, and at room temperature (~18°C) for 5 min. This solution can be stored at −20°C.

8. Never use the bovine serum albumin (BSA) that accompanies the commercially available restriction enzymes, as it could be contaminated with traces of DNA fragments of *E. coli* or of the bovine genome.

9. Mix the following oligonucleotides in equal molar ratios to yield 0.35 μg/μL in the annealing buffer (10 mM Tris–HCl [pH 7.5], 1 mM EDTA, and 10 mM $MgCl_2$): *Bam*HI(*Bgl*II)-*Sma*I d(GAT CCC CGG G) and p*Sma*I linker d(pCCC GGG). Incubate at 65°C for 2 min, at 37°C for 10 min, and

at room temperature (~18°C) for 5 min. This solution can be stored at –20°C.

10. The collected sample volume may be more than 20 μL due to the residual TE solution on the wet filter. Thus, it is recommended to thaw three tubes of ElectroMAX DH12S cells at this step.

Acknowledgments

We are obliged to Dr. Daisuke Okuzaki for technical advice and Dr. Patrick Hughes for critical reading of the manuscript. This work was supported in part by a Grant-in-Aid for Scientific Research (S) from the Ministry of Education, Culture, Sports, Science and Technology of Japan, and a Grant-in-Aid from the Regional Research and Development Resources Utilization Program from the Japan Science and Technology Agency to HN.

References

1. Thompson, J.R., Marcelino, L.A. and Polz, M.F. (2002) Heteroduplexes in mixed-template amplifications: formation, consequence and elimination by 'reconditioning PCR'. *Nucleic Acids Res.* **30**, 2083–2088.
2. Van-Gelder, R.N., Von-Zastrow, M.E., Yool, A., Dement, W.C., Barchas, J.D. and Eberwine, J.H. (1990) Amplified RNA synthesized from limited quantities of heterogeneous cDNA. *Proc. Natl Acad. Sci. USA* **87**, 1663–1667.
3. Phillips, J. and Eberwine, J.H. (1996) Antisense RNA amplification: a linear amplification method for analyzing the mRNA population from single living cells. *Methods* **10**, 283–288.
4. Dafforn, A., Chen, P., Deng, G., Herrler, M., Iglehart, D., Koritala, S., Lato, S., Pillarisetty, S., Purohit, R., Wang, M. *et al.* (2004) Linear mRNA amplification from as little as 5 ng total RNA for global gene expression analysis. *Biotechniques* **37**, 854–857.
5. Moll, P.R., Duschl, J. and Richter, K. (2004) Optimized RNA amplification using T7-RNA-polymerase based *in vitro* transcription. *Anal. Biochem.* **334**, 164–174.
6. Kurimoto, K., Yabuta, Y., Ohinata, Y., Ono, Y., Uno, K.D., Yamada, R.G., Ueda, H.R. and Saitou, M. (2006) An improved single-cell cDNA amplification method for efficient high-density oligonucleotide microarray analysis. *Nucleic Acids Res.* **34**, e42.
7. Kurimoto, K., Yabuta, Y., Ohinata, Y. and Saitou, M. (2007) Global single-cell cDNA amplification to provide a template for representative high-density oligonucleotide microarray analysis. *Nat. Protoc.* **2**, 739–752.
8. Saitou, M., Yabuta, Y. and Kurimoto, K. (2008) Single-cell cDNA high-density oligonucleotide microarray analysis: detection of individual cell types and properties in complex biological processes. *Reprod. Biomed. Online* **16**, 26–40.
9. Eberwine, J., Yeh, H., Miyashiro, K., Cao, Y., Nair, S., Finnell, R., Zettel, M. and Coleman, P. (1992) Analysis of gene expression in single live neurons. *Proc. Natl Acad. Sci. USA* **89**, 3010–3014.
10. Tietjen, I., Rihel, J.M., Cao, Y., Koentges, G., Zakhary, L. and Dulac, C. (2003) Single-cell transcriptional analysis of neuronal progenitors. *Neuron* **38**, 161–175.
11. Osawa, M., Egawa, G., Mak, S.S., Moriyama, M., Freter, R., Yonetani, S., Beermann, F. and Nishikawa, S. (2005) Molecular characterization of melanocyte stem cells in their niche. *Development* **132**, 5589–5599.
12. Jensen, K.B. and Watt, F.M. (2006) Single-cell expression profiling of human epidermal

stem and transit-amplifying cells: Lrig1 is a regulator of stem cell quiescence. *Proc. Natl Acad. Sci. USA* **103**, 11958–11963.

13. Baugh, L.R., Hill, A.A., Brown, E.L. and Hunter, C.P. (2001) Quantitative analysis of mRNA amplification by *in vitro* transcription. *Nucleic Acids Res.* **29**, e29.
14. Iscove, N.N., Barbara, M., Gu, M., Gibson, M., Modi, C. and Winegarden, N. (2002) Representation is faithfully preserved in global cDNA amplified exponentially from sub-picogram quantities of mRNA. *Nat. Biotechnol.* **20**, 940–943.
15. Tougan, T., Okuzaki, D. and Nojima, H. (2008) Chum-RNA allows preparation of a high-quality cDNA library from a single-cell quantity of mRNA without PCR amplification. *Nucleic Acids Res.* **36**, e92.

Chapter 3

Construction of a Full-Length cDNA Library from Castor Endosperm for High-Throughput Functional Screening

Chaofu Lu, James G. Wallis, and John Browse

Abstract

It is desirable to produce high homogeneity of novel fatty acids in oilseeds through genetic engineering to meet increasing demands by the oleo-chemical industry. However, expression of key enzymes for biosynthesis of industrial fatty acids usually results in low levels of desired fatty acids in transgenic oilseeds. The abundance of unusual fatty acids in their natural species suggests that additional genes are needed for high production in transgenic plants. We used the model oilseed plant *Arabidopsis thaliana* expressing a castor fatty acid hydroxylase (FAH12) to identify genes that can boost hydroxy fatty acid accumulation in transgenic seeds. We described previously a high-throughput approach that in principle can allow testing of the entire transcriptome of developing castor seed endosperm by shotgun transforming a full-length cDNA library into a FAH12-expressing *Arabidopsis* line. The resulting transgenic seeds can be screened by high-throughput gas chromatography. The most critical step of the approach is the construction of a full-length cDNA library. In this chapter, we describe in detail the construction of the cloning vectors and a full-length cDNA library from developing castor seed endosperms. The approach we describe has broad applicability in many areas of biology.

Key words: High-throughput, Full-length cDNA library, Cap-trapping, Gateway® compatible vectors

1. Introduction

Plant oils in most oilseed crops are comprised primarily of only a few fatty acids, such as palmitic, stearic, oleic, linoleic, and linolenic acids. However, certain plant species produce fatty acids with special chemical structures such as alterations in fatty acyl chain length, double-bond positions, and oxygenated functional groups (1). Physical or chemical properties make some of these unusual fatty acids valuable for their applications in numerous industrial products. In recent years, considerable efforts have been made to genetically engineer oilseed crops for production of such unusual

Chaofu Lu et al. (eds.), *cDNA Libraries: Methods and Applications*, Methods in Molecular Biology, vol. 729,
DOI 10.1007/978-1-61779-065-2_3, © Springer Science+Business Media, LLC 2011

fatty acids, and thus provide a low-cost feedstock resource for the oleo-chemical industry (2–5). However, the results of these experiments have typically been disappointing, producing, in most cases, plant lines with very low yields of the desired fatty acids (2–4, 6). For example, ricinoleic acid (12-hydroxyoctadec-*cis*-9-enoic acid; 18:1-OH) biosynthesis in castor (*Ricinus communis* L.) is catalyzed by the oleate Δ12-hydroxylase (FAH12) (7). Heterologous expression of FAH12 in the model oilseed plant *Arabidopsis* produced only up to 17% hydroxy fatty acids in seed oils. This level is much lower than that found in castor oil, where ricinoleate constitutes ~90% of the total fatty acids. Other efforts to produce unusual fatty acids, such as acetylenic, monoenoic, eleostearic, and parinaric acids, have likewise found that accumulation of unusual fatty acids in transgenic seeds is far below the proportion found in their respective natural sources (3, 4, 6). These results show that expressing the single catalytic enzymes required for unusual fatty acid biosynthesis is insufficient to create transgenic plants producing large amounts of these fatty acids in seed storage oil. Since the unusual fatty acids occur in abundance in their natural sources, we believe that additional necessary components for increased accumulation of unusual fatty acids in transgenic plants can be obtained from the source species. Although we describe our approach for a specific application, the methods we have developed are applicable to many questions on biotechnology and functional genomics research.

We described a high-throughput approach to screen genes from castor that may boost hydroxy fatty acids accumulation in seed oils of transgenic *Arabidopsis* (8). The procedure is comprised of five major steps: (a) construction of a full-length cDNA library in a high-throughput λ phage vector from developing endosperm of castor seeds, and introduction of the cDNA library into a high-throughput plant expression binary vector, (b) transformation of the full-length cDNA library into the FAH12 transgenic line, and production of seeds from individual transgenic plants, (c) high-throughput screening of fatty acid composition of transgenic seeds by gas chromatography, and (d) confirmation of putative lines with increased hydroxy fatty acids accumulation by analyzing fatty acids composition in the next generation and retransformation of the identified cDNA. The most important step is the use of a full-length cDNA library representing the entire transcriptome from castor endosperm. This was achieved by synthesizing full-length first-strand cDNAs at a high temperature (55°C) using the SuperScript III reverse transcriptase (Invitrogen). The first-strand cDNAs were then recovered by a biotinylated cap-trapping approach (9), and the second-strand cDNA was synthesized by the single-strand linker ligation method (10). To facilitate shotgun cloning and avoid potential bias against long cDNAs, we created a λ_{GW} vector which was developed by incorporating the *attB* sites of the Gateway® cloning system (www.Invitrogen.com/gateway) into the λ_{ZAPII} vector acquired from Stratagene. In this chapter, we describe in detail the construction of the cloning vectors and a full-length cDNA library from developing castor seed endosperms.

2. Materials

2.1. Vector Construction

1. Vector λ_{ZAPII} (Stratagene).
2. Restriction enzymes *Sst*I and *Xho*I (New England Biolabs).
3. pDEST-C (Invitrogen).
4. Pfu Turbo DNA polymerase (Stratagene).
5. Four primers:

   ccdBfor1

   *GACAAGTTGTACAAAAAAGCAGGCT*GAGCTCAGTATGCGTATTTGCGCGCTG

   ccdBfor2

   AGCT*GACAAGTTTGTACAAAAAAGCAGGCT*GAGCTC

   ccdBrev1

   *CACCACTTTGTACAAGAAAGCTGGGT*CTCGAGTACGCTAGTGTCATAGTCCTG

   ccdBrev2

   TCGAC*ACCACTTTGTACAAGAAAGCTGGGT*CTCGAG
6. T4 DNA ligase (New England Biolabs).
7. λ-Phage packaging reaction (MaxPlax lambda packaging extract; Epicentre).
8. ccdB-permissive *Escherichia coli* strain C600 (Invitrogen).

2.2. Preparation of Total RNA

1. RNase-free water (see Note 1).
2. Liquid nitrogen.
3. Chloroform:isoamyl alcohol (IAA) (24:1, v:v).
4. Total RNA extraction kit, e.g., Midi RNAqueous Kit supplied with LiCl solution (Ambion, Austin, TX).
5. Plant RNA Isolation Aid (Ambion) if using the Ambion kit.
6. mRNA purification kit, e.g., the oligotex kit (Qiagen, Valencia, CA, USA).

2.3. cDNA Synthesis

1. SuperScript III reverse transcriptase supplied with First-Strand Buffer (Invitrogen).
2. RNase inhibitor, e.g., RNase OUT Recombinant RNase Inhibitor (Invitrogen, 40 U/μL).

3. First-strand primer (HPLC purified) containing an *Xho*I site (underlined): 5′-GAGAGAGAGAGAGAGAGAGGATCCA <u>CTCGAG</u>TTTTTTTTTTTTTTTTTVN-3′. Primers are resuspended in water at a high concentration (12 μM).
4. dNTP mix: 10 mM each dATP, dGTP, dTTP, and 5-methyl-dCTP at neutral pH.
5. Proteinase K: Dissolve in water at 10 μg/μL and store in small aliquots at –20°C.
6. DNase- and RNase-free glycogen.
7. EDTA: 0.5 M, pH 8.0.
8. Tris–HCl: 1 M, pH 7.5.
9. TE buffer: Use 1 mL 1 M Tris–HCl (pH 8.0) and 0.2 mL EDTA (0.5 M), and make up with double-distilled water up to 100 mL.
10. Sodium dodecyl sulfate (SDS): 10% in water.
11. Ammonium acetate: 10 M.
12. Phenol: pH 4.0, water equilibrated (not older than a few weeks).
13. Chloroform.
14. Absolute ethanol.
15. Biotin hydrazide long arm (MW 371.51) (Invitrogen).
16. Sodium acetate buffer: 1 M, pH 4.5.
17. Sodium acetate buffer: 1 M, pH 6.1.
18. $NaIO_4$ (MW213.9): 250 mM.
19. NaCl: 5 M.
20. Isopropanol.
21. DNA-free tRNA: 10–50 and 50 μg/mL.
22. Streptavidin-coated MPG beads (Dynabeads® MyOne™ Streptavidin C1: Invitrogen).
23. Beads washing/binding solution I: 2 M NaCl, 50 mM EDTA (pH 8.0).
24. Beads washing solution II: 10 mM Tris–HCl (pH 7.5), 0.2 mM EDTA, 40 μg/mL tRNA, 10 mM NaCl, and 20% glycerol.
25. cDNA Release Buffer: 50 mM NaOH and 5 mM EDTA.
26. RNase I (10 U/μL) (Promega, Madison, WI) with buffer.
27. DNA fractionation column (e.g., CL-4B).
28. GN5-A nucleotides with *Sst*I and *Sal*I sites: 100 μM (see Note 2).

   5′-GAGAGAGAGAGCAC<u>GAGCTCGTCGAC</u>TAGTG ACACTATAGAACCAGNNNNN-3′

29. GN5-B nucleotides: 100 μM.

    5′-TGGTTCTATAGTGTCACTAGTCGACGAGCTC GTGCTCTCTCTCTC-3′

30. N6-C nucleotides with *Sac*I and *Sal*I sites: 100 μM.

    5′-GAGAGAGAGAGCACGAGCTCGTCGACTAGT GACACTATAGAACCANNNNNN-3′

31. DNA Ligation Kit Ver. 2.1 (Takara) with solution I and solution II.
32. ExTaq DNA polymerase (Takara).
33. *Sst*I (10 U/μL).
34. *Xho*I (10 U/μL).
35. ^{32}P-dGTP (see Note 3).

2.4. Construction of a cDNA Library

1. MaxPlax lambda packaging extract (Epicentre, Madison, WI).
2. SM buffer: 100 mM sodium chloride, 10 mM magnesium sulfate (heptahydrate), 50 mM Tris–HCl, pH 7.5, 0.01% (w/v) gelatin, sterile solution.

3. Methods

3.1. Vector Construction

1. Phage DNA digestion

   Combine in a microfuge tube:

   1 μg of vector λ_{ZAPII}

   0.1 μL 100× BSA (Supplied with the enzymes)

   2 μL NEB buffer 4 (Supplied with the enzymes)

   Nuclease-free water to a final volume of 18 μL

   1 μL *Sst*I

   1 μL *Xho*I

   Mix thoroughly but gently

   Incubate at 37°C for 2 h

   Heat inactivate by incubating at 65°C for 20 min and store at 4°C

2. Sticky-end DNA amplification

   (a) Dilute the primers

   Using the details of synthesis provided by the primer manufacturer, dissolve the primers in nuclease-free water to a final concentration of 10 μM.

(b) Prepare the amplification master mix

Combine in a 250-μL microfuge tube kept on ice:

0.1-μg pDEST-C vector

2-μL Pfu Turbo enzyme

10-μL 10× Pfu Turbo buffer

0.8 μL of the 25 mM of each dNTP mix supplied with the enzyme

Nuclease-free water to a final volume of 96 μL

(c) Prepare the final amplification reactions
Divide the prepared reaction mix into two 48-μL aliquots in 250-μL thermocycler tubes kept on ice.

Add to one tube

1 μL 10 μM ccdBfor1

1 μL 10 μM ccdBrev1

Add to the second tube

1 μL 10 μM ccdBfor2

1 μL 10 μM ccdBrev2

(d) Amplify using the following thermocycler program

Denature at 94°C for 2 min

Denature at 94°C for 20 s

Anneal at (Tm−5°C) for 20 s

Extend at 72°C for 90 s

Repeat 29 times

Chill to 4°C

3. Sticky-end ligation

(a) Prepare insert
Combine 20 μL of each of the two amplification reactions in a single tube and mix briefly. Heat the combined sample to 94°C for 4 min in the thermocycler, and then allow it to cool slowly at room temperature for 15 min. One-fourth of the resulting DNA molecules have the correct extensions at each end (11).

(b) Ligation reaction (see Note 4)

Combine in a microfuge tube

5 μL of the vector prepared by restriction in step 1

5 μL of the sticky-end insert prepared in step 2

1-μL NEB ligase buffer (includes rATP)

1-μL NEB T4 DNA ligase

Nuclease-free water to a final volume of 20 μL

Incubate at 16°C for 2–18 h

4. Phage packaging and analysis
   (a) Heat 5 μL of the ligation mix to 65°C for 5 min to denature the ligase, then cool on ice.
   (b) To the ligation mix (5 μL), add 25 μL MaxPlax lambda packaging extract (Epicentre).
   (c) Incubate at 30°C for 90 min.
   (d) Add 500-μL SM buffer and 25-μL chloroform, mix. Store at 4°C.
   (e) Infect the C600 ccdB-permissive *E. coli* strain with a small sample of the packaged phage.
   (f) Test the resulting plaques for insertion of the ccdB cassette by amplification with ccdBfor1 and ccdBrev2 using the same polymerase chain reaction (PCR) thermocycling program described in the synthesis steps. Correct amplification products will be the size of the ccdB cassette, 1.7 kb.

3.2. Total RNA Extraction

The following protocol is for ~1.0 g of tissue (fresh or frozen castor endosperm).

1. Grind tissue into fine powder in liquid nitrogen.
2. Homogenize in 8 mL of lysis-binding buffer of the RNAqueous kit and 1 mL Plant RNA Isolation Aid (Ambion) in a glass Dounce tissue grinder.
3. Add 9 mL chloroform/IAA (24:1), vortex 1–2 min (see Note 5).
4. Centrifuge at 10,000 × *g* for 10 min.
5. Transfer the upper aqueous phase into a new tube by pipetting, taking care to avoid the precipitated material at the layer between the two phases.
6. (Optional) Back extract with 2 mL lysis-binding buffer.
7. Do RNA extraction and precipitate total RNA using LiCl following instructions of the Midi RNAqueous Kit (Ambion) (see Note 6).
8. Measure the RNA purity by spectrophotometry. The desired OD ratios are 230/260 < 0.5; 260/280 > 1.8 or ~2.0.
9. Prepare mRNA using oligotex kit (Qiagen).
10. Measure the mRNA purity and estimate yield.
11. Store mRNA at −20 to −80°C.

3.3. First-Strand cDNA Synthesis

Full-length first-strand cDNAs were precipitated and captured based on protocols described by Carninci and Hayashizaki (12).

1. In each of 10× microcentrifuge tubes (200 μL if using thermal cycler), assemble the following:

First-strand primer (12 μM)	1 μL
mRNA	500 ng–1.0 μg
dNTP Mix	1 μL

2. Heat mixture to 65°C for 5 min and incubate on ice for at least 1 min.
3. Collect the contents of the tube by brief centrifugation and in each tube add:

   4 μL 5× First-strand buffer (Invitrogen)

   1 μL 0.1 M DTT

   1 μL RNaseOUT Recombinant RNase inhibitor (Invitrogen, 40 U/μL)

   1 μL of SuperScript III reverse transcriptase (200 U/μL)
4. Mix by pipetting gently up and down.
5. Incubate at 55°C for 60 min. This step and the next step can be done using a thermal cycler. Avoid contamination of radioactive materials and use proper protection.
6. Inactivate the reaction by heating at 70°C for 15 min.
7. Take 0.5 μL from radiolabeled tube and run electrophoresis on 0.8% agarose gel along with a 1-kb DNA ladder. Evaluate the first-strand cDNA by phosphorimager or radioautography. A satisfactory result is shown in Fig. 1, where the length of the majority of cDNA should be in the 2–3 kb range.
8. Store the cDNA synthesis reaction at –20°C, or proceed to the following steps (see Note 7).

3.4. Organic Phase Extraction and cDNA Precipitation

1. Transfer all cDNA reactions (~200 μL) into a 1.5-mL microcentrifuge tube.
2. Add:

Water	37.5 μL (to final 250 μL)
EDTA (0.5 M) to final 10 mM	5 μL
SDS (10%) to a final 0.2%	5 μL
Proteinase K to a final 100 ng/μL	2.5 μL

3. Incubate at 45°C for >15 min.
4. Add, in a final 400 μL:

Water	30 μL
Ammonium acetate (10 M)	120 μL

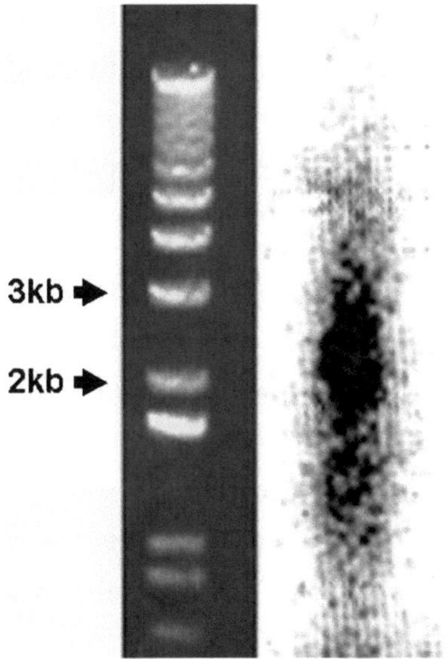

Fig. 1. Full-length, first-strand cDNA synthesis by reverse transcriptase reaction. Lane 1, molecular weight markers; lane 2, (α-^{32}P)dATP radiolabeled cDNA.

5. Perform phenol–chloroform and chloroform extraction:
   (a) Add 1.0 volume (400 μL) phenol/chloroform (1:1).
   (b) Vortex moderately until phases mix.
   (c) Leave on ice for 1–2 min.
   (d) Centrifuge for 2 min at 10,000 × *g*.
   (e) Carefully remove upper phase to a fresh 1.5-mL tube. Keep the tube for back extraction.
   (f) Mix cDNA with 200-μL chloroform, vortex gently.
   (g) Centrifuge for 2 min at 10,000 × *g*.
   (h) Transfer the aqueous phase to a new tube. Keep the tube with chloroform for the back extraction.
6. Perform back extraction using chloroform:
   (a) Add 100-μL water to the phenol–chloroform tube from point (e) of step 5, vortex gently.
   (b) Centrifuge for 2 min at 10,000 × *g*.
   (c) Transfer upper phase to the chloroform tube from point (h) of step 5 above, discard organic phase.
   (d) Vortex and centrifuge for 2 min at 10,000 × *g*.
   (e) Transfer the aqueous phase to the previously extracted cDNA (point (h) of step 5 above).
   (f) Split the samples in two 1.5-mL Eppendorf tubes (~250 μL each).

7. Precipitate cDNA with 2.5 volumes absolute ethanol:
   (a) In each tube, add 625-μL ethanol.
   (b) Incubate at −80°C for >20 min.
8. Prepare 10 mM biotin hydrazide long arm (MW 371.51) in water (see Note 8).

3.5. Capture Full-Length cDNA by Cap-Trapping

3.5.1. Biotinylation of cDNA Samples

1. Centrifuge at 10,000 × *g* for 15 min to precipitate cDNA.
2. Wash two times with 800 μL 80% ethanol.
3. Centrifuge for 2–3 min at 15,000 × *g*.
4. Combine two tubes and resuspend in 47-μL water.
5. Add 3.3-μL sodium acetate buffer (1 M, pH 4.5).
6. Add 1 μL *freshly prepared* $NaIO_4$ to a final 5 mM.
7. Wrap the tube with aluminum foil, incubate on ice for 45 min.
8. Add the following and incubate on ice for 45 min (or −20 to −80°C for 30 min) in the dark:

SDS (10%)	0.5 μL
NaCl (5 M)	11 μL
Isopropanol	61 μL

9. Centrifuge for 10 min at 10,000 × *g*.
10. Rinse two times with 70% ethanol, centrifuge for 2 min at 10,000 × *g*.
11. Resuspend cDNA pellet in 50-μL water.
12. To the cDNA add the following and incubate overnight (10–16 h) at room temperature:

Sodium acetate buffer, pH 6.1 (1 M)	5 μL
SDS (10%)	5 μL
Biotin hydrazide long arm from step 8 of Subheading 3.4	150 μL

13. To precipitate cDNA, add:

Sodium acetate, pH 6.1 (1 M)	75 μL
NaCl (5 M)	5 μL
Absolute ethanol	750 μL

14. Incubate for 1 h on ice or −80°C for 30 min.
15. Centrifuge for 10 min at 10,000 × *g*.
16. Wash once with 70% ethanol, once with 80% ethanol.
17. Redissolve sample in 70 μL of 0.1× TE.

18. To the cDNA sample add:

RNase I buffer (Promega)	20 μL
RNase I (10 U/μL, Promega)	20 μL
Water	90 μL

19. Incubate at 37°C for 15 min.
20. To stop reaction, put on ice and add 100 μg of tRNA and 100 μL of 5 M NaCl.

3.5.2. Capture of Full-Length cDNA

1. Add 100 μg of DNA-free tRNA to 500-μL streptavidin-coated MPG beads.
2. Incubate on ice for 30 min with occasional mixing.
3. Separate beads with a magnetic stand (for 3 min) and remove supernatant.
4. Wash three times (gently) with 500-μL washing/binding solution.
5. Resuspend the beads in 400 μL of the washing/binding solution.
6. Transfer beads into cDNA sample, after mixing, gently rotate the tube for 30 min at room temperature.
7. Separate beads on a magnetic stand. Discard the supernatant.
8. Wash beads twice with washing/binding solution I.
9. Wash once with 0.4% SDS, 50-μg/mL tRNA.
10. Wash once with beads washing solution II.
11. Wash once with 50-μg/mL tRNA in water.
12. To the beads, add 50-μL cDNA Release Buffer, briefly stir, and incubate 10 min at room temperature with occasional mixing.
13. Separate the beads, transfer cDNA into a tube on ice containing 50 μL of 1 M Tris–HCl, pH 7.5.
14. Repeat twice of steps 12–13, pool the eluted solutions.
15. To the cDNA elute, add 4 μL 10% SDS, 4 μL 0.5 M EDTA, 2-μL proteinase K (10 mg/mL). Incubate at 45°C for 15 min.
16. Add 200-μL phenol/chloroform/IAA (pH 7.9), vortex, and put on ice for 1–2 min.
17. Centrifuge for 2 min and transfer upper phase (80–90%) to a new tube, back extract the rest, and pool the extractions.
18. Add NaCl to a final 0.2 M, and 3 μg of glycogen.
19. Precipitate cDNA with one volume of isopropanol, wash with 70% ethanol.
20. Resuspend pellet in 50 μL of 0.1× TE.

3.5.3. Fractionation of cDNA

1. Perform cDNA fractionation following the column (e.g., CL-4B) manufacturer's instructions.
2. Precipitate the first four fractions by adding 0.2 M NaCl, 1-μg tRNA, and one volume of isopropanol.
3. Wash pellet with 70% ethanol.
4. Redissolve cDNA in 5-μL H_2O.

3.6. Synthesis of Second-Strand cDNA

The second-strand cDNA may be synthesized by a single-strand linker ligation method (10).

1. Prepare the GN5 and N6 linkers: In tubes (1) and (2), add the following:

(1) GN5 linker

GN5-A	5 μL
GN5-B	5 μL
0.5 M NaCl	4 μL (final 100 mM)
H_2O	6 μL

(2) N6 linker

N6-C	5 μL
GN5-B	5 μL
0.5 M NaCl	4 μL
H_2O	6 μL

2. Incubate at

65°C for 5 min

45°C for 5 min

37°C for 10 min

25°C for 10 min

3. Store the linkers at –20°C.
4. Linker-cDNA ligation: Ligate cap-trapper cDNA with mixed linkers (molar ratio N6:GN5 = 1:4) using DNA Ligation Kit Ver. 2 (Takara).

cDNA	4.8 μL
GN5 linker	2.16 μL
N6 linker	0.54 μL
Solution II	7.5 μL
Solution I	15 μL

5. Incubate at 15°C overnight.
6. Next morning, heat samples at 70°C for 2 min to inactivate ligase.
7. Perform proteinase K–phenol/chloroform extraction.
8. Precipitate and wash pellet, dissolve in 45-μL H_2O.
9. Add:

10× ExTaq buffer	6 μL
2.5 mM dNTPs	6 μL
H_2O	15 μL
ExTaq	3 μL

10. Incubate at 65°C for 5 min, 68°C for 30 min, and 72°C for 10 min.
11. Incubate sample with 0.2 mg/mL proteinase K in 10 mM EDTA/0.2% SDS at 45°C for 15 min.
12. Perform phenol/chloroform extraction and back extraction.
13. Precipitate cDNA with 0.2 M NaCl, 1-μg glycogen and one volume isopropanol.
14. Wash pellet with 70% ethanol.
15. Dissolve cDNA in 40 μL 0.1× TE (see Note 9).

3.7. Cleaving cDNA

1. To the cDNA sample, add:

10× ReAct2 buffer (Invitrogen)	5 μL
*Sst*I (10 U)	1 μL
*Xho*I (10 U)	0.5 μL
H_2O	To 50 μL

2. Incubate at 37°C for ~1 h.
3. Perform proteinase K–phenol/chloroform/IAA extraction, back extraction.
4. Fractionation of cDNA by CL-4B spun column.
5. Pool the first three fractions.
6. Precipitate by 0.2 M NaCl and 2.5 volumes ethanol.
7. Wash pellet with 70% ethanol.
8. Dissolve in 7-μL H_2O.

3.8. Ligation and Library Packaging

1. Ligate 1 μL λ_{GW} (100 ng) vector with 1–3 μL (10–20 ng) of cDNA in 5 μL.
2. Incubate at 15°C overnight.
3. Heat at 65°C for 5 min, cool on ice.

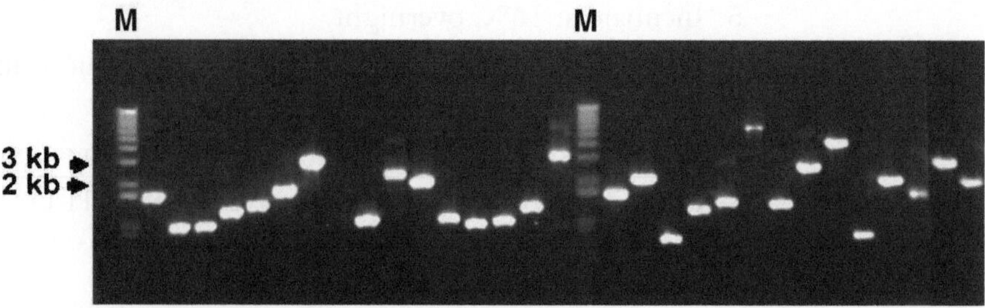

Fig. 2. Evaluation of a castor full-length cDNA library. Size distribution of cDNA inserts of randomly picked clones from the cDNA library was examined by plaque PCR using T3 and T7 primers. M, molecular weight markers.

4. To the ligation mix (5 µL), add 25-µL MaxPlax lambda packaging extract (Epicentre).
5. Incubate at 30°C for ≥90 min.
6. Add 500-µL SM buffer and 25-µL chloroform, mix.
7. Store at 4°C.

To evaluate the quality of this cDNA library, one may perform plaque PCR and sequencing of the cDNA inserts. As shown in Fig. 2, the insert size of the cDNA ranged from ~700 bp to over 6 kb, and the majority of the clones have inserts of 2–3 kb. We sequenced 140 inserts from the 5′ ends. When compared to sequences by BLAST, the results revealed that ~93% of the cDNA encoded an open reading frame that included the translational start codon, ATG. Since an oligo (dT) primer corresponding to the poly A signal sequence of the mRNA was used to synthesize the first-strand cDNA, the 3′ ends of the cDNA clones should include intact carboxyl termini of proteins. Therefore, this cDNA library has ~93% clones that encode full-length proteins from developing castor endosperms. Furthermore, by PCR analysis, we also determined that many important genes involved in lipid biosynthesis are present in this cDNA library, such as those genes for long-chain acyl synthetases, lysophosphatidic acyltransferases, diacylglycerol acyltransferases (DGAT1 and DGAT2), and phospholipid:diacylglycerol acyltransferases (data not shown).

4. Notes

1. To prepare RNase-free water, autoclaved double-distilled water or water of higher quality that has a resistivity of 18.2 MΩ cm and total organic content of less than five parts per billion, without DEPC treatment, is usually acceptable.

2. Restriction sites (*Sst*I, *Sal*I, and *Xho*I) are incorporated into the oligos to facilitate subsequent cloning by methylation-sensitive restriction cleaving of cDNA. These enzymes do not cut the hemimethylated sites of cDNA that is synthesized using ^{5m}C instead of C in the dNTP.
3. Radioactive ^{32}P-dGTP may be used to monitor the efficiency of full-length first-strand cDNA synthesis (Subheading 3.3). One may choose not to include the radiolabeled tube after step 1, Subheading 3.4. However, using radiolabeled cDNA samples throughout the procedures will help in monitoring recovery of full-length cDNA sample, which is usually very small amount (~10–100 ng) after being through steps before Subheading 3.5.2.
4. Ligation success can be improved at least five-fold by first combining the separate DNA preparations and co-precipitating them with 95% ethanol and 0.3 M sodium acetate.
5. Other commercial kits for RNA extraction may also work. This step is to get rid of oils in tissues such as castor endosperm in this chapter.
6. LiCl precipitation offers major advantages over other RNA precipitation methods in that it does not efficiently precipitate DNA, protein, or carbohydrate.
7. Any first-strand reaction mixture that is not used right away should be placed at –20°C. First-strand cDNA can be stored at –20°C for up to 1 month.
8. Always prepare fresh solution of biotin hydrazide. Long time and extensive mixing is needed for complete solubilization.
9. This amplified cDNA can be stored at –20°C for up to 3 months.

References

1. van de Loo, F. J., Fox, B. G., and Somerville, C. (1993) Unusual fatty acids, in *Lipid metabolism in plants* (Moore, T. S. J., Ed.), pp. 91–126, CRC Press, Boca Raton.
2. Broun, P. and Somerville, C. (1997) Accumulation of ricinoleic, lesquerolic, and densipolic acids in seeds of transgenic *Arabidopsis* plants that express a fatty acyl hydroxylase cDNA from castor bean. *Plant Physiol.* **113**, 933–942.
3. Lee, M., Lenman, M., Banas, A., Bafor, M., Singh, S., Schweizer, M., Nilsson, R., Liljenberg, C., Dahlqvist, A., Gummeson, P. O., Sjodahl, S., Green, A., and Stymne, S. (1998) Identification of non-heme diiron proteins that catalyze triple bond and epoxy group formation. *Science* **280**, 915–918.
4. Suh, M. C., Schultz, D. J., and Ohlrogge, J. B. (2002) What limits production of unusual monoenoic fatty acids in transgenic plants? *Planta* **215**, 584–595.
5. Smith, M. A., Moon, H., Chowrira, G., and Kunst, L. (2003) Heterologous expression of a fatty acid hydroxylase gene in developing seeds of *Arabidopsis thaliana*. *Planta* **217**, 507–516.
6. Cahoon, E. B., Carlson, T. J., Ripp, K. G., Schweiger, B. J., Cook, G. A., Hall, S. E., and Kinney, A. J. (1999) Biosynthetic origin of conjugated double bonds: Production of fatty acid components of high-value drying oils in transgenic soybean embryos. *Proc. Natl Acad. Sci. USA* **96**, 12935–12940.

7. van de Loo, F. J., Broun, P., Turner, S., and Somerville, C. (1995) An oleate 12-hydroxylase from *Ricinus communis* L is a fatty acyl desaturase homolog. *Proc. Natl Acad. Sci. USA* **92**, 6743–6747.
8. Lu, C., Fulda, M., Wallis, J. G., and Browse, J. (2006) A high-throughput screen for genes from castor that boost hydroxy fatty acid accumulation in seed oils of transgenic *Arabidopsis. Plant J.* **45**, 847–856.
9. Carninci, P., Shibata, Y., Hayatsu, N., Sugahara, Y., Shibata, K., Itoh, M., Konno, H., Okazaki, Y., Muramatsu, M., and Hayashizaki, Y. (2000) Normalization and subtraction of cap-trapper-selected cDNAs to prepare full-length cDNA libraries for rapid discovery of new genes. *Genome Res.* **10**, 1617–1630.
10. Shibata, Y., Carninci, P., Watahiki, A., Shiraki, T., Konno, H., Muramatsu, M., and Hayashizaki, Y. (2001) Cloning full-length, cap-trapper-selected cDNAs by using the single-strand linker ligation method. *Biotechniques* **30**, 1250–1254.
11. Zeng, G. (1998) Sticky-end PCR: new method for subcloning. *Biotechniques* **25**(2), 206–208.
12. Carninci, P. and Hayashizaki, Y. (1999) High efficiency full-length cDNA cloning. *Methods Enzymol.* **303**, 19–44.

Chapter 4

Full-Length Transcriptome Analysis Using a Bias-Free cDNA Library Prepared with the Vector-Capping Method

Seishi Kato, Mio Oshikawa, and Kuniyo Ohtoko

Abstract

Full-length complementary DNAs (cDNAs) are an essential resource for functional genomics. Recently, we have developed a simple and efficient method for preparing a full-length cDNA library from a small amount of total RNA, named the "vector-capping" method. The biggest advantage of this method is that the intactness of the cDNA can be assured by the presence of dG at the 5′ end of the full-length cDNA. Furthermore, the cDNA library represents the mRNA population in the cell owing to a bias-free procedure. In this chapter, we describe not only the protocol for preparing the library but also the points for analyzing the 5′-end sequence of the obtained cDNA.

Key words: Transcriptome, Full-length cDNA, Vector primer, Vector-capping method, Expression profile, Transcriptional start site, Splicing isoform, Long-sized cDNA

1. Introduction

The genome encodes information on an entire set of genes and on the regulation of expression of these genes. The information of the gene is once transcribed to mRNA, and then mRNA is translated to protein on the ribosome. Thus, the analysis of all mRNA molecules expressed in the given cell, so-called transcriptome, enables us to obtain information on the primary structure of all proteins constituting the cell and on the transcriptional start site of each gene on the genome. The discovery of reverse transcriptase enabled us to convert an unstable mRNA molecule into a stable complementary DNA (cDNA) molecule. Expressed sequence tag (EST) analyses using these cDNAs have produced a huge amount of information on gene expression profiles. In these analyses, a full-length cDNA clone containing an entire region from the cap site to the poly(A) tail is indispensable to obtain

Chaofu Lu et al. (eds.), *cDNA Libraries: Methods and Applications*, Methods in Molecular Biology, vol. 729,
DOI 10.1007/978-1-61779-065-2_4, © Springer Science+Business Media, LLC 2011

information on the primary structure of the protein encoded by the gene and on the transcriptional start site of the gene.

Okayama and Berg (1, 2) established the method for synthesizing the full-length cDNA by which a lot of genes were cloned as a form of full-length cDNA. We also developed a chimeric oligo-capping method by combining the oligo-capping method (3) and the Okayama–Berg method, and proposed the construction of a human full-length cDNA bank (4). Subsequently, the oligo-capping method (5) and the Cap-trapper method (6) were developed for synthesizing the full-length cDNA, and have been applied to large-scale analyses of various full-length cDNA libraries prepared from human and mouse tissues. Both the oligo-capping and Cap-trapper methods intended to utilize the cap structure that exists at the 5′ end of an intact mRNA molecule and to selectively synthesize the full-length cDNA starting from the cap site. Although these methods can enable us to synthesize the full-length cDNA, they require a lot of starting material mRNA and many steps in the procedure. The biggest issue is that it is difficult to judge whether the cDNA is full-length or not. The conventional methods have other problems such as the possibility of deletion or mutation caused by using a restriction enzyme digestion step or a PCR amplification step.

Recently, we have developed a novel method for preparing a full-length cDNA library, named the "vector-capping" method, solving above all problems (7). The characteristics of the vector-capping method are as follows: (1) it consists of a few steps, (2) it contains neither a PCR amplification step nor restriction enzyme treatment of the cDNA, (3) the intactness of the cDNA can be assured, (4) the full-length content is more than 95%, (5) several microgram of total RNA is sufficient to construct a cDNA library consisting of 10^6 independent clones, and (6) the procedure is free from biases by the size and expression level of mRNA.

The development of this method is attributed to the following three discoveries. (1) The cap-dependent dC addition to the 3′ end of the first-strand cDNA occurs owing to the terminal deoxynucleotidyl transferase activity of reverse transcriptase (8). As a result, the second-strand cDNA possesses an additional dG at its 5′ end. Thus, we can validate the full-length cDNA by the presence of this additional dG. (2) T4 RNA ligase can catalyze the ligation between the 5′ end of the double-stranded DNA and the 3′ end of the first-strand cDNA in the mRNA:cDNA hybrid. This leads to reduce the number of steps in the cDNA synthesis procedure. (3) Most mRNA molecules extracted from cultured cells or fresh tissues have an intact form, and total RNA can be used as a template without purifying mRNA. This means that it is not necessary to select full-length cDNAs if the reverse transcriptase reaction is complete, and that 1/50th–1/100th of the amount of total RNA necessary for conventional methods is sufficient to construct the same size of the cDNA library.

In this chapter, we describe not only the protocol for preparing the library, but also the points for analyzing the 5′-end sequences obtained by single-pass sequencing of the cDNA clones isolated from the library.

1.1. Principle of the Vector-Capping Method

Figure 1 shows the schematic procedure of the vector-capping method. The first step is annealing of a vector primer, a linear plasmid DNA with a poly(T) tail at one end, to a poly(A)+ RNA molecule. Instead of an oligo(dT) primer used in the conventional methods, the vector primer is adopted according to the Okayama–Berg method (1, 2). The use of the vector primer has two advantages: (1) since the vector primer scarcely primes to an intrinsic A stretch in mRNA, the synthesized cDNA frees from the deletion

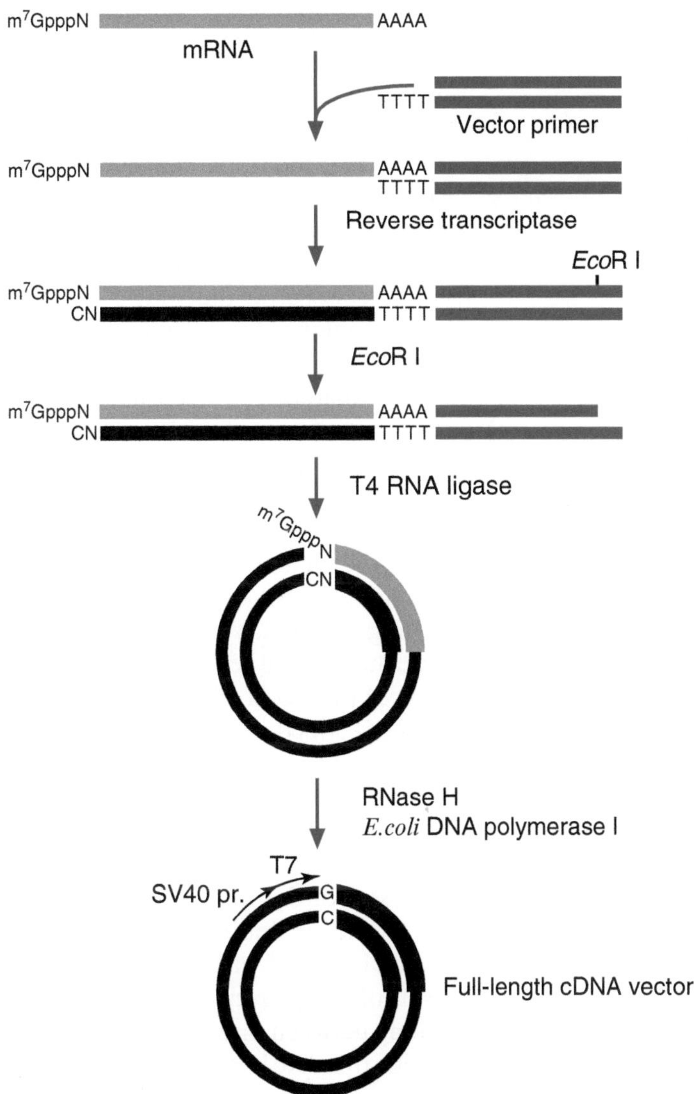

Fig. 1. Schematic procedure for synthesis of full-length cDNA.

of the 3′-end sequence downstream of the A stretch (see Note 1) and (2) since the insertion of cDNA into the vector is unidirectional, it is easy to identify an antisense mRNA-derived cDNA.

The conventional methods utilized poly(A)$^+$ RNA as a template, but our method used total RNA without purifying poly(A)$^+$ RNA (see Note 2). The merits of using total RNA are as follows: (1) the troublesome steps for purifying poly(A)$^+$ RNA can be omitted, (2) the degradation of mRNA during a poly(A)$^+$ RNA purification process can be avoided, (3) the amount of starting total RNA is 1/50th–1/100th of the amount necessary for the conventional methods, and (4) the use of the small amount of the template mRNA results in the effective synthesis of rare cDNA and long-sized cDNA, because the consumption of reverse transcriptase and substrate nucleotides by abundant mRNAs is suppressed.

The second step is a reverse transcriptase reaction. The reaction conditions of reverse transcriptase are the same as used in the conventional methods. Any reverse transcriptase commercially available can be used, but intrinsic RNase H-free one is necessary. We used SuperScript II or SuperScript III purchased from Invitrogen, and both enzymes gave long-sized full-length cDNAs with an insert size of >10 kbp. The termination of reverse transcriptase on the way may have merely occurred because we could not observe the size bias up to 10 kbp. Even if the reverse transcriptase stops on the way, cyclization may not occur owing to steric hindrance by the 5′ end of untranscribed mRNA, resulting in the selection of full-length cDNAs.

By analyzing the terminal sequence of the anchor-ligated cDNA, we found that the terminal transferase activity of reverse transcriptase caused a cap-dependent addition of dC to the 3′ end of the first-strand cDNA (ref. 8, see Note 3). As a result, the full-length cDNA has an additional dG at its 5′ end, and thus we can assure the intactness of the cDNA by the presence of this dG. The degraded mRNA starting with G also gave the truncated cDNA starting with dG. When the high quality of mRNA was used as a template, the full-length content was more than 95% (7, 9). Thus, most of the dG at the 5′ end of the cDNA can be assigned as an extra dG added in a cap-dependent manner.

The third step is an *Eco*RI digestion. Although this step can be omitted, this treatment is effective to reduce the amount of byproducts generated by priming of the incomplete vector primer (9).

The fourth step is a cyclization with T4 RNA ligase, which is a key step of this method. We found that T4 RNA ligase can join the 5′ end of double-stranded DNA to the 3′ end of the first-strand DNA of mRNA:cDNA hybrid. This discovery was the result of serendipity. T4 RNA ligase has been known to catalyze joining between single-stranded RNA molecules and also between single-stranded DNA molecules. The mechanism of the ligation between double-stranded DNA and mRNA:cDNA hybrid is unclear.

Since the head of cDNA looks like being capped by a vector, this method was named the "vector-capping" method. In the case of the full-length cDNA, one dG is inserted between the vector sequence and the 5′-end sequence of the cDNA. The T4 RNA ligase reaction requires high concentration of polyethylene glycol and incubation for a long time as well as the ligation between single-stranded DNAs.

The fifth step is a replacement of mRNA by DNA. If the ligation product is immediately used for transformation, this step can be omitted, because the replacement could occur in the *Escherichia coli* cells. However, this step is recommended to store the reaction product.

1.2. Preparation of the Vector Primer

The quality of the vector primer holds the key to the success of this method. Figure 2 shows the preparation method of the vector primer, which is basically the same as described in the Okayama–Berg method. Any vector can be used as long as it carries the following restriction enzyme sites: a 3′-protruding site for poly(dT) tailing (*Kpn*I in pGCAP10) and its proximal site for removing the poly(dT) tail of one end (*Eco*RV in pGCAP10). The vector pGCAP10 described in this chapter was constructed based on the multifunctional phagemid vector pKA1 that enables us to prepare a single-stranded sense cDNA, to prepare a sense or an antisense RNA, to perform *in vitro* transcription/translation, and to express

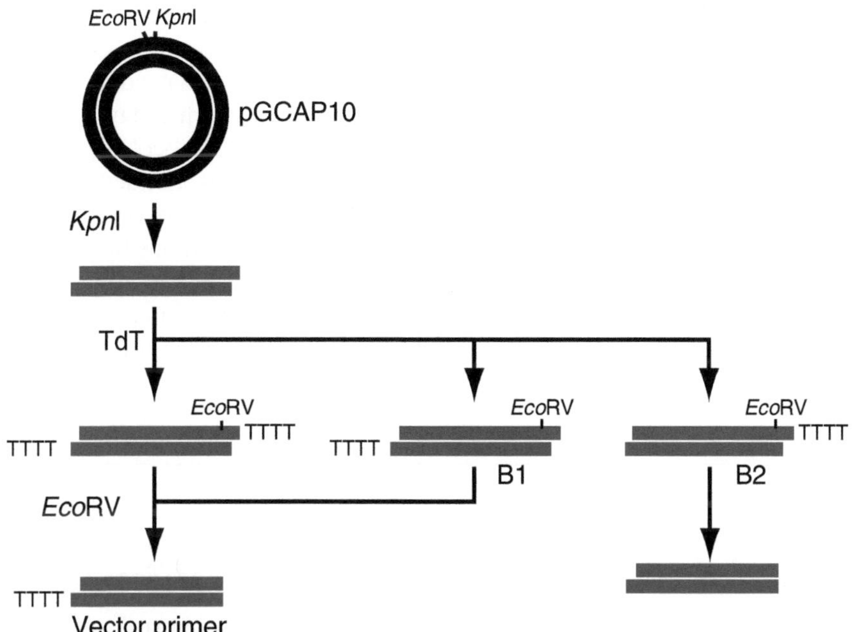

Fig. 2. Preparation of vector primer. *TdT* represents terminal deoxynucleotidyl transferase. **B1** and **B2** are byproducts generated by incomplete tailing.

the cDNA in mammalian cells (4, 9). The point in the process for preparing the vector primer is to avoid the contamination of uncut vectors and untailed vectors. The uncut vectors due to incomplete digestion with restriction enzymes become the background noise, causing the decrease of the cDNA insert content. Since the presence of the uncut vectors is inevitable, we perform the separation of the reaction products by agarose gel electrophoresis after each reaction step to remove the uncut vectors.

The number of dT per tail was adopted to be about 60 according to the Okayama–Berg method (1, 2). The number of the added dT can be controlled by changing substrate concentrations or reaction time. We used the optimum concentration of dTTP at the constant reaction time, which was determined with a pilot-scale experiment. The incomplete tailing reaction causes byproducts having only one tail (**B1** and **B2** described in Fig. 2).

The *Eco*RV digestion cuts off the poly(dT) tail from one end. The short tail fragment is removed by agarose gel electrophoresis. Okayama et al. further purified the poly(dT)-tailed vector using oligo(dA) cellulose column chromatography. At present, oligo(dA) cellulose is commercially not available. In our experience, purification by agarose gel electrophoresis is enough to obtain a functional vector primer. However, incomplete *Eco*RV digestion leaves **B1** and **B2** uncut. Since the sizes of **B1** and **B2** are similar to that of the vector primer, it is impossible to separate them by agarose gel electrophoresis. These byproducts can prime to mRNA and result in the production of the artifacts: **B1** causes an addition of the extra vector-derived sequence at the 5′ end of the cDNA and **B2** causes the opposite-directional insertion of the cDNA by which the cDNA starting from poly(dT) is produced. The production of these byproducts is remarkably reduced by adding an *Eco*RI digestion step in the cDNA synthesis protocol (9).

The quality of the vector primer is assessed by synthesizing the first-strand cDNA from total RNA derived from any eukaryotic cell source and by analyzing the products by agarose gel electrophoresis. If a smear band is visible at the upper region of the vector primer and a band of a vector primer is not detected, it indicates that almost all vector primers are used for synthesizing the cDNA, that is, the vector primer works well.

2. Materials

2.1. Preparation of the Vector Primer

1. Vector (pGCAP10, GenBank # AB371573).
2. *Kpn*I (40 U/μL, Takara Bio, Otsu, Japan).
3. 10× Low buffer: 100 mM Tris–HCl (pH 7.5), 100 mM $MgCl_2$, 10 mM dithiothreitol (DTT).
4. Phenol:chloroform:isoamyl alcohol (25:24:1, PCI).

5. 3 M Sodium acetate (pH 5.2).
6. Ethanol.
7. 80% (v/v) Ethanol.
8. TE: 10 mM Tris–HCl (pH 8.0), 1 mM EDTA.
9. Agarose.
10. TAE: 40 mM Tris–acetate (pH 8.0), 1 mM EDTA.
11. 0.5 μg/mL Ethidium bromide solution.
12. GENECLEAN II Kit (Qbiogene, Carlsbad, CA).
13. 5× Tailing buffer: 700 mM sodium cacodylate, 150 mM Tris–HCl (pH 6.8).
14. 1 mM Dithiothreitol (DTT).
15. 0.1 mM dTTP.
16. H_2O (ribonuclease free).
17. 10 mM $CoCl_2$.
18. Terminal deoxynucleotidyl transferase (TdT) (13 U/μL, Takara Bio).
19. 0.5 M EDTA (pH 8.0).
20. *Eco*RV (50 U/μL, Takara Bio).
21. 10× High buffer: 500 mM Tris–HCl (pH 7.5), 100 mM $MgCl_2$, 10 mM DTT, 1 M NaCl.
22. 1 M Tris–HCl (pH 8.0).
23. 10% SDS.
24. Proteinase K (20 mg/mL, Ambion).
25. RNase A (10 μg/mL, NIPPON GENE, Tokyo, Japan).

2.2. Preparation of Total RNA

1. Isogen (Nippon Gene).
2. Chloroform.
3. Isopropanol.

2.3. cDNA Synthesis

1. dNTP solution containing dATP, dCTP, dGTP, and dTTP, each at 5 mM.
2. Ribonuclease inhibitor (40 U/μL, Takara Bio).
3. Superscript II (200 U/μL, Invitrogen, Carlsbad, CA).
4. 5× Reaction buffer (Invitrogen).
5. *Eco*RI (8 U/μL, Takara Bio).
6. 10× Ligation buffer: 250 mM Tris–HCl (pH 7.5), 50 mM $MgCl_2$, and 2-mercaptoethanol.
7. 5 mM ATP.
8. 0.1 M DTT.
9. 10× T4 RNA ligase (40 U/μL, Takara Bio).

10. 50% (w/v) PEG 6000 (Nacalai tesque, Kyoto, Japan).
11. 5× Second-strand buffer: 100 mM Tris–HCl (pH 7.5), 20 mM $MgCl_2$, 50 mM $(NH_4)_2SO_4$, 500 mM KCl, and 250 μg/mL BSA.
12. *Escherichia coli* DNA polymerase I (10 U/μL, New England Biolabs, Ipswich, MA).
13. RNase H (60 U/μL, Invitrogen).

2.4. Construction of the cDNA Library

1. ElectroMAX™ DH10B cells (Invitrogen).
2. SOC medium.
3. LB agar plates containing ampicillin (50 μg/mL).
4. LB containing ampicillin (50 μg/mL).
5. 50% (v/v) Glycerol.

3. Methods

3.1. Preparation of the Vector Primer

3.1.1. KpnI Digestion

1. The plasmid vector pGCAP10 of 40 μg in 178 μL was mixed with 20 μL of 10× Low buffer, 2 μL of *Kpn*I (40 U/μL) (see Note 4).
2. Incubate at 37°C for 2 h.
3. Add an equal volume of phenol:chloroform:isoamyl alcohol (25:24:1, PCI) and mix well. Centrifuge at 12,000×*g* for 10 min at 4°C. Transfer the upper aqueous layer to the fresh tube. (Hereafter, this step was described as the sentence "Extract with PCI.")
4. Add 20 μL of 3 M sodium acetate (pH 5.2) and 500 μL of 100% ethanol. Centrifuge at 12,000×*g* for 10 min at 4°C. Remove the supernatant and rinse the pellet with 800 μL of 80% (v/v) ethanol. (Hereafter, this step was described as the sentence "Ethanol precipitate.")
5. Dissolve the pellet in 20 μL of TE.
6. Electrophorese the *Kpn*I-digested product in a 0.8% (w/v) agarose gel (10×10×0.6 cm) in TAE.
7. The gel was stained in a 0.5 μg/mL ethidium bromide solution.
8. Excise the DNA band containing the vector of 3.4 kbp with a razor blade under long wave UV light (see Note 5).
9. Transfer the gel slice to a 50-mL tube and determine the approximate volume of the gel slice by weighing.
10. Recover the DNA fragment from the crushed gel using a kit "GENECLEAN II" as follows. Refer to the manufacturer's protocol.

11. Add three volumes of a NaI solution and dissolve the gel.
12. Add GLASSMILK of 40 μL to the solution, mix, and incubate at room temperature for 5 min.
13. Spin the tube in a centrifuge for a minute to pellet.
14. Wash the pellet three times with 500 μL of prepared NEW Wash and dry the pellet under vacuum for 5 min.
15. Resuspend the pellet with TE in a volume equal to the volume of GLASSMILK.
16. Centrifuge for 30 s, remove the supernatant containing the eluted DNA and place in a new tube.
17. Measure the OD_{260} and calculate the concentration of the DNA.

3.1.2. Poly(dT) Tailing

1. The *Kpn*I-digested pGCAP10 of 20 μg in 32.3 μL was mixed with 20 μL of 5× tailing buffer, 10 μL of 1 mM DTT, 20 μL of 0.1 mM dTTP, 10 μL of 10 mM $CoCl_2$, and 7.7 μL of TdT (13 U/μL).
2. Incubate at 37°C for 1 h.
3. Add 5 μL of 0.5 M EDTA (pH 8.0).
4. Extract with PCI.
5. Ethanol precipitate.
6. Dissolve in 50 μL of TE.
7. Electrophorese the poly(dT)-tailed product in a 0.8% (w/v) agarose gel (10×10×0.6 cm) in TAE (see Note 6).
8. The gel was stained in a 0.5 μg/mL ethidium bromide solution.
9. Excise the DNA band containing the poly(dT)-tailed vector of ~3.5 kbp with a razor blade using long wave UV light.
10. Transfer the gel slice to a 50 mL tube and determine the approximate volume of the gel slice by weighing.
11. Recover the DNA fragment from the crushed gel using a kit "GENECLEAN II" (as described in Subheading 3.1.1, steps 11–17).

3.1.3. EcoRV Digestion

1. The poly(dT)-tailed vector in 89 μL was mixed with 10 μL of 10× High buffer, and 1 μL of *Eco*RV (50 U/μL).
2. Incubate at 37°C for 1 h.
3. Extract with PCI.
4. Ethanol precipitate.
5. Dissolve in 50 μL of TE.
6. Electrophorese the product in a 0.8% (w/v) agarose gel (10×10×0.6 cm) in TAE.

7. The gel was stained in a 0.5 μg/mL ethidium bromide solution.
8. Excise the DNA band containing the poly(dT)-tailed vector of ~3.5 kbp with a razor blade using long wave UV light.
9. Transfer the gel slice to a 50 mL tube and determine the approximate volume of the gel slice by weighing.
10. Recover the DNA fragment from the crushed gel using a kit "GENECLEAN II" (as described in Subheading 3.1.1, steps 11–16).
11. Ethanol precipitate.
12. Dissolve in 50 μL of H_2O.
13. Measure the OD_{260} and calculate the concentration of the DNA.

3.1.4. Proteinase K Treatment (see Note 7)

1. The vector primer in 91 μL was mixed with 1 μL of 1 M Tris–HCl (pH 8.0), 2 μL of 0.5 M EDTA, 5 μL of 10% SDS, and 1 μL of proteinase K (50 μg/μL).
2. Incubate at 40°C for 1 h.
3. Extract with PCI.
4. Ethanol precipitate.
5. Dissolve in 20 μL of H_2O.
6. Measure the OD_{260} and calculate the concentration of the DNA.
7. Adjust the concentration of the DNA to be 300 ng/μL by adding H_2O (see Note 8).

3.1.5. Quality Assessment of the Vector Primer

1. The total RNA of 10 μg in 15 μL is mixed with the vector primer of 0.3 μg in 1 μL, and 10 μL of the four dNTPs at 5 mM each (see Note 9).
2. Incubate at 65°C for 5 min, and then place on ice for 2 min.
3. Add 8 μL of 5× reaction buffer, 2 μL of 0.1 M DTT, 2 μL of Ribonuclease Inhibitor (40 U/μL), and 2 μL of Superscript II (200 U/μL).
4. Incubate at 43°C for 3 h.
5. Take 1 μL of the reaction product.
6. Mix with 8 μL of H_2O and 1 μL of RNase A (10 μg/mL) (see Note 10).
7. Incubate at 37°C for 5 min.
8. Electrophorese in a 0.8% agarose gel in TE.
9. Confirm that the smear band of cDNA is observed and the band of the vector primer is not detected, suggesting that the product works well as a vector primer.

3.2. Preparation of Total RNA (see Note 11)

1. Add 1 mL of ISOGEN to the pellet of 10^7 cells in a 1.5-mL tube and dissolve by passing through a 21-gage needle (see Note 12).
2. Let it stand for 5 min at room temperature.
3. Add 0.2 mL of chloroform and shake vigorously for 15 s.
4. Let it stand for 3 min at room temperature.
5. Centrifuge at 12,000 × *g* for 15 min at 4°C.
6. Transfer the aqueous phase to a new tube.
7. Add 0.5 mL of isopropanol and let it stand for 10 min at room temperature.
8. Centrifuge at 12,000 × *g* for 10 min at 4°C.
9. Wash the precipitate with 80% (v/v) ethanol.
10. Dissolve the pellet in 50 μL of H_2O.
11. Add 1 mL of ISOGEN and repeat the procedure from step 2 to step 9.
12. Dissolve the pellet in 50 μL of H_2O.
13. To check the quality of the total RNA, take 1 μL of the solution and electrophorese in a 1% agarose gel in TAE (see Note 13).
14. Add 200 μL of 100% ethanol and store at –80°C.

3.3. cDNA Synthesis

3.3.1. First-Strand cDNA Synthesis

1. Put the solution containing 10 μg of the total RNA in a tube.
2. Ethanol precipitate.
3. Dissolve in 15 μL of H_2O.
4. Add the vector primer of 0.3 μg in 1 μL, and 10 μL of the four dNTPs at 5 mM each.
5. Incubate at 65°C for 5 min, and then place on ice for 2 min.
6. Add 8 μL of 5× reaction buffer, 2 μL of 0.1 M DTT, 2 μL of Ribonuclease Inhibitor (40 U/μL), and 2 μL of Superscript II (100 U/μL).
7. Incubate at 43°C for 3 h.
8. Take 1 μL of the reaction product and confirm the extension of the first-strand cDNA by agarose gel electrophoresis according to Subheading 3.1.5.
9. Add 60 μL of H_2O.
10. Extract with PCI.
11. Ethanol precipitate.
12. Dissolve the pellet in 175 μL of H_2O.

3.3.2. EcoRI Digestion

1. Add 20 μL of 10× High buffer and 5 μL of *Eco*RI (8 U/μL).
2. Incubate at 37°C for 1 h.
3. Extract with PCI.
4. Ethanol precipitate.
5. Dissolve in 24 μL of H_2O.

3.3.3. Cyclization Using T4 RNA Ligase

1. Add 10 μL of 10× ligation buffer, 10 μL of 5 mM ATP, 2 μL of 0.1 M DTT, 1 μL of Ribonuclease Inhibitor (40 U/μL), 3 μL of T4 RNA ligase (40 U/μL), and 50 μL of 50% (w/v) PEG 6000.
2. Incubate at 20°C overnight.
3. Add 100 μL of H_2O.
4. Extract with PCI.
5. Ethanol precipitate.
6. Dissolve the pellet in 64 μL of H_2O.

3.3.4. Replacement of RNA by DNA

1. Add 20 μL of 5× second-strand buffer, 10 μL of the four dNTPs at 5 mM each, 5 μL of *E. coli* DNA polymerase I (10 U/μL), and 1 μL of RNase H (60 U/μL) (see Note 14).
2. Incubate at 12°C for 6 h.
3. Extract with PCI.
4. Ethanol precipitate.
5. Dissolve the pellet in 50 μL of TE.

3.4. Construction of cDNA Library

3.4.1. Transformation of E. coli Cells

1. 1 μL of the cDNA solution is mixed with 50 μL of ElectroMAX™ DH10B electro-competent cells.
2. The transformation is carried out by electroporation according to the manufacturer's protocol. In order to estimate the transformation efficiency, 1 μL of pUC19 (10 pg/μL) is used for transformation as a control.
3. Add the cells into 1 mL of SOC media.
4. Incubate at 37°C for 1 h.
5. Spread the culture media of 10 and 50 μL on LB agar plates containing ampicillin.
6. Incubate at 37°C overnight.
7. Count the number of colonies and estimate the size of the library (see Note 15).

3.4.2. Quality Assessment of the Library

1. Pick up the 96 colonies at random and suspend in 100 μL of LB containing ampicillin in a 96-well microtiter plate.
2. Incubate at 37°C overnight.

3. Add 50% (v/v) glycerol of 40 μL, mix well, and store at −80°C.
4. Isolate the plasmid from each sample.
5. Digest with *Swa*I and *Eco*RI, electrophorese in a 1% agarose gel, and measure the size of the cDNA insert (see Note 16).
6. Sequence the 5′ end of each cDNA.

3.4.3. Preparation of the cDNA Library

1. The remaining cDNA solution is used for transformation in the same scale as described in Subheading 3.4.1.
2. Spread the culture media on LB agar plates containing ampicillin.
3. Incubate at 37°C overnight.
4. Pick up all colonies and suspend in 60 μL of LB containing ampicillin in a 384-well microtiter plate (see Note 17).
5. Incubate at 37°C overnight.
6. Add 50% (v/v) glycerol of 40 μL, mix well, and store at −80°C.

3.5. Sequencing Analysis

3.5.1. Single-Pass Sequencing of the 5′ End of cDNA

The cDNA library prepared by the vector-capping method is so high in both insert and full-length contents that all clones isolated from the library are worth sequencing. Thus, single-pass sequencing of the 5′ end of each cDNA insert is the most effective way for identifying the gene encoded by the cDNA. The plasmid isolation and sequencing reaction are performed in 96-well or 384-well microtiter plates. The sequencing is performed using the automated capillary DNA sequencer.

3.5.2. Classification of the 5′-End Sequences

The sequence output from the DNA sequencer can be analyzed by commercially available software. First, we remove the clones lacking sequence information of the cDNA insert, including (1) a clone showing ambiguous sequence due to failure of a sequencing reaction or due to deletion of a sequencing primer site, (2) an uncut vector, (3) a vector without a cDNA insert, and (4) a deleted vector. The remaining clones have a cDNA insert. A clone starting with a poly(A) tail is classified as a truncated cDNA, because it may not be generated by cyclization of an unprimed vector primer, but by priming of the vector primer to a degraded poly(A) fragment. Most of truncated cDNA clones carry a short 3′-untranslated region followed by a poly(A) tail, which is derived from a degraded mRNA. When this kind of short cDNA starts with an additional dG not existing in the genome, this should be regarded as a full-length cDNA derived from a short intact mRNA with the cap structure.

If the genome sequence is known like human or mouse, the query sequence can be easily mapped on the genome by the BLAST search. Mitochondria genome is also included as a target for search.

In most cases of the full-length cDNA, the query sequence overlaps with the first exon of the known gene or hits the upstream region of the gene. If the sequence starts from the middle of the exon or the intron and has an additional dG at the 5′ end, this starting point is also considered as a real transcriptional start site. The sequence overlapping with the opposite strand of the first exon of the known gene corresponds to its antisense gene. When the sequence does not overlap with the known gene but with EST sequences, the sequence refers to the full-length cDNA corresponding to the ESTs. When there is no gene around the mapped region, the cDNA may originate from a novel gene. The clone with a repetitive sequence at the 5′ end is difficult to be mapped, but the full sequence of the clone can be uniquely mapped.

In the case of the species whose genome sequence has not been available, the obtained sequences are classified by clustering. The gene having multiple promoters produces the different 5′-end sequences, so that the number of clusters is greater than the number of genes. In order to assess whether the two different clones originate from the same gene, it is necessary to compare the full sequences of the two clones or to wait for the determination of the genome sequence of the species.

3.5.3. Evaluation of an Additional Sequence at the 5′ End of cDNA

When the 5′-end sequence of the cDNA is mapped on the genome using BLAST search, almost all sequences have an additional sequence that does not exist in the genome sequence. Most of them are one nucleotide dG that is added to the 5′ end of a full-length cDNA in a cap-dependent manner, but some clones show further addition of nucleotides such as GG, AG, TG, etc. as shown in Fig. 3c. The nucleotide dT is preferentially added like (T)nG, and it should be kept in mind that the clone starting with (T)nG is possibly full-length. In the case of abundant clones, the presence of the additional nucleotide is obvious by aligning the 5′-end sequences. The rare clones should be compared with the genome sequence. It should be noted that some clones have no additional dG in spite of the full-length cDNA (9). These clones may be derived from naturally occurred cap-free mRNA or cap-removed mRNA.

Some clones have a short sequence derived from the vector primer that escaped from restriction enzyme digestion: *Eco*RV digestion in the process of the vector primer preparation and *Eco*RI digestion in the process of cDNA synthesis as shown in Fig. 3d and e, respectively. Usually, the additional vector sequence follows the full-length cDNA starting with an additional dG. To avoid missing these full-length clones, we should keep in mind the possibility of the presence of the extra vector-derived sequence at the 5′ end of the cDNA.

```
a  pGCAP10
tatagggaatttaaatgaattcggccggccgatatcctggtaccgcggccgcggatctccctttagtgag
                EcoRI         EcoRV   KpnI

b  GAPDH genome
cccccggtttctataaattgagcccgcagcctcccgcttcGCTCTCTGCTCCTCCTGTTCGACAGTCAGC

c  GAPDH cDNA
tatagggaatttaaatgaatt------------------GGCTCTCTGCTCCTCCTGTTCGACAGTCAGC
tatagggaatttaaatgaatt-----------------GGGCTCTCTGCTCCTCCTGTTCGACAGTCAGC
tatagggaatttaaatgaatt-----------------AGGCTCTCTGCTCCTCCTGTTCGACAGTCAGC
tatagggaatttaaatgaatt-----------------TGGCTCTCTGCTCCTCCTGTTCGACAGTCAGC
tatagggaatttaaatgaatt--------------TTTTGGCTCTCTGCTCCTCCTGTTCGACAGTCAGC

d  GAPDH cDNA
tatagggaatttaaatgaattcggccggccgat------GGCTCTCTGCTCCTCCTGTTCGACAGTCAGC
tatagggaatttaaatgaattcggccggccgat----TTGGCTCTCTGCTCCTCCTGTTCGACAGTCAGC

e  GAPDH cDNA
tatagggaatttaaatgaattcggccggccgatatcctgGGCTCTCTGCTCCTCCTGTTCGACAGTCAGC
```

Fig. 3. Alignment of the 5′-end sequences of full-length cDNAs encoding human glyceraldehyde-3-phosphate dehydrogenase (GAPDH). (**a**) The sequence of the cloning site of the vector pGCAP10. (**b**) The genome sequence of the transcriptional start site of *GAPDH* locus. The *uppercase letters* represent the first exon sequence. (**c**) Examples of the 5′-end sequences of GAPDH full-length cDNAs (*uppercase letters*). An extra sequence (*underlined*) was inserted. (**d**) Insertion of a vector-derived sequence due to incomplete digestion with *Eco*RI. (**e**) Insertion of a vector-derived sequence due to incomplete digestion with *Eco*RV and *Eco*RI.

3.5.4. Full-Sequencing

Having the same 5′-end sequences does not mean that the clones have an identical sequence of the entire region to the poly(A) tail. Most of human genes, especially long-sized or rare genes, have different downstream sequences because of alternative splicing and/or alternative polyadenylation. The reverse is also true. The cDNA clones with the different 5′-end sequence are often originated from the same gene owing to alternative transcriptional start sites. As a result, they have a different open reading frame, and encode proteins with a different amino acid sequence. When multiple clones are obtained for a target gene, one should fully sequence all clones obtained. This is indispensable for the functional analysis of the protein encoded by the cDNA clones.

Full-sequencing of a novel gene without its sequence information is carried out using a primer walking. The clone corresponding to the RefSeq can be full-sequenced using primer sets synthesized based on the sequence of RefSeq. However, some splicing isoforms may have an unreadable region by these primer sets. In that case, the primer walking is used. Recently, transcriptome analysis using the next-generation DNA sequencer attracts the attention of researchers. However, we cannot distinguish splicing isoforms from these sequence data because of fragment sequencing. In order to determine the precise sequence of alternative splicing isoforms, the full-sequencing of the full-length cDNA is indispensable.

4. Notes

1. In the clones whose full sequences we determined, there was no clone lacking the 3′ end sequence due to priming to the intrinsic A stretch. We found that some sequences registered in GenBank stopped at the intrinsic A stretch.
2. Poly(A)$^+$ RNA also can be used in this method, but our experience showed that the use of poly(A)$^+$ RNA caused a reduction in the number of rare clones and the number of long-sized clones.
3. When *in vitro* transcribed mRNA possessing an A cap instead of a m^{7}G cap was used as a template, the nucleotide added to the 3′ end of the first-strand cDNA was dT, suggesting that the additional nucleotide is complementary to the nucleotide in the cap structure (7, 8). When cap-free mRNA was used, no nucleotide was added.
4. The vector DNA should have a supercoiled form, because the end at a nick serves as a primer for TdT. We prepared a vector DNA using a plasmid isolation kit purchased from Qiagen.
5. You can distinguish the bands of cut and uncut vectors by electrophoresing for a long time in a large-sized agarose gel. When the gel slice is cut, you should take care not to contaminate with an uncut plasmid band.
6. Before doing a preparative scale experiment, a small portion of the poly(dT)-tailed product is mixed with an equal amount of *Kpn*I-digested vector, and then electrophoresed. The band of poly(dT)-tailed vector can be distinguished from the untailed one. If two bands cannot be separated, it means the failure of the tailing.
7. This step is recommended to degrade contaminated RNases.
8. The vector primer solution is divided into small portions and stored at −20°C.
9. We usually prepared a large amount of total RNA from a cultured cell line and used it for assessing the vector primer and for determining the optimum ratio of the vector primer to the total RNA in the step of the first-strand cDNA synthesis.
10. The use of RNase should be done at the restricted area separated from the room where the cDNA synthesis is carried out.
11. The vector-capping method does not include any selection process for intact mRNA or full-length cDNA. Thus, the full-length content depends on the intactness of mRNA. The most important thing for preparing a high-quality cDNA library is to use intact mRNA samples and to avoid the

degradation of mRNA during experimental procedures. To avoid the contamination of RNases into the sample, all experiments in handling RNA samples are carried out in the clean bench.

12. Here, we show the preparation method of total RNA from cultured cells as an example.
13. It is necessary that the clear bands of 28S and 18S rRNAs can be seen and that the fragmented RNAs do not exist near the migration front on the gel.
14. The use of *E. coli* DNA ligase is omitted from the step of the second-strand cDNA synthesis to avoid cyclization of a contaminated incomplete vector primer and ligation between cDNA-carrying vectors.
15. The library composed of 10^5–10^6 independent clones is obtained using the scale described in this protocol.
16. Since the cDNA synthesized using the present method is guaranteed to have a poly(A) tail, comparing the 5′-end sequence with the RefSeq database enables us to identify the gene, that is, to estimate the size of the cDNA insert from the database. However, if the RefSeq has not been available, it is necessary to measure the size of the cDNA insert by agarose gel electrophoresis after digesting with restriction enzymes.
17. We used a colony picker when more than 10,000 colonies were picked up.

Acknowledgments

This work was supported by a grant from the Ministry of Health, Labor, and Welfare of Japan.

References

1. Okayama, H. and Berg, P. (1982) High-efficiency cloning of full-length cDNA. *Mol. Cell. Biol.* **2**, 161–170.
2. Okayama, H., Kawaichi, M., Brownstein, M., Lee, F., Yokota, T. and Arai, K. (1987) High-efficiency cloning of full-length cDNA; construction and screening of cDNA expression libraries for mammalian cells. *Methods Enzymol.* **154**, 3–28.
3. Maruyama, K. and Sugano, S. (1994) Oligo-capping: a simple method to replace the cap structure of eukaryotic mRNAs with oligoribonucleotides. *Gene* **138**, 171–174.
4. Kato, S., Sekine, S., Oh, S.-W., Kim, N.-S., Umezawa, Y., Abe, N., Yokoyama-Kobayashi, M. and Aoki, T. (1994) Construction of a human full-length cDNA bank. *Gene* **150**, 243–250.
5. Suzuki, Y., Yoshitomo-Nakagawa, K., Maruyama, K., Suyama, A. and Sugano, S. (1997) Construction and characterization of a full length-enriched and a 5′-end-enriched cDNA library. *Gene* **200**, 149–156.
6. Carninci, P., Kvam, C., Kitamura, A., Ohsumi, T., Okazaki, Y., Itoh, M., Kamiya, M., Shibata, K., Sasaki, N., Izawa, M., Muramatsu, M., Hayashizaki, Y. and Schneider, C. (1996) High-efficiency full-length cDNA cloning by biotinylated CAP trapper. *Genomics* **37**, 327–336.

7. Kato, S., Ohtoko, K., Ohtake, H. and Kimura, T. (2005) Vector-capping: a simple method for preparing a high-quality full-length cDNA library. *DNA Res.* **12**, 53–62.
8. Ohtake, H., Ohtoko, K., Ishimaru, Y. and Kato, S. (2004) Determination of the capped site sequence of mRNA based on the detection of cap-dependent nucleotide addition using an anchor ligation method. *DNA Res.* **11**, 305–309.
9. Oshikawa, M., Sugai, Y., Usami, R., Ohtoko, K., Toyama, S. and Kato, S. (2008) Fine expression profiling of full-length transcripts using a size-unbiased cDNA library prepared with the vector-capping method. *DNA Res.* **15**, 123–136.

Chapter 5

Construction of Improved Yeast Two-Hybrid Libraries

Richard H. Maier, Christina J. Maier, and Kamil Önder

Abstract

The Yeast Two-Hybrid (Y2H) system is the most frequently used method for identifying protein–protein interactions. The use of recombination-amenable Y2H vectors would reduce time and effort for cloning prey or bait vectors, and increase the quality of Y2H screenings due to the production of improved screening libraries. These libraries can heighten the amount of new candidates in Y2H screenings significantly by representing more correct candidate genes in frame and outperform a classical Y2H library. The described vectors can be used for the construction of genomic, peptide, or cDNA-based Y2H libraries. Furthermore, the compatibility to newer ORFeome libraries is also given. Here, we describe a vector system for site-specific recombination and for the construction of high-content Y2H libraries. In summary, we will describe the construction of these vectors and the production of Y2H screening libraries.

Key words: Yeast two-hybrid, Recombinational cloning, Protein interaction, Reading frame, High throughput, Screening library, ORFeome

1. Introduction

The Yeast Two Hybrid (Y2H) system invented by Fields and Song (1) is the most frequently used technique for identifying protein–protein interactions (PPIs), requiring minimal prior information of the putative interactors. In a classical Y2H screen, the protein of interest, the bait molecule, is normally screened against a prey library. Good screens are characterized through high-quality Y2H libraries, which normally should be highly representative of the investigated target cell/tissue's gene content. In the ideal case, the gene content in the library should not be biased by gene abundance, out-of-frame clones, or fragmented genes. Currently, existing libraries are either genomic DNA libraries or cDNA originated libraries, and often fall short in terms of quality of the insert. They are constructed using partial restriction endonuclease digestion of genomic DNA or from cDNA.

Chaofu Lu et al. (eds.), *cDNA Libraries: Methods and Applications*, Methods in Molecular Biology, vol. 729,
DOI 10.1007/978-1-61779-065-2_5, © Springer Science+Business Media, LLC 2011

Construction of both library types is accompanied by nonproductive cloning of genes, for example, gene fragments in the wrong reading frame or wrong orientation, gaps in the sequence, or truncated amino termini. cDNA libraries may not be fully representative of low abundance or conditionally expressed proteins, and may be no better at representing the hurdle than genomic libraries (2). In the recent years, a new type of screening library, the so-called ORFeomes, for genome-wide analysis of PPIs increased and ranged from bacteria (3) to human (4). The term "ORFeome" refers to the whole collection of all open reading frames of an organism present in "entry vectors" for recombinational cloning (5, 6) to enable large-scale, high-throughput "omics" applications. The great advantages of such ORFeomes cloned in those vectors are the use of full-length genes that are amenable to fast and high-efficiency cloning, flexibility due to many expression vectors suitable for recombinational cloning, and the maintenance of the orientation. The use of ORFeomes cloned gene by gene into Y2H vectors would be, for example, a systematic based approach like direct screening of a few thousand baits against the same amount of prey molecules performed automatic in a high-throughput manner. Finally, single clone ORFeome collections can be used to derive pooled libraries (e.g., equal amounts of all clones from a complete library are pooled and used for Y2H analysis) for classical Y2H screenings and are intrinsically normalized for gene abundance.

In this chapter, we will describe (a) the construction of a vector system that is amenable to recombinational or Gateway® cloning, and is reading frame independent. This vector system has the ability to clone in every reading frame without the need to know or maintain the reading frame of each gene. Our vector system enables any researcher to clone his/her own source of cDNA or genomic library cloned in entry vectors in the new vectors to take advantage of all possible reading frames, and will help to make Y2H libraries more representative. Moreover, we describe (b) the construction of an ORFeome-based Y2H library and a multi-frame Y2H library. An ORFeome-based Y2H library in one of the constructed bait or prey vectors (only one reading frame is required) will represent every nearby gene in the correct orientation as well as in the correct reading frame. Additionally, our vector system will increase the utility of already existing cDNA and genomic libraries cloned in an entry vector with a fixed reading frame.

2. Materials

2.1. Colony PCR/Adapter PCR

1. Adapter forward primer 5′-GGGGACAAGTTTGTACAAAAAAGCAGGCTTG-3′.

2. Adapter reverse primer 5′-GGGGACCACTTTGTACAA GAAAGCTGGGTA-3′.
3. Platinum® Pfx DNA polymerase (Invitrogen) in the concentration of 2.5 U/μL with the supplied 10× amplification buffer and 50 mM magnesium sulfate for the adapter PCR.
4. DONR™ vector-specific forward primer 5′-GTAAAAC GACGGCCAG-3′ and reverse primer 5′-CAGGAAAC AGCTATGAC-3′. The primers here are for the use with the pDONR™/Zeo (Invitrogen) vector.
5. pAD-Gate1–6 specific forward primer 5′-CTATTCGAT GATGAAGATACCCCA-3′ and reverse primer 5′-GTGAA CTTGCGGGGTTTTTCAG-3′.
6. pBD-Gate1–6 specific forward primer 5′-TCATCGGAA GAGAGTAGTAAC-3′ and reverse primer 5′-GAGTCAC TTTAAAATTTGTATACAC-3′.
7. Use a Taq polymerase for colony PCR, e.g., the BioThermRed™ polymerase (GenXpress; 5 U/μL) with the supplied 10× reaction buffer.
8. 100 mM dNTP set (Invitrogen) consists of dATP, dCTP, dTTP, and dGTP. Prepare a working solution with the concentration of 2 mM (each nucleotide).
9. Applied Biosystems 2720 thermal cycler was used in this protocol.

2.2. Transformation of Escherichia coli Competent Cells

1. Subcloning Efficiency™ DH5α™ competent cells (Invitrogen).
2. One Shot® ccdB Survival™ 2 T1R chemically competent cells (Invitrogen).
3. One Shot® TOP10 Electrocomp™ cells (Invitrogen) as super-competent cells for library transformation.
4. Water bath with temperature control.
5. Electroporator.
6. Electroporation Gene Pulser/MicroPulser Cuvettes, 0.1 cm gap (BioRad).
7. SOC Medium (Invitrogen): 2% tryptone, 0.5% yeast extract, 10 mM sodium chloride, 2.5 mM potassium chloride, 10 mM magnesium sulfate, and 20 mM glucose.
8. Thermomixer: shaking heat block with temperature control.
9. Luria broth plates: yeast extract 5 g/L, tryptone 10 g/L, sodium chloride 10 g/L, and agar 15 g/L containing the appropriate antibiotic; ampicillin (100 μg/mL for pAD-Gate1–6, respectively, pGADT7 plasmid selection); kanamycin (50 μg/mL for pBD-Gate1–6, respectively, pGBKT7 plasmid selection); or chloramphenicol (32 μg/mL).
10. Low-salt Luria broth plates: similar to normal Luria broth plates, but with 5 g/L sodium chloride with the antibiotic

Zeocin™ (Invitrogen; 50 μg/mL) for pDONR™/Zeo plasmid selection (see Note 1).

2.3. Cell Culture and Plasmid Preparation

1. Rotilabo® rubber wipers (Carl-Roth).
2. Liquid Luria broth: yeast extract 5 g/L, tryptone 10 g/L, and sodium chloride 10 g/L containing the appropriate antibiotic; ampicillin (100 μg/mL for pAD-Gate1–6, respectively, pGADT7 plasmid selection); kanamycin (50 μg/mL for pBD-Gate1–6, respectively, pGBKT7 plasmid selection); or chloramphenicol (32 μg/mL).
3. Low-salt Luria broth plates: similar to normal liquid Luria broth medium but with 5 g/L sodium chloride with the antibiotic Zeocin™ (Invitrogen; 50 μg/mL) for pDONR/Zeo plasmid selection (see Note 1).
4. GenElute™ HP Plasmid Maxiprep kit (Sigma–Aldrich) (see Note 2).
5. Incubator with temperature control and shaking platform.
6. Spectrophotometer.

2.4. DNA Cleavage by Restriction Endonucleases and Ligation

1. Vectors pGADT7 and pGBKT7 of the Matchmaker™ Yeast Two-Hybrid System (Clontech).
2. Restriction endonuclease SmaI (New England Biolabs; 20 U/μL) with the supplied reaction buffer.
3. T4 DNA ligase (Promega; 1–3 U/μL) with the supplied reaction buffer.
4. Gateway® reading frame cassettes rfA, rfB, and rfC.1 (Invitrogen).
5. DNA purification kit: Wizard® SV Gel and PCR Clean-Up System (Promega).

2.5. BP and LR Recombination Reactions

1. DONR Vector such as pDONR™/Zeo or pDONR™221 (both Invitrogen) (see Note 3).
2. Destination vectors pAD-Gate1–6 or pBD-Gate1–6 (2).
3. cDNA library in entry vector; e.g., SuperScript® Premade cDNA Library (Invitrogen).
4. Gateway® BP Clonase II enzyme mix (Invitrogen).
5. Gateway® LR Clonase II enzyme mix (Invitrogen).
6. Proteinase K solution (Invitrogen; 2 μg/μL).
7. TE buffer pH 8.0 (10 mM Tris–HCl; 1 mM EDTA).
8. Glycogen (Fermentas; 20 μg/μL).
9. 7.5 M Sodium acetate (Na_4OAc).
10. Ethanol: absolute and 70% (AppliChem).
11. Heraeus Microcentrifuge Fresco (ThermoScientific).
12. Water bath with temperature control.

3. Methods

3.1. Design of Recombination-Amenable Vectors

Gateway® cloning technology is based on recombinational cloning. The main advantages are fast and efficiency in cloning, maintenance of orientation, high-throughputness, and flexibility. In brief, for a single insert of interest, the DNA is flanked by recombination-specific sites, cloned into a DONR™ vector by BP-recombination cloning, selected for positive clones on appropriate antibiotics, *E. coli* propagated, and the insert then transferred by LR-recombination cloning into a destination vector of interest. While this procedure is easily achievable for a single or a few inserts, it would be a cumbersome and costly undertaking to clone a complete library into a DONR™ vector. For the production of Y2H libraries, there are two ways to start: either a library pre-cloned in a DONR™ vector is commercially obtained, or your own library has to be introduced into a DONR™ vector. To enable direct cloning of a library of fragmented or full-length inserts, we modified a DONR™ vector by introducing a blunt-end restriction site. Finally, a complete set of different reading frame destination vectors was produced.

3.1.1. DONR™/Entry Vectors

1. To use Gateway® terminology, the term DONR™ vector means a vector ready for a so-called BP reaction. In a cloning step, the to-be cloned gene, fragment, etc., are cloned into this vector by replacing the existing ccdB (7) gene (negative selection) and the information for chloramphenicol resistance. For more information, please read the Gateway® cloning instruction manual (Invitrogen). For terminology, it is important that the vector is called entry vector when recombination has occurred successfully. Here, we describe the introduction of a blunt-end restriction site into a DONR™ vector to obtain Gateway® flexibility with traditional restriction enzyme-based cloning.
2. Order an oligonucleotide with the restriction site of choice flanked with truncated attB1 + 2 attachment sites. To save costs, it is necessary to order only the sense oligonucleotide. Here, we show the oligonucleotide for the restriction site for ScaI (AGT/ACT):

   5′-AAAAAGCAGGCTTG<u>AGTACT</u>TACCCAGCT TTCTTGTAC-3′ (see Note 4).
3. The ordered oligonucleotide serves as a template for a standard PCR. An assay volume of 50 µL consisting of 1.25 U of proofreading polymerase, 1 mM of $MgSO_4$, dNTP mix (0.3 mM each), 0.4 µM each of forward and reverse adapter primers, 5 µL of 10× amplification buffer, and 10 pmol of template oligonucleotide. The 35 PCR cycles (94°C for 30 s, 48°C for 30 s, and 72°C for 1 min/kb) were preceded by

heating to 94°C for 5 min and were followed by a 7-min incubation at 72°C (see Note 5).

4. The Gateway®-compatible amplified product is used for recombination into the DONR™ vector. The sample should contain 2 μL of unpurified PCR product, 2 μL of BP Clonase™ II enzyme mix, 150 ng of DONR™ vector and up to 10 μL of TE buffer, pH 8.0. In this reaction, the attB-flanked PCR product undergoes recombination with the attP sites on the DONR™ vector, creating attL sites. After incubation overnight at 25°C or room temperature, the reaction is stopped by adding 1 μL of proteinase K and incubating at 37°C for 30 min. The BP reactions were directly used for bacterial transformation (see Note 6).
5. Transform the BP reaction into chemically competent *E. coli* DH5α cells. For this, thaw the cells on ice, add 2 μL of the BP reaction (do not pipette up and down), and incubate the cells for 30 min on ice. Heat-shock the cells for 30 s at 42°C in the water bath. Cool the cells on ice and add 150 μL of SOC medium. After incubation at 37°C for 1 h with gentle shaking, plate one half of the transforming reaction on LB plates containing the appropriate antibiotic for the selected DONR™ vector. Incubate the cells overnight at 37°C.
6. Pick up some colonies (up to 5) to check each in a colony PCR. Add 50 μL of the samples with 2.5 U of Taq polymerase, 0.3 μM each of DONR™ vector-specific forward and reverse primers, 0.2 mM of dNTP mix, and 5 μL of 10× reaction buffer. Colonies were picked with a sterile pipette tip and transferred to the PCR tubes. The 45 PCR cycles (94°C for 30 s, 55°C for 30 s, and 72°C for 1 min/kb) were preceded by heating to 94°C for 5 min and followed by a 7-min incubation at 72°C. Determine the sizes of the PCR products by agarose gel separation and ethidium bromide staining.
7. Inoculate clones harboring entry vectors with the correct size in liquid LB medium containing the appropriate antibiotic. The volume of the medium depends on the size of the desired plasmid purification kit. Incubate the culture overnight at 37°C with gentle shaking.
8. Purify the plasmid with a plasmid preparation kit following the manufacturer's instructions and determine the concentration of the prepared sample by absorbance measurement at 260 nm.
9. To verify the entry vector, sequence it with the same specific forward primer as that in the colony PCR. The resulting new entry vector combines a restriction site for classic cDNA or genomic library construction and the advantages of site-specific recombinational cloning. The new entry vector can be

used for library construction using restriction enzyme-based cloning (described elsewhere in the literature). Each DONR™ vector can be equipped in this way with several restriction sites (not only blunt end).

3.1.2. Reading Frame-Independent Y2H Destination Vectors

1. To use Gateway® terminology, the term "Destination" vector refers to a vector with attR attachment sites suitable for LR reactions. After successful recombination of an attL-site flanked piece in an entry vector in a LR reaction the Destination vector becomes an Expression vector with again attB sites.
2. Choose Y2H bait and prey vectors of choice for modifications. Here, the vectors pGADT7 and pGBKT7 were used. Select a blunt-end restriction site in the multiple cloning site, here SmaI (CCC/GGG), and cut both vectors with the respective restriction endonuclease. Place 2 μg of vector in a 20-μL sample together with 20 U of SmaI, 2 μL of 10× reaction buffer, and up to 20 μL of dH_2O. Mix the reaction sample well and centrifuge the sample briefly. Incubate the sample at 25°C or room temperature for 2 h.
3. Stop the reaction by purifying each sample using a DNA purification kit following the manufacturer's instructions, and determine the concentration of the purified linearized DNA by absorbance measurement at 260 nm (see Note 7).
4. In six different ligation reactions, the three Gateway® reading frames A, B, and C.1 are ligated into linearized pGADT7 and pGBKT7. In each 15 μL sample, add 200 ng of linearized vector, 10 ng of reading frame cassette, 1 μL of T4 DNA ligase, 1.5 μL of 10× reaction buffer, and up to 15 μL of dH_2O and incubate for at least 3 days at 4°C (see Note 8).
5. Stop the ligation reactions by purifying each sample using a DNA purification kit following the manufacturer's instructions.
6. Transform 5 μL of each purified ligation reaction into chemically competent CcdB survival cells as described in Subheading 3.1.1, step 5. The best way to test for positive ligation is to plate the cells on LB plates containing chloramphenicol. After positive ligation, the inserted reading frame cassettes contain the chloramphenicol resistance gene. Incubate the plates overnight at 37°C.
7. The next day pick up to ten colonies per transformation and inoculate each in liquid LB medium containing chloramphenicol for plasmid preparation as described in Subheading 3.1.1, steps 7 and 8.
8. Verify the orientation of the Gateway® cassettes in the new destination vectors through sequencing with the vector-specific forward primer and store one candidate of each

orientation of each cassette in each of the two vectors. Finally, a set of 12 Gateway® compatible prey and bait vectors for each possible reading frame is constructed. A vector map of the prey vectors pAD-Gate1–6 (derived from pGADT7) and bait vectors pBD-Gate1–6 (derived from pGBKT7) is shown in Fig. 1 (see Note 9).

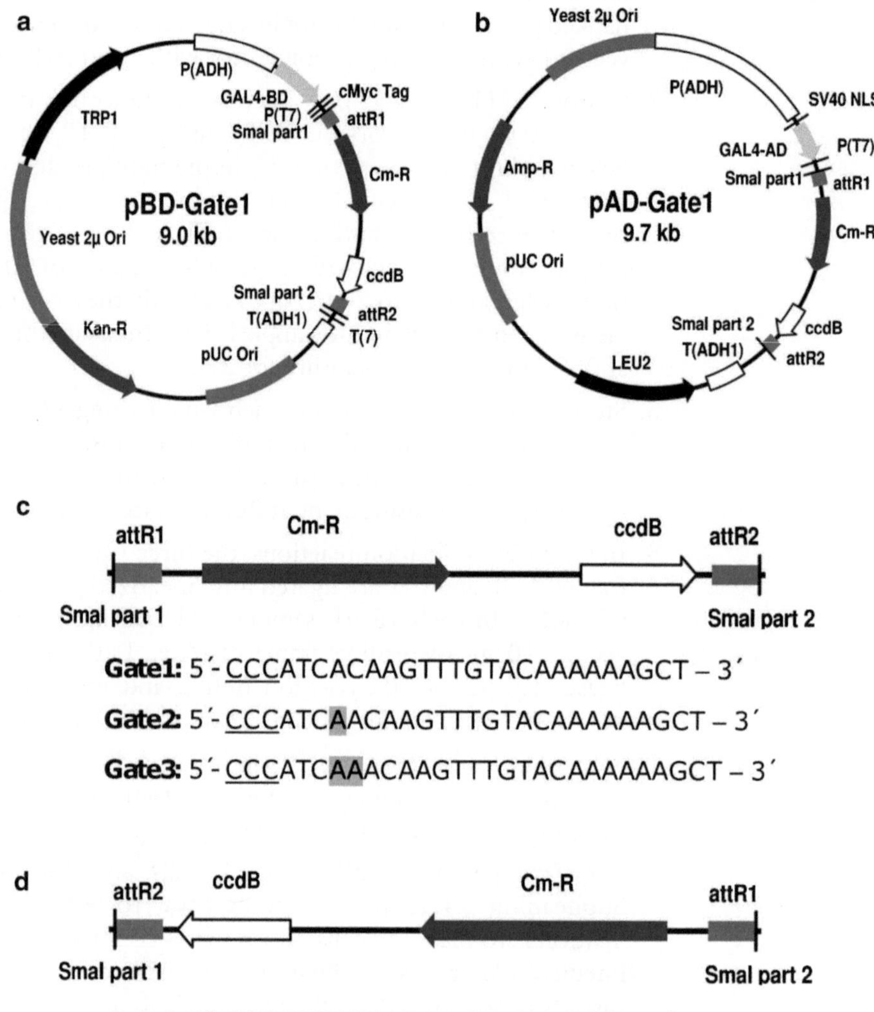

Fig. 1. Construction of the new vector system. Illustration and description of the set of the six Gateway®-compatible Y2H bait vectors pBD-Gate1–6 (**a**) and the six Gateway®-compatible Y2H prey vectors, pAD-Gate1–6 (**b**) These were constructed by blunt-end cloning the three Gateway® recombination cassettes in forward (**c**) and reverse (**d**) directions into the SmaI restriction site of the original Y2H vectors pGBKT7 and pGADT7. The sequences of the six different cassettes (Gate 1–6) are shown from the 5′ end. The first part of the residual SmaI restriction site is underlined, and the bases that were inserted to shift the reading frame are highlighted in *gray* (Copyright notice/credit © Biotechniques 2009 used by Permission).

3.2. Construction of an ORFeome-Based Y2H Library

In contrast to normal cDNA or genomic-derived libraries, an ORFeome-based Y2H library presents every gene of the chosen organism as full-length copy and in the right reading frame. For this, an ORFeome has to be constructed as described, for example, by Brandner et al. (3) with the human pathogen *Staphylococcus aureus*.

1. Each constructed ORFeome is constructed by PCR amplification of a full-length gene of interest, where a PCR product is 5-prime and 3-prime flanked with attB sites. In principle, this is done for several thousands of different genes (depends on the genome size of the studied organism) to produce a gene collection. While in the majority of ORFeome constructions the aim is to produce a nearly complete single gene collection of arrayed genes, it is also possible to use the PCR products directly for the production of Y2H libraries. Make a pool of these PCR products by pooling, for example, 2 μL of every PCR product in one tube.
2. Insert 2 μL of this pool of PCR products in a 10-μL assay containing 150 ng of DONR™ vector, 2 μL of BP Clonase™ II enzyme mix, and up to 10 μL of TE buffer, pH 8.0, and incubate the sample at 25°C or room temperature overnight. Because of the small amount of recoverable DNA, perform at least four identical samples. Stop the reactions next day by adding 1 μL of proteinase K and incubate at 37°C for 30 min.
3. Pool all duplicates of the BP reactions in one tube and add 60 μL of sterile dH_2O, 2 μL of glycogen, 50 μL of 7.5 M Na_4OAc, and 375 μL of absolute ethanol. Mix well by vortexing and place the tube on dry ice or at −80°C for 30 min. In the meantime, cool a microcentrifuge to 4°C (see Note 10).
4. Centrifuge the sample at 4°C for 30 min at 17,000 ×*g*. After centrifugation, a little DNA pellet is visible. Please mark it outside the tube. Discard the supernatant and without disturbing the pellet. Add 150 μL of cold 70% ethanol and centrifuge again at 4°C at 17,000 ×*g* for 5 min. Place the tube in the same position as before. Repeat this wash step and remove as much ethanol as possible after washing. Let the pellet dry for 15 min by keeping the lid of the tube open at room temperature till all ethanol is evaporated. Finally, dissolve the remaining pellet in 10 μL of TE buffer, pH 8.0, by pipetting up and down for at least 30 times.
5. Transform the precipitated BP reaction in supercompetent *E. coli* TOP10 cells. For library transformation, electrocompetent cells seem to be a better choice than chemically competent ones. The following transforming instructions assume the use of a BioRad MiniPulser Electroporation System. Thaw the cells on ice and pipette 3 μL into 50 μL of the competent

cells. Do not pipette up and down, or vortex. Take the entire ~53 µL and pipette it between the two electrodes of the electroporation cuvette. Place the cuvette in the MiniPulser, select the bacteria setting and pulse. Rinse the cells out of the cuvette with 1 mL of SOC media and put them in a sterile tube for incubation at 37°C for 1h with gentle shaking.

6. Plate each 100 µL of the transformation on 10 big (145 mm diameter) LB medium plates containing the appropriate antibiotic and incubate the plates overnight at 37°C.
7. Determine the amount of primary transformants the next day by counting the number of colonies. A good library should have at least 10^6 primary clones (see Note 11).
8. Scrape all colonies of each plate with a rubber wiper and collect the whole cell material in a plastic tube for plasmid preparation. Usually, the amount of cells obtained by this method is sufficient for a maxi plasmid preparation. Purify the plasmid DNA following the manufacturer's instructions and determine the concentration by absorbance measurement at 260 nm.
9. Due to the always same reading frame, because of the ORFeome construction strategy, there is only the need of one of the modified prey vectors. Normally, the to-be screened library in an Y2H screen is cloned into a prey vector, but it can switch off course. In the case of the ORFeome of *S. aureus* (3), the pAD-Gate 2 vector is the right choice, but this has to be determined in each case. Perform a LR reaction with 150 ng of library in entry vectors, 250 ng of Destination vector, 2 µL of LR clonase II enzyme mix, and up to 10 µL of TE buffer, pH 8.0. Make at least four duplicates, too and incubate the reaction mixture overnight at 25°C or room temperature.
10. Repeat the steps as described in Subheading 3.2, steps 3–8. The result is a Y2H library in a prey vector with nearly 100% cloning efficiency and full-length genes.
11. Finally, determine the average insert size by performing colony PCR of single colonies with vector-specific primers, and check for different inserts by sequencing plasmids from randomly chosen colonies.

3.3. Construction of a Multi-frame Y2H Library

In classic Y2H libraries, many candidates fail through incorrect orientation or wrong reading frame. Translation initiation generally scans the 5′ untranslated region till the first AUG is encountered. In one-frame libraries, two-thirds of these initial AUGs are out of frame (8). The constructed prey or bait vectors are able to clone in every reading frame without the need to know or maintain the reading frame. The six prey and bait vectors differ in the translated region in one or two nucleotides on both orientations, so that all possible reading frames for correct protein expression are available. In cases of Gateway® compatible cDNA libraries in entry vectors, only three reading frames are necessary for library construction, because the cDNA is normally cloned directionally.

1. The start point is either a researcher-constructed cDNA library in entry vectors suitable for site-specific recombination, or a commercial entry vector library is obtained. There are already many cDNA libraries available in Gateway® specific vectors.
2. Perform LR reactions with 150 ng of entry vector cDNA library, 250 ng of Destination vector (e.g., pAD-Gate 1–3), 2 μL of LR clonase II enzyme mix, and up to 10 μL of TE buffer, pH 8.0. Make four identical samples for each prey vector and incubate the reaction mix overnight at 25°C or room temperature. Stop the reactions next day by adding 1 μL of proteinase K and incubate at 37°C for 30 min (see Note 12).
3. Pool all identical LR reactions and repeat the steps described in Subheading 3.2, steps 3–8.
4. The results are three different Y2H libraries in prey vectors of the same source in three different reading frames, which have to be tested for insert sizes and cloning efficiency. For the final Y2H screen, there are two possibilities. First, perform three different screens with the same bait molecule together with each one of the three different reading frame libraries. Second, pool the same amount of each library together and perform one screen against one bait.
5. Alternatively, perform a LR reaction with 150 ng of entry vector cDNA library, 250 ng of Destination vector mix (equal amount of pAD-Gate 1–3), 2 μL of LR clonase II enzyme mix, and up to 10 μL of TE buffer, pH 8.0. Make again four identical samples and incubate the reaction mix overnight at 25°C or room temperature. Stop the reactions next day by adding 1 μL of proteinase K and incubate at 37°C for 30 min. Continue as described in Subheading 3.2, steps 3–8.
6. The produced libraries can each deliver positive Y2H colonies in a Y2H screen that overlap only to a small extend between the libraries, as reported by Maier et al. (2) in a test screen (Fig. 2). The constructed libraries were functional and contain equal amount of autoactivators. The overall amount of new found Y2H interacting candidates increases significantly compared to that with a single frame library.

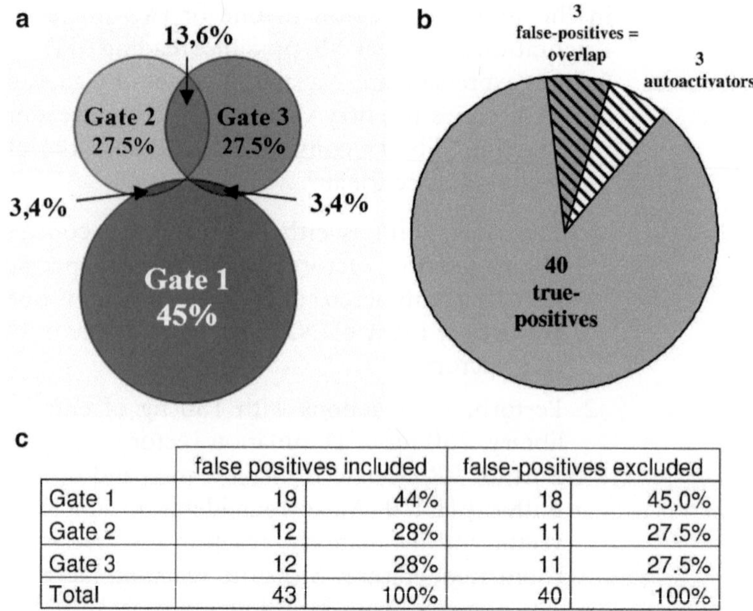

	false positives included		false-positives excluded	
Gate 1	19	44%	18	45,0%
Gate 2	12	28%	11	27.5%
Gate 3	12	28%	11	27.5%
Total	43	100%	40	100%

Fig. 2. Comparison of a three reading frame Y2H library screenings. In three individual Y2H screenings of a human cDNA library cloned in pAD-Gate 1, 2, and 3 against a human Vitamin D receptor (VDR) bait molecule, a total of 43 candidate clones were detected, sequenced, and tested for autoactivity. The analysis showed a significant increase in found interactors compared to a single reading frame screen. (**a**) Illustration of the recovered VDR interactors and their distribution over the different Gate vector libraries. (**b**) Overall false-positive and true-positive content of the Y2H library screening. (**c**) Detailed analysis of the screenings (Copyright notice/credit © Biotechniques 2009 used by Permission).

4. Notes

1. Zeocin™ selection reagent is light sensitive. Store all contents containing the drug in the dark. Also, strong acidity or basicity inhibits the activity of Zeocin™. So keep the pH of the media to 7.5.
2. Many vendors of diverse kits exist; the articles listed in Subheading 2 are only examples.
3. The vector pDONR™/Zeo is well adaptable because there are no Destination vectors with Zeocin™ resistance. If the vector pDONR 221 is used, there are problems with the vectors pBD-Gate1–6, for example, because they have the same selection marker, kanamycin.
4. Pay attention that there are no additional restriction sites of the chosen enzyme in the DONR™ vector. But there can be one in the area between the two attP sites in the DONR™ vector because this section is replaced by the PCR product.
5. The use of a proofreading polymerase is recommended because there are no mutations incorporated during PCR.

If a normal Taq polymerase is used, reduce the amount of PCR cycles down to 15–25. Our own experiments has shown that even PCR products that are not detectable on agarose gel with ethidium bromide staining deliver enough material for successful BP recombination.

6. In most cases, it is sufficient to use unpurified PCR products. If primers bind unspecifically and deliver multiple PCR products, purify the PCR product of appropriate size from the agarose gel using a DNA purification kit.
7. Alternatively, the enzymes can be inactivated by heat inactivation. Refer to the enzyme product manual. In our own studies, we got better results by simply purifying the samples with DNA purification kits without heat inactivation.
8. From our own observations, cloning efficiency is much higher in long ligations at 4°C compared to that in overnight ligations at 4°C (up to 95%). For standard, not library, ligation reactions, 2 h at 16°C is sufficient.
9. The functionality of new vectors has to be proved by testing known interactions with these vectors. Also, tests for autoactivation of reporter genes have to be done (see Maier et al.) (2). The vectors pAD-Gate1–6 and the pBD-Gate series are all fully functional with the yeast strain AH109 of the Matchmaker™ Y2H system.
10. The precipitation samples can be stored overnight at –20°C in cases of time constraints.
11. For counting cells, it is better to plate a series of dilutions on LB agar plates with the appropriate antibiotic, e.g., 1:10, 1:100, and $1:10^3$.
12. For directionally cloned cDNA libraries, the reverse-orientated prey vectors pAD-Gate 4–6 are not useful. In cases of genomic libraries in entry vectors where it is not clear if the inserts are wrong or correctly orientated, all possible reading frame vectors have to be inserted in the LR reaction.

References

1. Fields, S., and Song, O. (1989) A novel genetic system to detect protein-protein interactions. *Nature* **340**, 245–6.
2. Maier, R., Brandner, C., Hintner, H., Bauer, J., and Onder, K. (2008) Construction of a reading frame-independent yeast two-hybrid vector system for site-specific recombinational cloning and protein interaction screening. *Biotechniques* **45**, 235–44.
3. Brandner, C. J., Maier, R. H., Henderson, D. S., Hintner, H., Bauer, J. W., and Onder, K. (2008) The ORFeome of *Staphylococcus aureus* v 1.1. *BMC Genomics* **9**, 321.
4. Rual, J. F., Hirozane-Kishikawa, T., Hao, T., Bertin, N., Li, S., Dricot, A., Li, N., Rosenberg, J., Lamesch, P., Vidalain, P. O., Clingingsmith, T. R., Hartley, J. L., Esposito, D., Cheo, D., Moore, T., Simmons, B., Sequerra, R., Bosak, S., Doucette-Stamm, L., Le Peuch, C., Vandenhaute, J., Cusick, M. E., Albala, J. S., Hill, D. E., and Vidal, M. (2004)

Human ORFeome version 1.1: a platform for reverse proteomics. *Genome Res.* **14**, 2128–35.

5. Hartley, J. L., Temple, G. F., and Brasch, M. A. (2000) DNA cloning using *in vitro* site-specific recombination. *Genome Res.* **10**, 1788–95.
6. Walhout, A. J., Temple, G. F., Brasch, M. A., Hartley, J. L., Lorson, M. A., van den Heuvel, S., and Vidal, M. (2000) GATEWAY® recombinational cloning: application to the cloning of large numbers of open reading frames or ORFeomes. *Methods Enzymol.* **328**, 575–92.
7. Bernard, P., Kezdy, K. E., Van Melderen, L., Steyaert, J., Wyns, L., Pato, M. L., Higgins, P. N., and Couturier, M. (1993) The F plasmid CcdB protein induces efficient ATP-dependent DNA cleavage by gyrase. *J. Mol. Biol.* **234**, 534–41.
8. Lynch, M., Scofield, D. G., and Hong, X. (2005) The evolution of transcription-initiation sites. *Mol. Biol. Evol.* **22**, 1137–46.

Chapter 6

Normalization of Full-Length-Enriched cDNA

Ekaterina A. Bogdanova, Ekaterina V. Barsova, Irina A. Shagina, Alexander Scheglov, Veronika Anisimova, Laura L. Vagner, Sergey A. Lukyanov, and Dmitry A. Shagin

Abstract

A well-recognized obstacle to efficient high-throughput analysis of cDNA libraries is the differential abundance of various transcripts in any particular cell type. Decreasing the prevalence of clones representing abundant transcripts before sequencing, using cDNA normalization, may significantly increase the efficacy of random sequencing and is essential for rare gene discovery. Duplex-specific nuclease (DSN) normalization allows the generation of normalized full-length-enriched cDNA libraries to permit a high gene discovery rate. The method is based on the unique properties of DSN from the Kamchatka crab and involves denaturation–reassociation of cDNA, degradation of the ds-fraction formed by abundant transcripts by DSN, and PCR amplification of the remaining ss-DNA fraction. The method has been evaluated in various plant and animal models.

Key words: Duplex-specific nuclease, cDNA normalization, Full-length-enriched cDNA, cDNA library

1. Introduction

Whole transcriptome analysis is a general requirement before many basic biological questions can be addressed. One of the limitations of transcriptome sequencing is associated with significant fluctuations in the concentration of different transcripts within cells (1). Methods to decrease the prevalence of highly abundant transcripts and to equalize mRNA concentrations in a cDNA library are designated "cDNA normalization." Normalization brings the frequency of each transcript in the library within a narrow range and results in a substantial increase in the gene discovery rate (2–5).

Recently, we developed an effective cDNA normalization method especially optimized for full-length-enriched cDNA (6, 7).

Chaofu Lu et al. (eds.), *cDNA Libraries: Methods and Applications*, Methods in Molecular Biology, vol. 729,
DOI 10.1007/978-1-61779-065-2_6, © Springer Science+Business Media, LLC 2011

This method, termed duplex-specific nuclease (DSN) normalization, does not include physical separation steps and is based on selective hydrolysis of the ds-DNA fraction formed by abundant transcripts, using Kamchatka crab DSN. DSN exhibits a strong preference for ds-DNA as a substrate and is stable at elevated temperatures (8). This allows effective removal of ds-DNA from complex nucleic acids at hybridization temperatures.

DSN normalization has become a standard methodology because of its simplicity, applicability to total RNA, and availability. The efficacy of the method has been demonstrated using full-length-enriched library preparations as well as in sequencing projects employing the new parallel sequencers such as the 454 Life Sciences GS 20 model, using a variety of experimental organisms, including plants, insects, mollusks, nematodes, and fish (9–16).

DSN normalization includes cDNA denaturation followed by re-hybridization of denatured ds-cDNA. For each specific transcript, the hybridization rate is proportional to the square of the transcript concentration because nucleic acid hybridization is a second-order chemical reaction. Therefore, abundant transcripts convert to the ds-form more effectively than those that are less common (17), and the ss-cDNA fraction is equalized. After re-hybridization, ds-cDNA is hydrolyzed by DSN, whereas the ss-cDNA fraction remains unchanged. The latter is amplified by PCR and can be used for construction of a normalized cDNA library or immediate high-throughput sequencing.

The cDNA to be normalized should contain known adapter sequences at both ends. Adapter sequences can be introduced onto the ends of cDNA by various methods, e.g., by adapter ligation, or during cDNA synthesis employing the SMART approach (18). Our current protocol utilizes the SMART method as resulted cDNA is enriched with full-length sequences, can be obtained both from poly(A)+ and total RNA (even if only small amounts of starting material are available), and can be flanked by different adapter sequences. In such a protocol, we recommend adapters with asymmetric sites for the *Sfi*I restriction nuclease, to allow direct cloning of cDNA.

2. Materials

2.1. First-Strand cDNA Synthesis (see Note 1)

1. Isolated total or polyA(+) RNA. A number of methods are suitable for RNA isolation, yielding stable RNA preparations from most biological sources. Examples are the TRIzol method (GIBCO/Life Technologies) and the RNeasy kit (QIAGEN). RNA can also be isolated by the well-known method of Chomczynski and Sacchi (19), with one variation: all procedures re-performed at neutral pH instead of the acidic pH originally suggested.

2. 5′-oligonucleotide adapter for the template switching reaction:

   5′-AAGCAGTGGTATCAACGCAGAGTGGCCATTAC GGCCGGG-3′ (see Note 2).

3. CDS-3M adapter:

   5′-AAGCAGTGGTATCAACGCAGAGTGGCCGAGGCG GCC$(T)_{20}$VN-3′ (N = A, C, G, or T; V = A, G, or C).

4. MMLV reverse transcriptase with first-strand buffer (see Note 3).
5. dNTP mix (10 mM of each).
6. 20 mM $MnCl_2$.
7. 20 mM DTT.
8. Fresh MilliQ water.
9. RNase inhibitor (20 U/μL, Ambion).
10. Molecular biology-grade mineral oil.

2.2. PCR Amplification

1. Polymerase mix for long and accurate PCR with buffer (see Note 4).
2. dNTP mix (10 mM each).
3. PCR primer-M1 (10 μM): 5′-AAGCAGTGGTATCAACG CAGAGT-3′.
4. PCR primer-M2 (10 μM): 5′-AAGCAGTGGTATCAACG CAG-3′.
5. Fresh MilliQ water.
6. Molecular biology-grade mineral oil.

2.3. cDNA Purification

1. Commercial PCR purification kit such as QIAquick PCR Purification Kit (QIAGEN) or equivalent (see Note 5).
2. 3 M sodium acetate (NaAc), pH 4.8.
3. 98% (v/v) ethanol.
4. 80% (v/v) ethanol.
5. Columns for cDNA size selection equilibrated in TE buffer (e.g., CHROMA SPIN™-400 or 1000; Clontech).
6. Fresh MilliQ water.

2.4. Agarose Gel Electrophoresis

1. 1 kb DNA size markers.
2. 1.5% (w/v) agarose gel with EtBr.
3. 1× TAE buffer.

2.5. cDNA Hybridization and DSN Treatment (see Note 6)

1. 4× Hybridization buffer (20 mM HEPES, pH 7.5; 2 M NaCl).
2. Duplex-specific nuclease (Evrogen).
3. DSN storage buffer (50 mM Tris–HCl, pH 8.0; provided with the DSN enzyme).

4. 2× DSN master buffer (100 mM Tris–HCl, pH 8.0; 10 mM $MgCl_2$; 2 mM DTT; provided with the DSN enzyme).
5. DSN Stop solution (5 mM EDTA, provided with the DSN enzyme).
6. Glycerol.
7. Fresh MilliQ water.

3. Methods

3.1. cDNA Preparation

3.1.1. First-Strand cDNA Synthesis

1. Before taking an RNA aliquot, heat the total RNA sample at 65°C for 1–2 min and mix the contents by gently flicking the tube, to prevent RNA aggregation. Spin the tube briefly in a microcentrifuge.
2. To 3 μL RNA solution in MilliQ water [0.5–2 μg of total or polyA(+) RNA], add 1 μL of 10 μM CDS-3M adapter and 1 μL of 10 μM 5′-oligonucleotide adapter. Incubate at 72°C for 2 min, and place the tube on ice for 2 min. Spin the tube briefly in a microcentrifuge to collect the contents at the bottom.
3. Add 2 μL of 5× first-strand buffer, 1 μL of 20 mM DTT, 1 μL of dNTP mix (10 mM of each dNTP), and 1 μL reverse transcriptase. Also, 0.5 μL RNase inhibitor should be added to prevent RNA degradation during cDNA synthesis. Mix the contents by gentle pipetting and spin the tube briefly in a microcentrifuge. If a water bath or thermal cycler is used for incubation, cover the reaction mixture with one drop of mineral oil to prevent the loss of volume caused by evaporation.
4. Incubate the tube at 42°C for 30 min, add 1 μL of 20 mM $MnCl_2$, and incubate for 1.5 h at 42°C. After incubation, place the tube on ice to terminate first-strand synthesis. First-strand cDNA can be stored at −20°C for up to 1 month.

3.1.2. cDNA Amplification

5. Prepare the PCR mixture as follows: 80 μL of MilliQ water, 10 μL of 10× PCR buffer (provided with the polymerase mix), 2 μL of dNTP mix (10 mM of each dNTP), 4 μL of PCR primer-M1, 2 μL of first-strand cDNA solution (from step 4), and 2 μL of polymerase mix (see Note 7). Mix the contents by gently flicking the tube. Spin the tube briefly in a microcentrifuge. If the thermal cycler used is not equipped with a heated lid, overlay each reaction with a drop of mineral oil.
6. Subject the tube to PCR cycling using the following program: initial denaturation at 95°C for 1 min; a variable number (*N*) of PCR cycles at 95°C for 15 s, 66°C for 20 s, and 72°C for 3 min; and a final extension at 66°C for 20 s and 72°C for 3 min. Use Table 1 to determine the approximate number of

Table 1
Recommended number of PCR cycles for a given amount of RNA

Total RNA (μg)	Poly(A)+ RNA (μg)	Number of PCR cycles (*N*)
1.0–2.0	0.5–1	14–15
0.5–1.0	0.1–0.5	15–16
0.25–0.5	0.1 or rather less	17–18

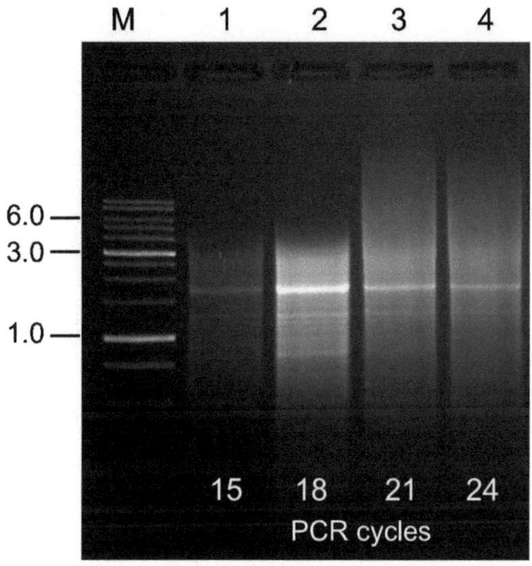

Fig. 1. Agarose gel electrophoresis of amplified human brain cDNA after different numbers of PCR cycles. The number of PCR cycles performed is indicated at the bottom. Lane M: 1 kb DNA size markers (SibEnzyme, Russia), 0.1 μg. 1 μg of total human brain RNA was used for cDNA synthesis. PCR products (4 μL per lane) after 15, 18, 21, and 24 cycles were analyzed on a 1.5% (w/v) agarose/EtBr gel in 1× TAE buffer alongside 0.1 μg of 1 kb DNA size markers. After 21 cycles, a smear appeared in the high-molecular-weight region of the gel, indicating that the reaction was overcycled. Because the plateau was reached after 19–20 cycles, the optimal cycle number for this experiment was 18.

PCR cycles (*N*) required for a given amount of total or poly(A)+ RNA used for first-strand cDNA synthesis (see Notes 8 and 9).

7. When cycling is completed, analyze 5 μL of the PCR product alongside 0.1 μg of 1 kb DNA size markers on a 1.5% (w/v) agarose gel, with EtBr staining, run in 1× TAE buffer. For comparison, Fig. 1 shows a characteristic gel profile of ds-cDNA synthesized from total human brain RNA. In the

case of PCR undercycling, subject the PCR to two more cycles and recheck the product (see Note 9). cDNA can be stored at –20°C for up to 3 months.

3.1.3. cDNA Purification

8. Purify the amplified cDNA to remove primer excess, dNTPs, and salts, using a commercial PCR purification kit. Be sure that the kit used effectively removes excess primer.
9. Transfer the cDNA solution, containing about 700–1, 300 ng of purified cDNA, into a novel sterile tube. Add 0.1 volume of 3 M NaAc, pH 4.8, and 2.5 volumes of 98% (v/v) ethanol. Do not use any co-precipitant in the cDNA precipitation procedure. Vortex the mixture thoroughly and centrifuge the tube for 15 min at maximum speed in a microcentrifuge at room temperature. Remove the supernatant carefully.
10. Gently overlay the pellets with 100 μL of 80% (v/v) ethanol. Centrifuge the tube for 5 min at maximum speed in a microcentrifuge at room temperature. Carefully remove the supernatant. Repeat this step.
11. Air-dry pellet for 10–15 min at room temperature. Be sure that pellet dry completely.
12. Dissolve the pellet in MilliQ water to a final cDNA concentration of about 100–150 ng/μL. To check cDNA quality and concentration, analyze 1 μL of cDNA solution alongside 0.1 μg of 1 kb DNA size markers on a 1.5% (w/v) agarose/EtBr gel in 1× TAE buffer. The cDNA can be stored at –20°C for up to 3 months and normalized afterwards.

3.2. cDNA Normalization

3.2.1. cDNA Hybridization

13. Warm the 4× hybridization buffer at 37°C for 10 min to dissolve any precipitate. Be sure that there is no visible pellet or precipitate in the buffer before use.
14. To a cDNA aliquot comprising 600–1,200 ng of cDNA in MilliQ water (4–12 μL of the cDNA solution from step 12), add 4 μL of 4× hybridization buffer, and MilliQ water, to a total volume of 16 μL (see Notes 10 and 11). Mix the contents and spin the tube briefly in a microcentrifuge.
15. Aliquot 4 μL of the reaction mixture into each of the four appropriately labeled (e.g., see Table 2) sterile PCR tubes. Overlay the reaction mixture in each tube with a drop of mineral oil and centrifuge the tubes at maximum speed in a microcentrifuge for 2 min.
16. Incubate the tubes in a thermal cycler at 98°C for 2 min.
17. Incubate the tubes at 68°C for 5–6 h and proceed immediately to DSN treatment. Do not remove the samples from the thermal cycler before DSN treatment.

Table 2
Setting up for DSN treatment

	Experimental tubes			Control tube
DSN solution	Tube 1 (S1 DSN1)	Tube 2 (S1 DSN1/2)	Tube 3 (S1 DSN1/4)	Tube 4 (S1 Control)
DSN stock solution (1 U/μL)	1 μL	–	–	–
"1/2 DSN" solution (0.5 U/μL)	–	1 μL	–	–
"1/4 DSN" solution (0.25 U/μL)	–	–	1 μL	–
DSN master buffer	–	–	–	1 μL

S<Number>, cDNA sample specification

3.2.2. DSN Treatment

18. Shortly before the end of the hybridization procedure, prepare the following dilutions of DSN enzyme in two sterile tubes:
    (a) Add 1 μL DSN storage buffer and 1 μL DSN stock solution (1 U/μL, see Note 12) to the first tube. Mix by gently pipetting the reaction mixture up and down. Label the tube as "1/2 DSN." Place the tube on ice.
    (b) Add 3 μL DSN storage buffer and 1 μL DSN stock solution (1 U/μL, see Note 12) to the second tube. Mix by gently pipetting the reaction mixture up and down. Label the tube as "1/4 DSN." Place the tube on ice.
19. Preheat the DSN master buffer at 68°C. Add 5 μL of the preheated DSN master buffer to each tube containing hybridized cDNA (from step 17), spin each tube briefly in a microcentrifuge, and return the tubes to the thermal cycler. Do not remove the tubes from the thermal cycler except for the time necessary to add preheated DSN master buffer.
20. Incubate the tubes at 68°C for 10 min.
21. Add DSN enzyme as specified in Table 2. After DSN addition, return the tubes immediately to the thermal cycler. Do not remove the tubes from the thermal cycler except for the time necessary to add DSN enzyme. If a tube is left at room temperature after DSN addition, nonspecific digestion of secondary structures formed by ss-DNA may occur, thus decreasing normalization efficiency.
22. Incubate the tubes in the thermal cycler at 68°C for 25 min; and next add 10 μL of DSN stop solution to each tube. Mix the tube contents and spin the tubes briefly in a microcentrifuge.

23. Incubate the tubes in the thermal cycler at 68°C for 5 min. Next, place the tubes on ice.
24. Add 20 μL of MilliQ water to each tube. Mix the contents and spin the tubes briefly in a microcentrifuge. Place the tubes on ice. The samples can be stored at −20°C for up to 2 weeks and used afterwards to amplify normalized cDNA.

3.2.3. First PCR Amplification

25. For each reaction tube from step 24, prepare a PCR mixture as follows: 40.5 μL of MilliQ water, 5 μL of 10× PCR buffer (provided with the polymerase mix), 1 μL of dNTP mix (10 mM of each dNTP), 1.5 μL of PCR primer-M1, 1 μL of normalized cDNA (from step 24), and 1 μL of polymerase mix (see Note 7). Mix the tube contents by gentle flicking. Spin each tube briefly in a microcentrifuge. If the thermal cycler used is not equipped with a heated lid, overlay each reaction with a drop of mineral oil.
26. Subject the tube to PCR cycling using the following program: initial denaturation at 95°C for 1 min; seven PCR cycles at 95°C for 15 s, 66°C for 20 s, and 72°C for 3 min (see Note 8).
27. After seven PCR cycles, use the control tube (see Table 2) to determine the optimal number of PCR cycles, using the procedure described in steps 28–29. Store the other tubes on ice.
28. Collect 10 μL from the seven-cycle PCR control tube into a clean tube for further agarose gel electrophoresis, and subject the remaining PCR mixture to further PCR cycling with the collection of 12 μL aliquots after 9, 11, 13, and 15 PCR cycles.
29. Analyze 5 μL amounts of the aliquots of each PCR (from step 28) alongside 0.1 μg of 1 kb DNA size markers on a 1.5% (w/v) agarose/EtBr gel, run in 1× TAE buffer. Determine the optimal number of cycles required for amplification of the control DNA (see Note 9 and Fig. 2). Store the remaining materials on ice.
30. Retrieve the seven-PCR experimental tubes from ice, return them to the thermal cycler, and if necessary, subject them to additional PCR cycles, until the optimal number as indicated by the control cDNA is completed. Next, immediately subject the tubes to additional nine cycles (see Note 13).
31. Analyze 5 μL of each PCR alongside a 5 μL aliquot from the control PCR tube (with the optimal PCR cycle number) and 0.1 μg of 1 kb DNA size markers on a 1.5% (w/v) agarose/EtBr gel run in 1× TAE buffer. Select the tube(s) showing efficient normalization (see Note 14). For comparison, Fig. 3 shows a characteristic gel profile of normalized human placental cDNA. If the cDNA from two or more tubes seems well normalized, combine the contents of these tubes into one sterile tube, mix well by vortexing, and spin the tube

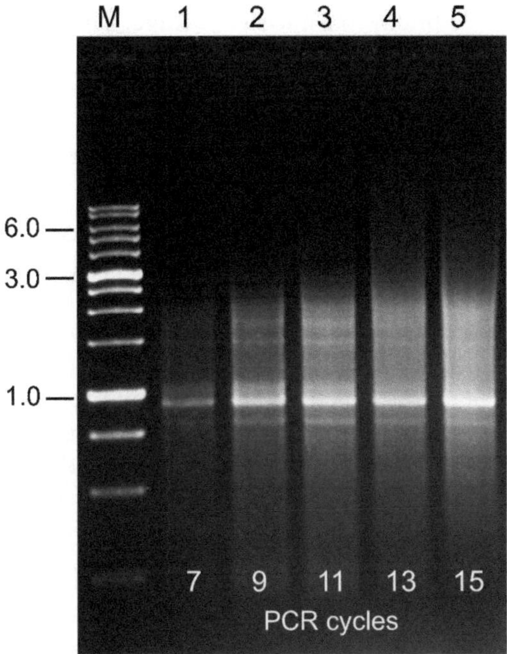

Fig. 2. Agarose gel electrophoresis of amplified human placental cDNA from step 29. The number of PCR cycles performed is indicated at the bottom. Lane M: 1 kb DNA size markers (SibEnzyme), 0.1 μg. 5 μL of each aliquot from the control tube (see step 28) was analyzed on a 1.5% (w/v) agarose/EtBr gel in 1× TAE buffer following the indicated number of PCR cycles. The optimal number of cycles determined in this experiment was nine.

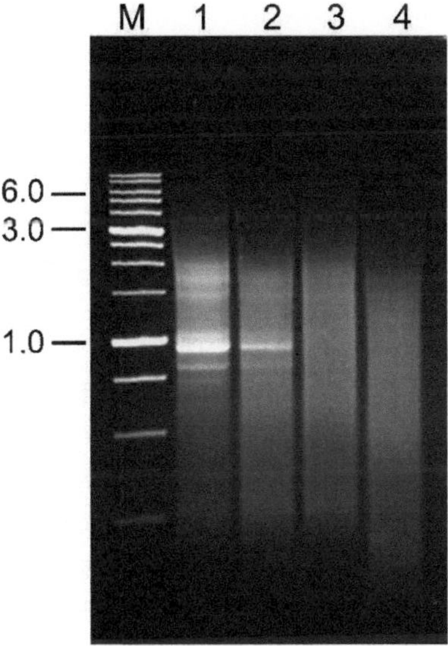

Fig. 3. Analysis of cDNA normalization results. Five-microliter aliquots of the PCR products from step 31 were loaded on a 1.5% (w/v) agarose/EtBr gel. Lane M: 1 kb DNA size markers (SibEnzyme), 0.1 μg. Lane 1: cDNA from the control tube. Lane 2: cDNA from the S1_DSN1/4 tube. Lane 3: cDNA from the S1_DSN1/2 tube. Lane 4: cDNA from the S1_DSN1 tube. In this experiment, efficient normalization was achieved in the S1_DSN1/2 tube (lane 3). In the S1_DSN1/4 tube (lane 1), normalization was not completed, and in the S1_DSN1 tube (lane 4) DSN treatment was excessive, resulting in partial cDNA degradation.

briefly in a microcentrifuge. This amplified normalized cDNA can be stored at –20°C for up to 1 month and used afterwards to prepare more normalized cDNA.

3.2.4. Second PCR Amplification

32. Dilute 2 μL of the normalized cDNA from step 31 in 20 μL MilliQ water.
33. Prepare a PCR mixture as follows: 80 μL of MilliQ water, 10 μL of 10× PCR buffer (provided with the polymerase mix), 2 μL of dNTP mix (10 mM of each dNTP), 4 μL of PCR primer-M2, 2 μL of diluted normalized cDNA (from step 32), and 2 μL of polymerase mix (see Note 7). Mix the tube contents by gentle flicking. Spin the tube briefly in a microcentrifuge. If the thermal cycler used is not equipped with a heated lid, overlay each reaction with a drop of mineral oil.
34. Subject the tube to PCR cycling using the following program: initial denaturation at 95°C for 1 min; 12 PCR cycles at 95°C for 15 s, 64°C for 20 s, and 72°C for 3 min; and a final extension at 64°C for 15 s and 72°C for 3 min (see Note 8).
35. When cycling is completed, analyze 5 μL of the PCR products by electrophoresis alongside 0.1 μg of 1 kb DNA size markers on a 1.5% (w/v) agarose/EtBr gel run in 1× TAE buffer to check PCR quality and DNA concentration. If necessary, subject the sample to 1–2 additional PCR cycles. Amplified normalized cDNA can be stored at –20°C for up to 1 month.

4. Notes

1. cDNA synthesis can be performed using a Mint kit (Evrogen) according to the manufacturer's instructions. Alternatively, the SMART™ cDNA Library Construction Kit (Clontech) or the Creator™ SMART™ cDNA Library Construction Kit (Clontech) can be used, except for the CDS primer. The CDS-3M adapter defined in Subheading 2 must be used instead of the CDS primer provided in the Clontech kit.
2. A 5′-oligonucleotide adapter serves as a second template for the template switching reaction and should contain three riboG nucleotides at the 5′-end to allow effective template switching by reverse transcriptase. This adapter (SMART IV Oligonucleotide) is commercially available in the SMART™ cDNA Library Construction Kit (Clontech) and the Creator™ SMART™ cDNA Library Construction Kit (Clontech). Alternatively, the Mint kit (Evrogen) provides instructions on how to use an adapter containing only deoxyribonucleotides (the adapter has a 3′-end nucleotide with a blocked 3′-OH group but otherwise has the same sequence). Please note

that, under standard conditions, the deoxyribonucleotide adapter is not involved in the template switching reaction.

3. MMLV reverse transcriptase capable of template switching must be used. It has been shown that some mutants of MMLV reverse transcriptase add several nontemplate deoxycytidines to the 3′-ends of newly synthesized first-strand cDNA (20). These deoxycytidines serve as an annealing site for the 5′-oligonucleotide adapter. Reverse transcriptase identifies the 5′-oligonucleotide adapter as an extra part of the RNA template and continues first-strand cDNA synthesis to the end of the oligonucleotide, thus incorporating the adapter sequence into the 5′-end of cDNA (21). For efficient template switching, 1× first-strand cDNA buffer should contain 2.5–3.5 mM $MgCl_2$. Suitable enzymes include Superscript II (Invitrogen), Mint (Evrogen), and SMARTScribe (Clontech) reverse transcriptases. Superscript III (Invitrogen) does not perform effective template switching.
4. Long and accurate PCR is achieved by combining a highly processive thermostable DNA polymerase with a second thermostable polymerase with proofreading (3′→5′ exonuclease) activity. This combination dramatically increases the length of the PCR product (22). In addition, a hot start must be used to reduce nonspecific DNA synthesis during PCR steps. Therefore, PCR kits allowing amplification of long cDNA samples and including automatic hot start are recommended. These include the Encyclo PCR Kit (Evrogen) and the Advantage™ 2 PCR Kit (Clontech).
5. The PCR Purification Kit should efficiently remove primer excess, dNTPs, and salts.
6. All these reagents are provided in the Trimmer and Trimmer-Direct kits (Evrogen).
7. Please use the polymerase mix at the concentration recommended by the manufacturer.
8. Cycling parameters in this protocol have been optimized for an MJ Research PTC-200 DNA machine and the reagents provided in the Encyclo PCR Kit (Evrogen) and the Advantage™ 2 PCR Kit (Clontech). Optimal parameters may vary with different thermal cyclers, polymerases, and templates.
9. Use of the optimal number of PCR cycles ensures that the ds-cDNA remains in the exponential phase of amplification. PCR overcycling is extremely undesirable as this yields nonspecific PCR products. Therefore, it is better to use fewer cycles than too many. The optimal number of PCR cycles must be determined individually for each experimental sample. When the yield of PCR products stops increasing with an additional cycle, the reaction has reached a plateau.

The optimal number of cycles should be one or two cycles less than that needed to reach the plateau. A typical electrophoresis result, indicative of an optimal number of PCR cycles, should appear as a moderately strong cDNA smear of the expected size distribution with several bright bands corresponding to abundant transcripts. For cDNA prepared from most mammalian RNAs, the overall signal intensity (relative to the 1 kb DNA ladder size markers, with 0.1 μg run on the same gel) should be roughly similar to that shown for the experiments of Fig. 1, lane 2 and Fig. 2, lane 2. If the cDNA smear appears in the high-molecular-weight region of the gel (e.g., as in lane 4 in Figs. 1 and 2), especially if no bright bands are distinguishable, this indicates that too many amplification cycles have been employed. If the smear is much fainter (lane 1), this indicates too few PCR cycles. If the size distribution of cDNA is generally less than expected (e.g., less than 2 kb for cDNA from mammalian sources), this could indicate that the initial RNA is of poor quality or has been degraded during storage/synthesis. If the optimal number of PCR cycles is more than 25–26, this indicates that only a few target DNA molecules were amplified; the resulting amplified cDNA thus probably does not contain rare transcripts.

10. When integrity of very long transcripts (more than 5 kb) is crucial, 1 μL of the thermostable single-stranded DNA binding (SSB) protein (with concentration 1–5 μM) can also be added to the reaction mixture. In this case, less PCR cycles are required to amplify normalized cDNA (step 30 and Note 12).

11. Occasionally, the undesired prevalence of several specific sequences occurs in normalized libraries. Failure of the normalization procedure may be attributed to ineffective hybridization of specific sequences due to high TA content or formation of secondary structures. To overcome this problem, DSN normalization may be accompanied by subtraction of undesired known transcripts (14, 23). For subtraction, 1 μL of the driver DNA should be added to the hybridization mixture at step 14. For driver preparation, perform PCR amplification of fragments (about 100 bp in length) of genes to be eliminated using gene-specific primers, purify PCR products using any commercially available PCR purification kit, and mix the fragments together to a final concentration of each fragment 10 ng/μL. Please use plasmid DNA with cloned gene fragments as PCR template.

12. To prepare the DSN stock solution, lyophilized DSN is diluted in DSN storage buffer with 50% (v/v) glycerol as described in the manufacturer's instructions. The final DSN concentration in the stock solution should be 1 U/μL. DSN stock solution should be stored at –20°C, whereas lyophilized DSN is stored at +4°C.

13. If SSB protein is used during DSN treatment (see step 14 and Note 10), less PCR cycles are required to amplify normalized cDNA, e.g., in this case, seven-PCR experimental tubes should be subjected to additional PCR cycles, until the optimal number as indicated by the control cDNA is completed, and to additional six cycles.
14. A typical result, indicative of efficient normalization, should have the following characteristics:
    (a) The overall signal intensity of PCR products from experimental tubes should be similar to the signal intensity of control PCR products. A smear from the experimental tubes that is much fainter than that shown by the control indicates PCR undercycling. In such a case, subject the experimental tubes to two or three additional PCR cycles and repeat the electrophoresis. If the overall signal intensity of PCR products from experimental tubes is much stronger than that of the control, especially if the bright bands are distinguishable, this may indicate that the normalization process was not successful, possibly because DNS has become inactivated during storage.
    (b) The pattern of PCR products from experimental tube(s) containing efficiently normalized cDNA appears as a smear without clear bands, whereas a number of distinct bands are usually present in the pattern of PCR products from the nonnormalized control tube.
    (c) The average length of PCR products from the experimental tube(s) containing efficiently normalized cDNA is congruous with the average length of PCR products from the nonnormalized control tube.

Acknowledgments

This work was supported by Evrogen JSC (Moscow, Russia) and by the program "State Support of the Leading Scientific Schools" (NS-5638.2010.4).

References

1. Alberts, B., Bray, D., Lewis, J., Raff, M., Roberts, K., and Watson, J. D. (1994) Molecular biology of the cell, 3rd ed., Garland Publishing, New York.
2. Soares, M., Bonaldo, M., Jelene, P., Su, L., Lawton, L., and Efstratiadis, A. (1994) Construction and characterization of a normalized cDNA library. *Proc Natl Acad Sci USA* **91**, 9228–32.
3. Carninci, P., Kvam, C., Kitamura, A., Ohsumi, T., Okazaki, Y., Itoh, M., Kamiya, M., Shibata, K., Sasaki, N., Izawa, M., Muramatsu, M., Hayashizaki, Y., and Schneider, C. (1996) High-efficiency full-length cDNA cloning by biotinylated CAP trapper. *Genomics* **37**, 327–36.
4. Luk'ianov, K. A., Gurskaia, N. G., Matts, M. V., Khaspekov, G. L., D'iachenko, L. B.,

Chenchik, A. A., Il'evich-Stuchkov, S. G., and Luk'ianov, S. A. (1996) A method for obtaining the normalized cDNA libraries based on the effect of suppression of polymerase chain reaction. *Bioorg Khim* **22**, 686–90.

5. Bogdanova, E. A., Shagin, D. A., and Lukyanov, S. A. (2008) Normalization of full-length enriched cDNA. *Mol Biosyst* **4**, 205–12.
6. Zhulidov, P. A., Bogdanova, E. A., Shcheglov, A. S., Vagner, L. L., Khaspekov, G. L., Kozhemyako, V. B., Matz, M. V., Meleshkevitch, E., Moroz, L. L., Lukyanov, S. A., and Shagin, D. A. (2004) Simple cDNA normalization using kamchatka crab duplex-specific nuclease. *Nucleic Acid Res* **32**, e37.
7. Zhulidov, P. A., Bogdanova, E. A., Shcheglov, A. S., Shagina, I. A., Wagner, L. L., Khazpekov, G. L., Kozhemyako, V. V., Lukyanov, S. A., and Shagin, D. A. (2005) A method for the preparation of normalized cDNA libraries enriched with full-length sequences. *Russ J Bioorganic Chem* **31**, 170–7.
8. Shagin, D. A., Rebrikov, D. V., Kozhemyako, V. B., Altshuler, I. M., Shcheglov, A. S., Zhulidov, P. A., Bogdanova, E. A., Staroverov, D. B., Rasskazov, V. A., and Lukyanov, S. (2002) A novel method for SNP detection using a new duplex-specific nuclease from crab hepatopancreas. *Genome Res* **12**, 1935–42.
9. Cheung, F., Haas, B. J., Goldberg, S. M., May, G. D., Xiao, Y., and Town, C. D. (2006) Sequencing *Medicago truncatula* expressed sequenced tags using 454 Life Sciences technology. *BMC Genomics* 7, 272.
10. Moroz, L. L., Edwards, J. R., Puthanveettil, S. V., Kohn, A. B., Ha, T., Heyland, A., Knudsen, B., Sahni, A., Yu, F., Liu, L., Jezzini, S., Lovell, P., Iannucculli, W., Chen, M., Nguyen, T., Sheng, H., Shaw, R., Kalachikov, S., Panchin, Y. V., Farmerie, W., Russo, J. J., Ju, J., and Kandel, E. R. (2006) Neuronal transcriptome of aplysia, neuronal compartments and circuitry. *Cell* **127**(7), 1453–67.
11. Simon, A., Glöckner, G., Felder, M., Melkonian, M., and Becker, B. (2006) EST analysis of the scaly green flagellate *Mesostigma viride* (Streptophyta), implications for the evolution of green plants (Viridiplantae). *BMC Plant Biol* **6**, 2.
12. Sandhu, S. K., Jagdale, G. B., Hogenhout, S. A., and Grewal, P. S. (2006) Comparative analysis of the expressed genome of the infective juvenile entomopathogenic nematode, *Heterorhabditis bacteriophora*. *Mol Biochem Parasitol* **145**(2), 239–44.
13. Danley, P. D., Mullen, S. P., Liu, F., Nene, V., Quackenbush, J., and Shaw, K. L. (2007) A cricket Gene Index, a genomic resource for studying neurobiology, speciation, and molecular evolution. *BMC Genomics* **8**, 109.
14. Quilang, J., Wang, S., Li, P., Abernathy, J., Peatman, E., Wang, Y., Wang, L., Shi, Y., Wallace, R., Guo, X., and Liu, Z. (2007) Generation and analysis of ESTs from the eastern oyster, *Crassostrea virginica* Gmelin and identification of microsatellite and SNP markers. *BMC Genomics* **8**, 157.
15. Wang, J., Jemielity, S., Uva, P., Wurm, Y., Gräff, J., and Keller, L. (2007) An annotated cDNA library and microarray for large-scale gene-expression studies in the ant *Solenopsis invicta*. *Genome Biol* **8**, R9.
16. Meyer, E., Aglyamova, G. V., Wang, S., Buchanan-Carter, J., Abrego, D., Colbourne, J. K., Willis, B. L., and Matz, M. V. (2009) Sequencing and de novo analysis of a coral larval transcriptome using 454 GSFlx. *BMC Genomics* **10**, 219.
17. Young, B. D. and Anderson, M. (1985) Quantitative analysis of solution hybridization. In Nucleic acids hybridisation, a practical approach (eds. Hames, B. D. and Higgins, S. J.), 47–71, IRL Press, Oxford/Washington, DC.
18. Zhu, Y. Y., Machleder, E. M., Chenchik, A., Li, R., and Siebert, P. D. (2001) Reverse transcriptase template switching: a SMART approach for full-length cDNA library construction. *Biotechniques* **30**, 892–7.
19. Chomczynski, P. and Sacchi, N. (1987) Single-step method of RNA isolation by acid guanidinium thiocyanate–phenol–chloroform extraction. *Anal Biochem* **162**, 156–9.
20. Schmidt, W. M. and Mueller, M. W. (1999) CapSelect: a highly sensitive method for 5′ CAP-dependent enrichment of full-length cDNA in PCR-mediated analysis of mRNAs. *Nucleic Acids Res* **27**(21), e31.
21. Matz, M., Shagin, D., Bogdanova, E., Britanova, O., Lukyanov, S., Diatchenko, L., and Chenchik, A. (1999) Amplification of cDNA ends based on template-switching effect and step-out PCR. *Nucleic Acids Res* **27**, 1558–60.
22. Barnes, W. M. (1994) PCR amplification of up to 35-kb DNA with high fidelity and high yield from lambda bacteriophage templates. *Proc Natl Acad Sci USA* **91**, 2216–20.
23. Bogdanova, E. A., Shagina, I. A., Mudrik, E., Ivanov, I., Amon, P., Vagner, L. L., Lukyanov, S. A., and Shagin, D. A. (2009) DSN depletion is a simple method to remove selected transcripts from cDNA populations. *Mol Biotechnol* **41**, 247–53.

Chapter 7

Bioinformatic Methods for Finding Differentially Expressed Genes in cDNA Libraries, Applied to the Identification of Tumour Vascular Targets

John M.J. Herbert, Dov J. Stekel, Manuela Mura, Michail Sychev, and Roy Bicknell

Abstract

The aim of this method is to guide a bench scientist to maximise cDNA library analyses to predict biologically relevant genes to pursue in the laboratory. Many groups have successfully utilised cDNA libraries to discover novel and/or differentially expressed genes in pathologies of interest. This is despite the high cost of cDNA library production using the Sanger method of sequencing, which produces modest numbers of expressed sequences compared to the total transcriptome. Both public and propriety cDNA libraries can be utilised in this way, and combining biologically relevant data can reveal biologically interesting genes. Pivotal to the quality of target identification are the selection of biologically relevant libraries, the accuracy of Expressed Sequence Tag to gene assignment, and the statistics used. The key steps, methods, and tools used to this end will be described using vascular targeting as an example. With the advent of next-generation sequencing, these or similar methods can be applied to find novel genes with this new source of data.

Key words: Gene expression, Candidate genes, Differential gene expression, cDNA libraries

1. Introduction

Originally, cDNA clone and library sequencing were performed to find the existence of genes in non-annotated and incomplete genomes (1–3); however, subsequently, they have been used to find differentially expressed genes as ultimately, gene expression determines a cell's phenotype and function (4, 5). Once generated and following publication, cDNA libraries are deposited in public sequence repositories at the National Center for Biotechnology Information (NCBI) (6) and Cancer Genome Anatomy Project (CGAP) (7). By combining biologically relevant libraries from these databases, differential gene expression analyses

Chaofu Lu et al. (eds.), *cDNA Libraries: Methods and Applications*, Methods in Molecular Biology, vol. 729,
DOI 10.1007/978-1-61779-065-2_7, © Springer Science+Business Media, LLC 2011

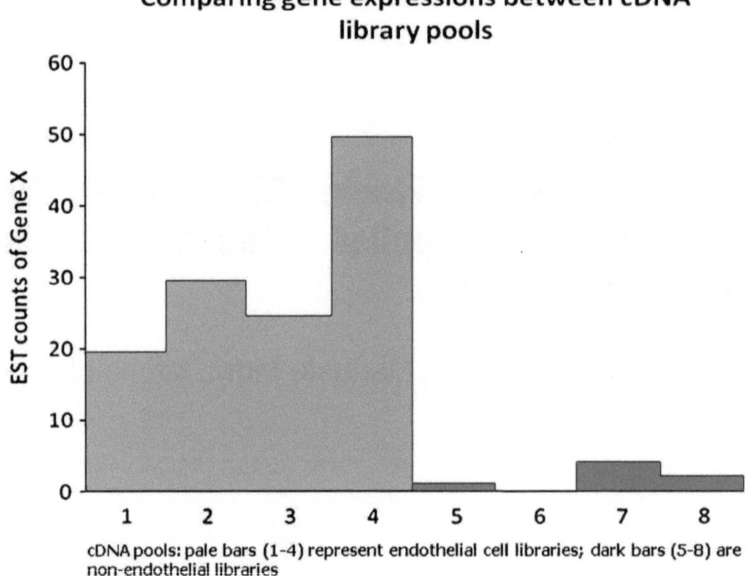

Fig. 1. This shows the simplest scenario of measuring differential expression between two pools of cDNA libraries. *Pale bars*, numbered 1–4, each represent the EST count of gene X in that particular endothelial cell cDNA library; and the *dark* (5–8), the counts of gene X in the non-endothelial cDNA libraries. With the assumption that the total EST count for all genes is the same for each cDNA library pool, gene X is preferentially expressed in endothelial cells.

lead to the identification of candidate genes that can be validated in the laboratory (4, 8, 9).

At the simplest level, measuring differential gene expression between two biologically different pools of cDNA libraries is portrayed in Fig. 1. Each bar in the chart corresponds to a count of Expressed Sequence Tags (ESTs) from a cDNA library, thus, the expression level of gene X in a particular cDNA library. As an example, the pale bars represent endothelial cell cDNA libraries and the dark, non-endothelial libraries. In this simple scenario, it is assumed that both pools contain the same total number of EST counts for all genes and, therefore, it is obvious that gene X is preferentially expressed in endothelial cells. There are three important aspects of cDNA library analyses: relevant cDNA library selection, accurate EST to gene assignment, and statistics.

1.1. Selecting Biologically Relevant cDNA Libraries

cDNA library analysis usually entails comparing one pool of libraries with another pool, each containing biologically relevant data. For this, two questions need to be answered: which cDNA libraries to select and where to find them quickly? cDNA libraries can be found using either the Unigene cDNA library browser (10) or the CGAP cDNA library finder (11). Both resources contain essentially the same data.

Before choosing which cDNA libraries to use in a differential gene expression analyses, a researcher has to think carefully about the biological aim and what constitutes a good library. For example, cDNA libraries can be normalised or subtracted during preparation to select for rare transcripts instead of those highly abundant. For measuring true statistically significant differential gene expression, these libraries should not be used as they bias results. However, if your aim is to find novel tissue-specific genes and the accuracy of the statistics is not so important, then it is good to include them. cDNA library preparation can be performed in a variety of ways and a good overview of the different methods used can be found at CGAP (12). Choosing appropriate libraries may seem obvious; to find vascular targets, a researcher could choose to compare all tumour libraries with all normal libraries. However, from an angiogenesis and vascular targeting perspective, normal libraries can also include placenta or embryonic libraries, all of which undergo active angiogenesis and may affect results.

Our own research has centred on the prediction and validation of endothelial and tumour endothelial genes (8, 9, 13, 14). The biological basis of our research is that bulk tumours need a blood supply to grow beyond 2 mm in diameter; in addition, all a blood vessels in the body, including tumour vessels, are lined with endothelial cells, and endothelial cell gene expression is dependent on its surrounding environment. Tumour vessels have an environment very different to that of normal vessels: tumour vessels are irregular and tortuous, starved of oxygen (1%) in the presence of high levels of vascular endothelial growth factor (VEGF), and starved of glucose at an acidic pH (pH 6.8) and under low shear stress or blood flow. Normal vessels, on the contrary, are regular, in an environment with glucose, higher pH (pH 7.4), and 4% oxygen, subject to a higher shear stress, and with a lower level of VEGF. Since endothelial cells in a tumour environment selectively express genes not present in a normal environment, therapeutic antibodies specific to these genes can destroy tumour vasculature on direct injection into the blood stream, without damage to normal vasculature. This is an attractive therapeutic strategy as there is little toxicity to patients.

In this article, the aim of biology is to find genes selectively expressed on tumour endothelial cells (tumour endothelial markers, TEMs). To predict TEMs, two differential gene expression analyses are needed: the first finds genes selectively expressed in endothelial cells and the second, genes specific to bulk tumour tissues. The intersection constitutes putative TEMs.

1.2. EST to Gene Assignment

Each EST represents a transcript of a gene, and accurate EST to gene assignment is imperative for quality cDNA library analyses. ESTs are between 100 and 600 bases of single pass sequencing from either the 3′ or the 5′ end of a gene and, therefore, can contain sequencing errors. Currently, there are over eight million human ESTs and some contain contaminants such as repetitive elements and bacterial/vector sequences. Unigene has an automated pipeline that constructs contigs of ESTs to assign each EST to a unique gene. This is based on sequence alignments and Boolean filtering, such as the presence of a poly A signal, for example (15). The original Unigene builds did not incorporate genome information and we attempted to improve on this by doing so. Our method involved a stringent process that first BLAST searched each EST against the reference sequence database of nucleotides for high-quality alignments and then compared that result with the genome position of both the EST and matched Refseq gene. If both results concurred, the EST was assigned that gene, otherwise it was rejected (see ref. 8). More recently, Unigene has also started to build EST contigs using genome information (16) but our method is simpler and more stringent, and is used in this work.

1.3. Statistics

A number of methods have been developed to test the significance of differentially expressed genes. The common assumption made by all methods is that each gene can be analysed independently of the others, so that the analysis is applied to one gene at a time. This assumption can be statistically reasonable provided that no one gene contributes too many sequences to a library. If a gene is contributing more than about 20% of a library's sequences, then the assumption may be problematic.

Every method considers the number of counts of the gene of interest in each library and compares them with the total number of counts. The classical statistical approach is to start with a "null hypothesis", namely, that the frequency of the gene should be the same in all libraries. This is analogous to the approach taken in other statistical tests such as the *t*-test. Based on the null hypothesis, it is possible to compute the expected number of counts of the gene in each library. If the numbers are close, then the null hypothesis is accepted, and the gene is considered not to be differentially expressed. If the numbers are very different, then the null hypothesis is rejected and the gene is considered to be differentially expressed.

In this context, there are three common ways of calculating whether these numbers are close or not. The simplest method, as described by Susko and Roger (17), is to use a chi-squared statistic. This is appropriate if the expected number of counts of the gene in each library under the null hypothesis is at least 5. However, when the expected number of transcripts is lower than 5 in one or more libraries, the chi-squared statistic does not give reliable results.

A common alternative, used by the NCBI's Digital Differential Display (DDD) (18), is Fisher's Exact Test (19). This is reliable for rare transcripts; however, as discussed by Claverie (20), the assumptions of Fisher's Exact Test are not really appropriate for this type of data because it assumes that the total number of counts for each gene is a known, fixed number, which is then partitioned between the libraries. This is clearly not the case as any number of transcripts can be sequenced.

Stekel et al. (21) proposed an alternative method that models the number of counts in each library as a Poisson distribution to compute an R-value statistic and uses a randomisation procedure based on these Poisson distributions to compute empirical p-values. This approach has the advantage of being able to be used even for rare transcripts, and, unlike the approaches described below, can be used to compare any number of libraries.

A very different statistical approach is the Bayesian framework that can incorporate "prior" information about gene expression or calculate a "posterior" probability for a particular level of differential gene expression. Audic and Claverie (22) describe a Bayesian method that can be applied for comparing two libraries for differentially expressed genes. One advantage of this approach is that, for example, when a gene is not observed in a library, the classical approaches assume that it is absent; the Bayesian approaches can include probability contributions from when it is rare but unobserved in this particular case. However, Audic and Claverie's approach can only be applied to comparing two libraries. The SAGEmap xprofiler (23) is another Bayesian approach that gives a very different output. Instead of returning a p-value of rejecting a null hypothesis, it computes a probability of the gene being more than x-fold differentially expressed, typically set by the user as twofold. However, this method can also only be used with two libraries (or groups of libraries), and does not produce p-values.

Once a method has been chosen, it would be applied to each gene, one at a time. There are typically hundreds or thousands of genes analysed and so a large list of p-values is generated. This leads to an important statistical consideration: the problem of multiple testing. A p-value is normally interpreted as the probability of seeing a result at least as disparate as the observed result if the null hypothesis were true and so the data truly random. This means, for example, that if we applied a test statistic to 1,000 randomly generated data sets with no true differential expression, we would expect to discover 50 "differentially expressed" genes with p-values less than 0.05. So any analysis of cDNA library data is highly prone to false positives, and statistical methods are needed to control them. An effective approach to counteract the multiple testing problem was developed by Benjamini and Hochberg (24). Termed the False Discovery Rate (FDR), this method generates a q-value (25) that takes into account multiple testing in a way that maximises the power of the test; that is, the ability to discover true positives.

All of the methods described above have their strengths and weaknesses. As described at the start of the section, they all rely on the assumption that the genes can be analysed one at a time. In truth, we know that many genes are co-regulated, by common transcription factors or because different transcription factors are modified by common metabolites. Analyses of this type of data that could take this into account have not yet been developed.

2. Materials

All that is needed are a scientist with a clear idea of the biology, a computer with access to the Internet, and a spreadsheet program such as MS Excel or OpenOffice Calc. A link to the accompanying website, containing software and example files, is found at http://sara.molbiol.ox.ac.uk/userweb/jherbert/tissue_diffex/tissue_index.html (26).

3. Methods

3.1. Preliminaries

Subheading 3 provides step-by-step instructions for a user to perform cDNA library analyses for the prediction of TEMs. It should be achievable by any scientist who has basic Internet browsing skills and experience of spreadsheets. This author used MS Office 2007 and the Google Chrome web browser (version 6.0.472.63) to perform the analyses outlined here. Advice is provided in Subheading 4, and two online files are available as extra examples.

1. Example_TEM_screen (27) is an example Excel file you will end up with if you follow these instructions and this can help a user who does not understand this guide. This file is located on the accompanying website (26).
2. SimpleGuide (28), a Word document, also found on the website (26), goes into more detail (as does in Subheading 4) for some steps and will help anyone who is not fully accustomed to Excel.

To find putative TEMs, perform the following steps.

3.2. Finding cDNA Libraries of Endothelial Cells

Importantly, in two ways, the environment of endothelial cells cultured *in vitro* mimic the environment of endothelial cells of vessels in tumours; the presence of growth factors and the absence of blood flow. Therefore, there is potential for TEMs to be expressed. For the endothelial screen, cDNA libraries from cell isolates are preferred as bulk tissue libraries comprise only ~2% endothelium and expression of genes from other cell types in the bulk tissue could confound results. However, fully understanding the context and the cDNA library origin, and checking for accurate annotation can alter which libraries are selected (see Note 1). Now, follow the instructions listed to create a list of endothelial cell cDNA libraries.

1. Choosing the endothelial pool: go to the cDNA library finder at CGAP, see ref. 11.
2. Fill out the form as follows: Organism = Homo sapiens, Library Group = All EST Libraries, Tissue Type = Any, Tissue Preparation = Any, Tissue Histology = Any, Library Protocol = Any, Library Name = leave empty, and Keyword = endot (a string that will match endothelium or endothelial, etc.). Click the "Submit Query" button. The result of this search, December 2009, produces a list of 34 libraries.
3. This may not be the final list of endothelial libraries to use, so it is useful to make a record of the libraries by storing them in a spreadsheet (Calc or Excel). To do this, there is a link at the top of the page "[Full Text]". Click on this and it will produce a text version of your cDNA library table. Do not be scared, the window will be full of text. Copy all of the text and paste it into an Excel or Calc spreadsheet, using the "paste special" option from the "Paste" drop-down menu and choosing "Text". This way the columns of data will be preserved. Note that there can be problems preserving the columns, see Note 2 or the SimpleGuide (28) if you have this problem as it is important. Save the file, naming it my_TEM_screen, and label the columns the same as in the original html table (Title, Tissue, Histology, Type, Protocol, and Keywords). Also, label the worksheet tab at the bottom, "EndoLibs". This spreadsheet is useful; it enables you to review and keep a record of the libraries you chose and a repository for the results. See the example Example_TEM_screen file (27) and compare it with your results here – it should look similar for this and subsequent steps.

   The list is reviewed to select a final set of endothelial libraries to use in a differential gene expression analysis. From step 2, it can be seen that the selection was liberal and the only specifications were that the libraries are human and contain the string "endot" in the keywords. It is possible that "endot" could match another word, so a review is made to check that all libraries are from endothelium.
4. Review the libraries by reading what is there. You will find that the THP-1 OligodT and THP-1 TAP Libraries are not endothelial cells. Mark them to be removed. Add an extra

column into the sheet (at column G), title it "use" and, by the libraries to be removed, type the string "no" (see ref. 27). Select all (by simultaneously pressing the "Ctrl" and "A" keys), then go to the data tab, and click the sort icon. Select "My data has headers" and sort by the "use" column. This will segregate the endothelial cell and the THP-1 libraries, and there should be a final list of 32 libraries from endothelium (December 2009). See ref. 29 for help on sorting. The two "no" libraries should be at the top of the worksheet. Check your results by comparing them with the example Excel file, sheet "EndoLibs" (27). Your results should be similar, although new data may have become available.

3.3. Finding Non-endothelial Cell cDNA Libraries

To find genes selectively expressed in endothelial cells and not in other cell types, a pool of non-endothelial cells has to be collected for comparison. Again, the biology has to be thought about and some libraries have to be excluded as they potentially contain endothelium or undergo active angiogenesis.

1. Choose the non-endothelial pool; return to the cDNA library finder start page at CGAP (11) and press the reset button of the form.
2. Fill out the form in the following manner: Organism = Homo sapiens, Library Group = All EST Libraries, Tissue Type = Any, Tissue Preparation = Cell Line, Tissue Histology = Normal, Library Protocol = Any, Library Name leave empty, and Keyword = adult. See Fig. 2, which shows the filled-out form. Press "Submit Query". The result of this search, conducted in December 2009, produces a list of 203 non-endothelial libraries.
3. Select "[Full Text]" again and copy this set of libraries into a new worksheet in the Excel/Calc file (copy, paste special, text; as before). Label the tab at the bottom "NonEndoLibs" and put the column titles in as before (see ref. 27). Although there are 203 libraries to review, to make the final list of endothelial genes as accurate as possible, it is worth spending time to look for possible contaminating libraries.
4. Filtering libraries (see Note 3); add a new column in G, titled "use" like in Subheading 3.2, step 4. Search for the word strings: "placenta", "endot", "foetus", "fetus", "foetal", "fetal", "embryo", "unknown developmental stage" and "vascular". Add in the word "no" next to libraries found with any of these words. From the 203 libraries, five are flagged and removed in this way. Some other libraries could be removed depending on how stringent the review is. For instance, there are several CD34 libraries that could be removed, as CD34-positive cells are vascular associated and highly expressed in

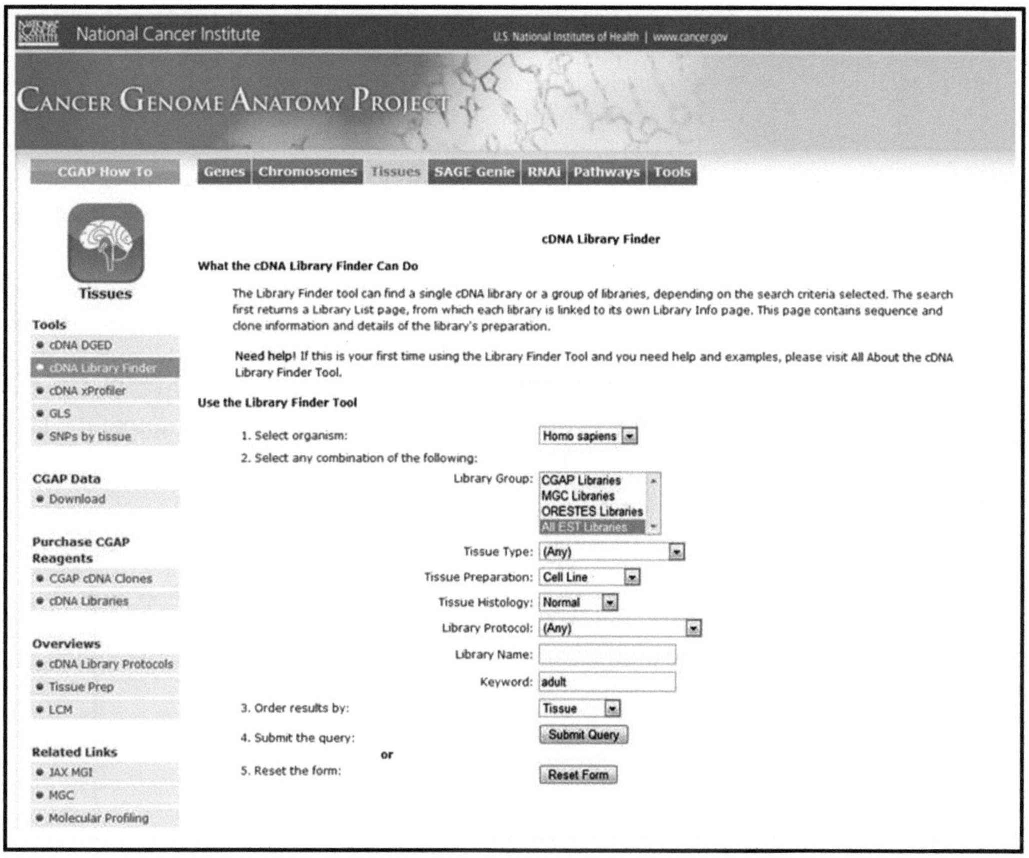

Fig. 2. This figure portrays the web interface to the Cancer Genome Anatomy Project cDNA library finder tool. It enables you to select cDNA libraries with specific attributes; making it a useful query tool in choosing libraries for differential expression analyses. The example form shown here is filled out for the normal, non-endothelial library pool selection.

placenta. Leave them in. You will have 198 non-endothelial cell libraries to use in the endothelial screen and these are listed in the example file, Example_TEM_screen (27). The five cDNA libraries removed include AHMSC2, D3OST2, N1ESE2, T1ESE2, and human aortic endothelium.

3.4. Finding Differentially Expressed Endothelial Genes

With both sets of cDNA libraries in place, the next step is to find differentially expressed genes. We have created a web form to perform differential gene expression analysis using the *R*-value statistic and stringent EST to gene mappings. This tool will be used to predict endothelial genes.

1. From a web browser, go to the Springer book chapter page (26), see Subheading 2, and choose the "Compare two pools of cDNA libraries".
2. You are presented with a web form that has two input windows and a series of statistical options underneath.

3. Simply copy and paste the Title column (containing cDNA library names) from the "EndoLibs" sheet into "pool A" minus the libraries flagged as "no" in Subheading 3.2, step 4 (32 library names). This is the list of endothelial cell libraries. Do the same for the non-endothelial libraries by copying and pasting them into "pool B", not including those labelled "no" in Subheading 3.3, step 4. The form will now look like Fig. 3.
4. Keep the default *R*-value statistic radio button checked and click the "Run Stats" button. The results are returned quickly. As done previous; copy, paste special (using the text option) the results into a new sheet called "EndoResults". Only select the table, from the column titled "Gene" and ignore the lines at the top (Total genes, etc.). See Note 4.

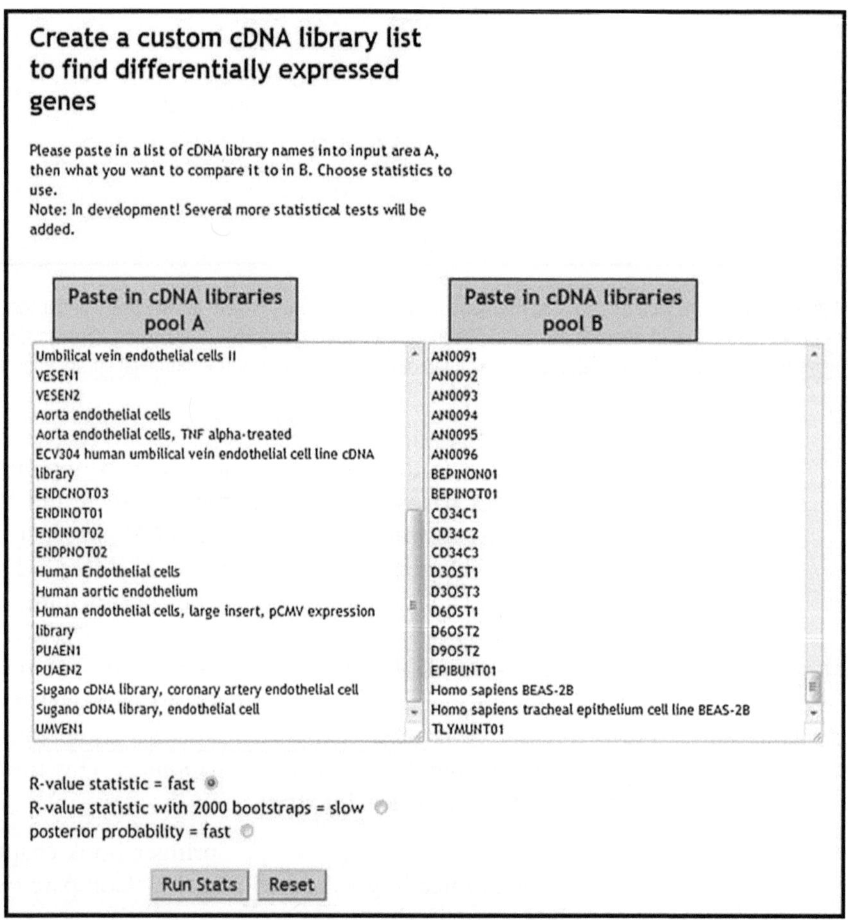

Fig. 3. The simple "diffex" web form is the interface to running differential gene expression analyses between two groups of cDNA libraries. The name of each pool is simply pasted into each of the input windows, pool A and pool B. A choice of statistics to run is listed and the default is set for the *R*-value statistic, which is recommended.

3.5. Endothelial Screen; Interpreting the Results

Figure 4 shows the results of the endothelial screen and the full set can be viewed in the supplementary Excel file on the accompanying website (see Subheading 2), in the "EndoResults" sheet. The results contain six columns of data:

- Gene; the gene symbol
- UP or down; whether a gene is up or downregulated in pool A
- *Q*-value; the False Discovery Rate *q*-value
- log2(fold_change); log2 of the fold change in expression
- TPM_GroupA; number of ESTs in pool A, displayed as transcripts per million
- TPM_GroupB; number of ESTs in pool B, displayed as transcripts per million

Segregate the up and downregulated genes by sorting the data in the following manner. Select all and sort using the following criteria, at the same time: the "Up or down" column in alphabetical descending order (Z to A), (by using "Add Level") in numerical

Results for your differential gene expression analysis

TotalGenes Found 9196
Total EST count Group A = 32374
Total EST count Group B = 49977

Only genes with a False Discovery Rate of 0.05 (Q-value) or less are shown

Gene	UP or down	Q-value	log2(fold_change)	TPM_GroupA	TPM_GroupB
GPNMB	down_groupA	0.0000	-8.34	30	10024
LOC100008589	down_groupA	0.0000	specific	0	34775
LOC100293593	down_groupA	0.0000	-4.44	525	11365
LOC100008588	down_groupA	0.0000	-9.02	30	16087
FUCA1	down_groupA	0.0000	specific	0	15567
ROBO4	UP_groupA	0.0000	specific	6455	0
MMP1	UP_groupA	0.0000	5.61	6826	140
TGM2	UP_groupA	0.0000	5.16	6455	180
MMP9	down_groupA	0.0000	specific	0	6342
MALAT1	down_groupA	0.0000	-5.97	92	5802
SRGN	UP_groupA	0.0000	5.35	4911	120
MCAM	UP_groupA	0.0000	5.10	4818	140
A2M	UP_groupA	0.0000	5.45	4386	100
ENG	UP_groupA	0.0000	4.53	4633	200
HDLBP	UP_groupA	0.0000	3.51	5251	460
SPARCL1	UP_groupA	0.0000	specific	2996	0
EFEMP1	UP_groupA	0.0000	5.81	3366	60
APLP2	UP_groupA	0.0000	2.14	8216	1860
CTGF	UP_groupA	0.0000	2.57	6177	1040
LGMN	down_groupA	0.0000	-4.97	123	3861
ITGA5	UP_groupA	0.0000	3.99	3490	220
VWF	UP_groupA	0.0000	specific	2254	0
BGN	UP_groupA	0.0000	5.42	2563	60
SGK1	UP_groupA	0.0000	3.46	3521	320
EMP1	UP_groupA	0.0000	2.76	4200	620
RPL13A	UP_groupA	0.0000	3.86	2903	200
PPIA	UP_groupA	0.0000	3.57	3088	260
P4HB	UP_groupA	0.0000	2.51	4448	780
CAV1	UP_groupA	0.0000	4.89	2378	80
PLS3	UP_groupA	0.0000	5.76	2162	40
DCTN1	UP_groupA	0.0000	specific	1884	0
RPL37A	UP_groupA	0.0000	specific	1884	0
EIF4G2	UP_groupA	0.0000	2.25	4849	1020
RPL26	UP_groupA	0.0000	4.86	2316	80
HHIP	UP_groupA	0.0000	specific	1853	0
FN1	down_groupA	0.0000	-1.84	1884	6763
LOC100293090	down_groupA	0.0000	-3.34	339	3441

Fig. 4. This figure shows the output of "diffex". It reports back several attributes that can be useful for result interpretation. The attributes are gene symbol, whether the gene was up- or downregulated, the *q*-value (measure of multiple testing significance), log2 fold change in gene expression (negative values mean downregulated), and normalised transcript counts for a gene in each pool (expressed as transcripts per million).

ascending order ("Smallest to Largest") for the "Q-val" column (remember to use "My data has headers"), and lastly, adding another level, "Largest to Smallest" log2 (fold_change). Add a new column G and label it "use" as before. Insert "no" next to all downregulated genes, those labelled "down_groupA" (see the example file from the website (Subheading 2)). This places the most endothelial-specific genes at the top of the list and all downregulated genes at the bottom, which will not be used further. The already validated endothelial genes: ROBO4, VWF, HHIP, MMRN1, CDH5, ANGPT2, SELE, TIE1, and PCDH12; are specific and near the top of the list. Therefore, the first step in TEM prediction appears reliable and is complete. A total of 542 genes were found to be upregulated or specific to endothelial cells with a q-value ≤ 0.05 (see Note 5).

3.6. Finding Tumour Genes

For the tumour screen, only bulk tumour tissue libraries are used as they contain ~2% endothelium, which is not present in cancer cell lines. TEMs can be identified by comparing the results of a tumour screen with that of the endothelial screen. The methods to search for uterine tumour genes are given here, which are almost identical to those used for the endothelial screen.

1. Find bulk tissue uterus cancer libraries; Use the cDNA library finder with the following options: Organism = Homo sapiens, Library Group = All EST Libraries, Tissue Type = Uterus, Tissue Preparation = Bulk, Tissue Histology = Cancer, Library Protocol = Any, and Library Name and Keyword leave empty. This finds 109 libraries. Copy and paste them into a new "UterusBulkCancer" worksheet of the Excel file. Check if all your libraries are from uterine cancers.
2. Find normal bulk libraries of all adult tissues. Reset the cDNA library finder and use the following options: Organism = Homo sapiens, Library Group = All EST Libraries, Tissue Type = Any, Tissue Preparation = Bulk, Tissue Histology = Normal, Library Protocol = Any, Library Name leave empty, and Keyword = adult. A total of 1,703 libraries are found. Copy and paste, as before, into a new sheet, labelled "AllBulkNorm". Add a "use" column and perform filtering by searching for the following words: "endot", "placenta" (there are many), "foetus", "fetus", "foetal", "fetal", "embryo", "unknown developmental stage", and "vascular". Mark any library with any of these strings as "no" again (see Note 6). A total of 359 cDNA libraries were tagged with "no" for removal. Select all and sort by the "use" column. The number of usable libraries is 1,344 (see the example file on the website, Subheading 2).

3. Carry out a differential gene expression analysis like before, using the cancer libraries (title column) as pool A and normal libraries (title column) as pool B. Do not use those marked with "no".
4. Copy the results to a new sheet, named "CancerScreenResults", pasting special as text, and sort as done in Subheading 3.5, step 1. This positions the tumour-specific genes at the top of the list. As before, add a "use" column and label all "down_groupA" genes "no". These genes will not be used further. Over 600 genes (q-value ≤ 0.05) are found to be upregulated or specific to uterine tumours when their expression is compared to normal tissue cDNA libraries.

3.7. Choosing TEMs; a Comparison of Endothelial and Tumour Genes

1. Create another new sheet in the Excel file, called "PutativeTEMs".
2. Copy the column titles, except "use", from "EndoResults" to "PutativeTEMs". Then copy all of the endothelial upregulated genes (labelled "UP_groupA") and data (except the use column) from the "EndoResults" sheet and paste them in the new "PutativeTEMs" sheet (see the example file). The column labelled "gene" must be in the first column (column A) of the "PutativeTEMs" sheet.
3. To find genes upregulated in both tumours and endothelial cells, a function called a VLOOKUP is used. VLOOKUP enables a cross-reference of data between two different worksheets in an Excel file. Add two new columns at G and H in the "PutativeTEMs" sheet and name them "TEMs vlookup from CancerScreenResults" and "putative TEM" respectively.
4. In the "TEMs vlookup from CancerScreenResults" column, the following function syntax is used: =VLOOKUP(A2,CancerScreenResults!$A$2:$B$650,2,FALSE). Copy this function, double click in cell G2, and paste. Important: the VLOOKUP function will only work if sheets and columns are labelled exactly as instructed in this guide. On pressing return, the string "#N/A" should appear next to the first gene in the list, assuming that the cDNA libraries have not changed (see Note 7).
5. Copy the VLOOKUP function all the way down column G ("TEMs vlookup from CancerScreenResults"). Then, sort the genes by the VLOOKUP result ("TEMs vlookup from CancerScreenResults", column G and alphabetically ascending). This will put putative TEMs at the top of the worksheet, which are labelled in column G as "UP_groupA" and not "#N/A".
6. Type the string "yes" into the "putative TEM" column of the genes labelled "UP_groupA" in the "TEMs vlookup from CancerScreenResults" column (column G). A total of 53 putative TEMs were found (see example file on website).

3.8. Interpreting Results

These analyses give a list of genes, some of which will be genuine TEMs and others not. Further data mining can also be performed to further prioritise a list of candidate genes by doing, for instance, a comprehensive literature search or a functional analyses with DAVID (30), or by using orthologue relationships to infer biological processes (31). This list of putative TEMs is a prediction based on expression profiles of public cDNA library data, and some predictions will be false because of low transcriptome coverage and experimental errors. However, mining cDNA libraries only costs a little time on a computer, which is cheaper than screening an entire transcriptome experimentally. It facilitates a quick list of candidate genes, whose expression profiles can be validated in the laboratory using techniques such as Quantitative PCR and/or immunohistochemistry/fluorescence.

From this quick screen, there is encouraging evidence that at least two TEMs have been predicted: ANXA2 and GBP4. Sharma et al. (32) found that an antibody to ANXA2 blocks plasmin-dependent invasion and migration in breast cancer and went on to suggest ANXA2 as a promising vascular target. In addition, from our work in 2008 (8), GBP4 was predicted as a TEM and it is also found among the 46 TEMs predicted here. Subsequent experimental validation with immunofluorescence by confocal laser scanning microscopy has confirmed this prediction. Figure 5 shows two tissue array sections of normal (Fig. 5a) and cancerous oesophagus (Fig. 5b). An antibody to a pan-endothelial marker, *Ulex europaeus* agglutinin-1 (UEA-I), which specifically binds to fucose residues on human endothelial cells of blood vessels, was used to identify vessels in the tissues (red staining). Both the cancer and normal tissues showed the presence of blood vessels (areas circled, Fig. 5a1 + b1). Green staining, by a GBP4-specific antibody, showed staining of blood vessels only in cancer and not the normal tissues (Fig. 5a2 + b2). Also, the GBP4 antibody co-localised with Ulex staining only on the cancer tissue, thus indicating GBP4 as a TEM (orange/yellow staining on Fig. 5b3). The fluorescence seen inside the vessels is background staining to red blood cells, which is confirmed by the lack of blue nucleoli staining (DAPI, a fluorescent nuclear stain) of these cells.

3.9. Summary

This method shows a researcher how to perform cDNA library analysis to predict valuable molecular targets, involved in a tissue or pathology of interest, for a very small amount of time and money. A similar approach can also be taken with microarray, serial analysis of gene expression (SAGE), and next generation sequencing (NGS) published data to find further candidate genes. Good luck and happy gene hunting.

3.10. Implementation of a Website

The website (see Subheading 2) includes implementations of several methods to determine differentially expressed genes in cDNA libraries; this allows a user to make comparisons. Particular

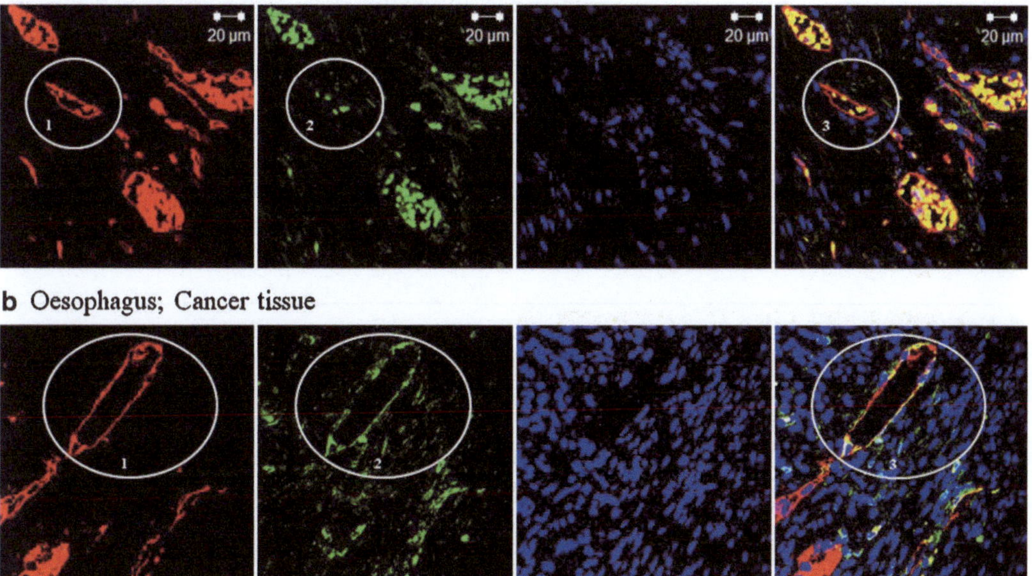

Fig. 5. Successful TEM prediction is confirmed by this figure displaying the immunofluorescence results of oesophagus tumour and normal tissues. An antibody to *Ulex europaeus* agglutinin-1 (UEA-I), which fluoresces *red* in the diagram, specifically binds to fucose residues on human endothelial cells of blood vessels and clearly shows their presence in both tissues (*circled* a1, b1, a3, and b3). A green fluorescent antibody, specific to GBP4, stains blood vessels only in the tumour and not in the normal tissue (a2, b2), thus showing a TEM-specific expression profile. Vessel expression is confirmed by the co-localisation of the *green* GBP4 antibody with the *red* endothelial specific antibody, giving a mixed fluorescence colour of *yellow* that is found only in the tumour tissue (a3, b3).

emphasis is placed on the methods described in the study by Stekel et al. (21) and Herbert et al. (8). In all cases, a Benjamini and Hochberg FDR correction has been implemented.

Four types of analysis can be carried out.

(a) One group of libraries in order to identify genes that are differentially expressed in one or more libraries relative to the others. For this analysis, the method described by Stekel et al. (21) can be used.

(b) Two groups of libraries in order to identify genes that are differentially expressed in one group of libraries relative to the other. For this analysis, the method described by Herbert et al. (8) can be used.

(c) Two groups of paired libraries, for example, a group of patients from whom a normal tissue library and a disease tissue library are both available. The aim is to find genes that are differentially expressed in one condition relative to the other, making use of the paired structure. For this analysis, we have generated a signed R-statistic as a measure of differential expression for each pair by combining the method described by Stekel et al. (21), with an assignment of a positive or negative value depending on whether the gene is up- or downregulated. A nonparametric Wilcoxon test using the signed R-statistics is then used to obtain a p-value.

(d) Two groups of paired libraries as in (c), in order to identify genes that are positively correlated. This analysis does not make use of R-statistics, but computes a one-sided p-value for the Spearman correlation coefficient based on the frequencies of the gene in each of the libraries.

4. Notes

1. Although some endothelial libraries are annotated as "bulk", it is apparent from the keywords that they are isolated endothelial cells. As an example, the "Umbilical vein endothelial cells II" library is annotated as "bulk" and, in theory, should not be used in the endothelial screen here. However, the authors' laboratory has extensive experience of isolating endothelial cells from umbilical cords, and it is evident that this library was produced in a similar way. In addition, the term "bulk endothelium" refers to endothelium being "the largest or principal portion" of a mass of tissue. Therefore, although ten libraries were classed as either "bulk" or "uncharacterised", there was sufficient evidence to satisfy the author that these libraries were principally endothelial cells and not of mixed bulk tissues.
2. Pasting text into Excel can be a problem as it could be set to a different delimiter. For instance, if the Excel is expecting comma-delimited data, lines of text not having any will cause the entire text to be placed into a single cell. This is not what is required, so a simple solution is to use the "Convert text to columns wizard" option in Excel, which enables changing the delimiter to a tab. A quick guide to this is as follows:
   - Delete all data from the Excel sheet and type a single word into a single cell.
   - Select this cell, then In the Data Tools group, click TEXT TO COLUMNS, which has an icon like this.

- Step 1; the text you typed into a cell will be displayed in the "Preview of selected data" section.

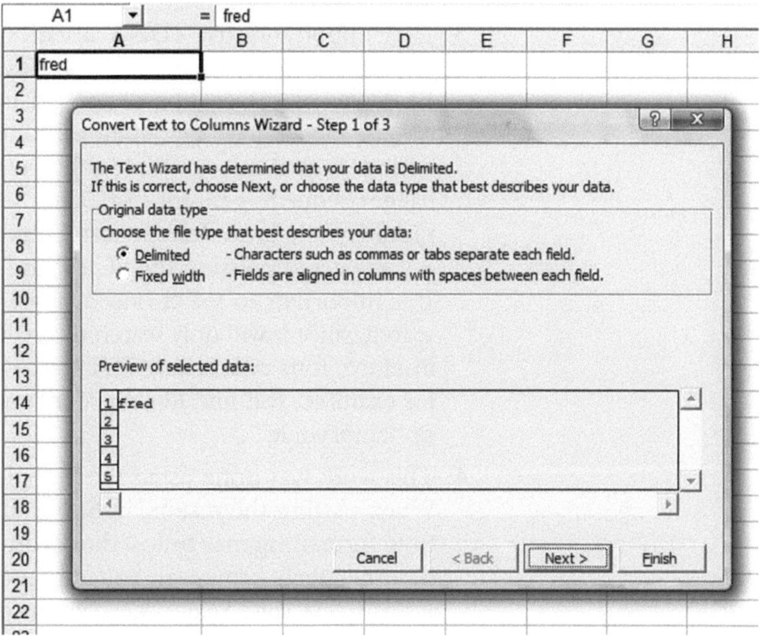

- Under the "Original data type" section, select the "Delimited" radio button and click next.
- Step 2; under "Delimiters", select the "Tab" delimiter option only, making sure all other delimiters are not selected and click next.

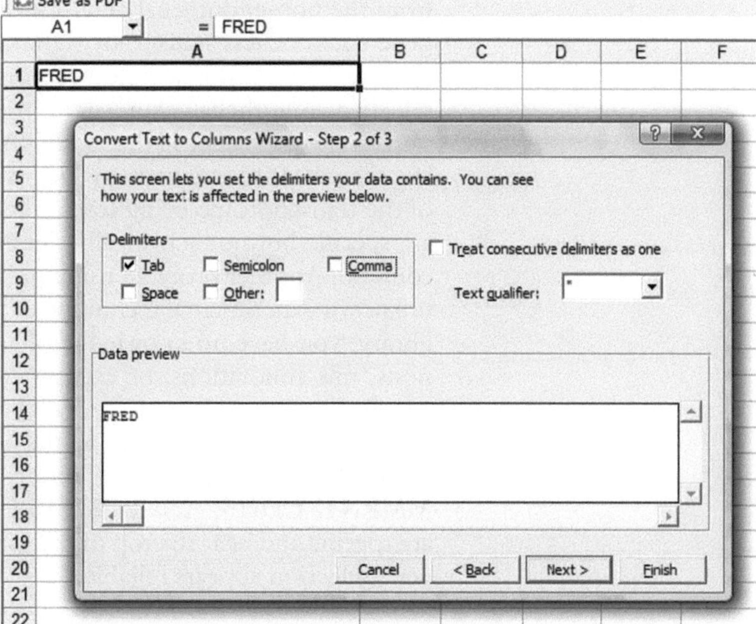

- Step 3; this step can be ignored and just click on the "Finish" button. Now pasting text (paste special text) from the cDNA library finder page should work correctly.

3. It is important that cDNA libraries to be subtracted from endothelial cells do not express vascular/angiogenic genes. Examples of cDNA libraries to exclude are bulk tissues, placenta, and embryonic libraries. Cancer cell lines are also excluded, as there is possible abhorrent expression of endothelial genes due to genome instability, re-arrangements, and copy number variations. Searches can be performed by selecting cell A1 alone and pressing the "Ctrl" and "f" keys simultaneously. It is important to select only a single cell in Excel before you search, since it will only search the cells selected if you have, for instance, four cells selected. A search term, such as "embryo" for example, will find libraries containing the terms "embryo" or "embryonic".
4. This work was done using the Google Chrome web browser. When pasting the results, in Subheading 3.4, step 4, the column formatting may be lost due to using MS Internet Explorer or other web browsers. You can use "Text to columns" in Excel to fix this. Select the column of results, then "Text to columns", and choose "next", then the "space" radio button. Select finish and it will format the data correctly. However, if you do this, you will have to set the delimiter back to tab for collecting the tumour/normal libraries in step 3.6.
5. It is important to note that genes can be either upregulated or specific to endothelial cells. If a gene has zero counts in the TMP_GroupB column, it is not expressed in any cell type from the non-endothelial cDNA library pool. Such a specific gene could be less statistically significant than an upregulated one but important in a biological context as the aim is to find selective endothelial expression. Another consideration of statistical significance is that a cDNA library usually contains less than 10,000 sequences, which is only a small proportion of the transcriptome being transcribed. Therefore, genes that are specific but not statistically significant should not be discounted. Another problem using public cDNA libraries is the unknown experimental technique of those who produced the library; you have no knowledge of any mistakes, contaminations, mis-annotations, or experimental errors. Therefore, a control in this screen is to see if established endothelial genes appear near or at the top of the upregulated genes. The already validated endothelial genes: ROBO4, VWF, HHIP, MMRN1, CDH5, ANGPT2, SELE, TIE1, and PCDH12, are specific and near the top of the list. Therefore, endothelial gene selection appears reliable.

6. It may speed up the process if you first sort the table by the "Tissue" column. It will be quicker to classify the unwanted 353 placenta libraries present by copying "no" next to all these libraries listed consecutively in the sheet.
7. About VLOOKUP; the first argument in the function uses the gene symbol in cell A2 and looks to see if this gene symbol exists in the first column of the "CancerScreenResults" sheet, referenced by the second VLOOKUP argument (CancerScreenResults!$A$2:$B$650). If the gene symbol does exist, the data stored in column 2 (the third argument) of the "CancerScreenResults" sheet, titled "UP or down", are returned and displayed. Now, genes with "Up" next to them are putative TEMs. The FALSE argument specifies exact matches when comparing gene symbols. Any gene not found in the "CancerScreenResults" sheet using this VLOOKUP function returns and displays the string "#N/A". For a more detailed description on how VLOOKUP's work, please see online refs. 33, 34 or refer to the Example_TEM_screen file (27).

Acknowledgments

John Herbert would like to specially thank Dov Stekel for his patient help and advice with the statistics and Professor Roy Bicknell for his expert biological advice. Thanks also to Dr. Manuela Mura and Michail Sychev for help with the immunofluorescence validation and website, and Mr. Stephen Taylor, Dr. Simon Mcgowan, and Dr. Zong-pei Han at the Computational Biology Research Group, who provided a Bioinformatic computing server. Thanks to all the "guinea pigs" from the Angiogenesis Group, Birmingham, who helped with testing the method. John M. J. Herbert was funded by Cancer Research UK, project Grant no. C4719/A6766.

References

1. Adams, M. D., Dubnick, M., Kerlavage, A. R., Moreno, R., Kelley, J. M., Utterback, T. R., Nagle, J. W., Fields, C., and Venter, J. C. (1992) Sequence identification of 2,375 human brain genes. *Nature* **355**, 632–634.
2. Adams, M. D., Kelley, J. M., Gocayne, J. D., Dubnick, M., Polymeropoulos, M. H., Xiao, H., Merril, C. R., Wu, A., Olde, B., Moreno, R. F., et al. (1991) Complementary DNA sequencing: expressed sequence tags and human genome project. *Science* **252**, 1651–1656.
3. Dugaiczyk, A., Haron, J. A., Stone, E. M., Dennison, O. E., Rothblum, K. N., and Schwartz, R. J. (1983) Cloning and sequencing of a deoxyribonucleic acid copy of glyceraldehyde-3-phosphate dehydrogenase messenger ribonucleic acid isolated from chicken muscle. *Biochemistry* **22**, 1605–1613.

4. Bortoluzzi, S., Bisognin, A., Romualdi, C., and Danieli, G. A. (2005) Novel genes, possibly relevant for molecular diagnosis or therapy of human rhabdomyosarcoma, detected by genomic expression profiling. *Gene* **348**, 65–71.
5. Itoh, K., Okubo, K., Utiyama, H., Hirano, T., Yoshii, J., and Matsubara, K. (1998) Expression profile of active genes in granulocytes. *Blood* **92**, 1432–1441.
6. NCBI. (2009) National Center for Biotechnology Information (http://www.ncbi.nlm.nih.gov/).
7. CGAP. (2009) Cancer Genome Anatomy Project (http://cgap.nci.nih.gov/cgap.html).
8. Herbert, J. M., Stekel, D., Sanderson, S., Heath, V. L., and Bicknell, R. (2008) A novel method of differential gene expression analysis using multiple cDNA libraries applied to the identification of tumour endothelial genes. *BMC Genomics* **9**, 153.
9. Huminiecki, L., and Bicknell, R. (2000) In silico cloning of novel endothelial-specific genes. *Genome Res.* **10**, 1796–1806.
10. NCBI. (2009) Unigenes cDNA library browser, p Unigenes cDNA library browser (http://www.ncbi.nlm.nih.gov/UniGene/lbrowse2.cgi).
11. CGAP. (2008) Cancer Genome Anatomy Project cDNA Library Finder, p The Cancer Genome Anatomy Project library finder; enables searches of a cDNA library database (http://cgap.nci.nih.gov/Tissues/LibraryFinder).
12. CGAP. (2009) Cancer Genome Anatomy Project: cDNA Library Protocols Overview (http://cgap.nci.nih.gov/Tissues/LibProtocols).
13. Armstrong, L. J., Heath, V. L., Sanderson, S., Kaur, S., Beesley, J. F., Herbert, J. M., Legg, J. A., Poulsom, R., and Bicknell, R. (2008) ECSM2, an endothelial specific filamin a binding protein that mediates chemotaxis. *Arterioscler. Thromb. Vasc. Biol.* **28**, 1640–1646.
14. Huminiecki, L., Gorn, M., Suchting, S., Poulsom, R., and Bicknell, R. (2002) Magic roundabout is a new member of the roundabout receptor family that is endothelial specific and expressed at sites of active angiogenesis. *Genomics* **79**, 547–552.
15. NCBI. (2009) Unigene clustering: build 1 (http://www.ncbi.nlm.nih.gov/UniGene/build1.html).
16. NCBI. (2009) Unigene clustering: build 2 (http://www.ncbi.nlm.nih.gov/UniGene/build2.html).
17. Susko, E., and Roger, A. J. (2004) Estimating and comparing the rates of gene discovery and expressed sequence tag (EST) frequencies in EST surveys. *Bioinformatics* **20**, 2279–2287.
18. NCBI. (2009) Digital Differential Display tool, Digital Differential Display tool ed (http://www.ncbi.nlm.nih.gov/UniGene/ddd.cgi).
19. Fisher, R. (1925) *Statistical Methods for Research Workers.* Oliver and Boyd, Edinburgh.
20. Claverie, J. M. (1999) Computational methods for the identification of differential and coordinated gene expression. *Hum. Mol. Genet.* **8**, 1821–1832.
21. Stekel, D. J., Git, Y., and Falciani, F. (2000) The comparison of gene expression from multiple cDNA libraries. *Genome Res.* **10**, 2055–2061.
22. Audic, S., and Claverie, J. M. (1997) The significance of digital gene expression profiles. *Genome Res.* 7, 986–995.
23. Lash, A. E., Tolstoshev, C. M., Wagner, L., Schuler, G. D., Strausberg, R. L., Riggins, G. J., and Altschul, S. F. (2000) SAGEmap: a public gene expression resource. *Genome Res.* **10**, 1051–1060.
24. Benjamini, Y., and Hochberg, Y. (1995) Controlling the false discovery rate: a practical and powerful approach to multiple testing. *J. R. Stat. Soc. B* **57**, 289–300.
25. Storey, J. D., and Tibshirani, R. (2003) Statistical significance for genomewide studies. *Proc. Natl. Acad. Sci. USA* **100**, 9440–9445.
26. Herbert, J. M. J. (2009) Springer chapter accompanying site; Bioinformatic methods for finding differentially expressed genes in cDNA libraries, applied to the identification of tumour vascular targets (http://sara.molbiol.ox.ac.uk/userweb/jherbert/tissue_diffex/tissue_index.html).
27. Herbert, J. M. J. (2009) Example results file; Example_TEM_screen.xls, p Example results file; Example_TEM_screen.xls (http://sara.molbiol.ox.ac.uk/userweb/jherbert/temp/Example_TEM_screen.xls).
28. Herbert, J. M. (2009) A simple guide to relevant Excel. An easy guide to some basic Excel needed to perform the data manipulation that comes with cDNA library analyses (http://sara.molbiol.ox.ac.uk/userweb/jherbert/temp/SimpleGuide.zip).
29. Microsoft. (2009) Excel help; a web page that can be searched for all versions of Excel for help on sorting (http://office.microsoft.com/en-gb/excel/default.aspx).
30. DAVID. (2009) The Database for Annotation, Visualization and Integrated Discovery (http://david.abcc.ncifcrf.gov/home.jsp).

31. Herbert, J. M., Buffa, F. M., Vorschmitt, H., Egginton, S., and Bicknell, R. (2009) A new procedure for determining the genetic basis of a physiological process in a non-model species, illustrated by cold induced angiogenesis in the carp. *BMC Genomics* **10**, 490.

32. Sharma, M. R., Koltowski, L., Ownbey, R. T., Tuszynski, G. P., and Sharma, M. C. (2006) Angiogenesis-associated protein annexin II in breast cancer: selective expression in invasive breast cancer and contribution to tumor invasion and progression. *Exp. Mol. Pathol.* **81**, 146–156.

33. Microsoft. (2009) VLOOKUP reference (http://office.microsoft.com/en-us/excel/HP052093351033.aspx).

34. Productivity Portfolio. (2009) Learning VLOOKUP in Excel (http://www.timeatlas.com/5_Minute_Tips/General/Learning_VLOOKUP_in_Excel).

Part II

Visionary Applications

Chapter 8

Enzymatic Production of RNAi Libraries from cDNAs and High-Throughput Selection of Effective shRNA Expression Constructs

Kohtaroh Sugao and Kenzo Hirose

Abstract

RNA interference (RNAi) using small interfering (siRNA) or short hairpin RNA (shRNA) has become the first choice for gene silencing maneuver in mammalian cells. Because different siRNAs of the same gene have variable silencing efficacy and only limited siRNAs are functional, many candidates are necessary to identify optimal siRNAs. We have previously reported a method named *e*nzymatic *p*roduction of *RNAi l*ibrary (EPRIL), by which a great variety of shRNA expression constructs (RNAi library) can be produced simultaneously from cDNAs of interest. Recently, we have improved this method and developed a more efficient method. We describe in this chapter detailed protocols for the improved version of EPRIL and high-throughput selection of effective shRNA expression constructs from an RNAi library.

Key words: RNAi, RNAi library, siRNA, shRNA, EPRIL, Enzymatic production of RNAi library

1. Introduction

RNA interference (RNAi) is a cellular phenomenon of double-stranded RNA (dsRNA) directed post-transcriptional gene silencing (1). In mammalian cells, RNAi can be artificially triggered by two different approaches: (1) 21–29 nt RNA duplex, called small interfering RNA (siRNA), or short hairpin RNA (shRNA), which is a precursor of siRNA, is prepared *in vitro* and introduced into cells (2–5) and (2) siRNA or shRNA is intracellularly produced from DNA-based expression construct (6–12). In cells, siRNA is incorporated into a large protein complex known as RNA-induced silencing complex then used as guide for selection and cleavage of complementary mRNA. Because of its high potency and specificity,

Chaofu Lu et al. (eds.), *cDNA Libraries: Methods and Applications*, Methods in Molecular Biology, vol. 729,
DOI 10.1007/978-1-61779-065-2_8,

siRNA and its expression construct have become a powerful tool for loss-of-function experiments, and are in clinical trials for use as drugs (13, 14).

However, several factors limit the use of siRNA-based RNAi. It is known that the potency of siRNA is strongly dependent on its sequence, and only a small fraction of siRNAs can induce highly effective gene silencing. Moreover, there are several evidences that siRNAs alter the expression level of unintended (off-target) mRNAs through partial sequence complementarity (15, 16). Therefore, the choice of optimal siRNA sequences is critical for successful gene silencing. To identify rules for optimization of siRNA sequences, much effort has been devoted and several siRNA design algorithms have been reported (17–19). However, even with the use of such algorithms, it is often required to prepare and test a number of candidates until optimal siRNA sequences are identified. In general, both siRNA and its expression constructs are prepared using chemically synthesized RNA or DNA oligonucleotides. For expression constructs, cloning of DNA oligonucleotides is also required. These preparation procedures are costly and time-consuming task.

To circumvent these problems, we and others have developed alternative approaches for preparation of siRNAs or shRNA expression constructs using enzymatic techniques (20–25). In our technology (23), named *e*nzymatic *p*roduction of *RNAi* *l*ibrary (EPRIL), cDNA of interest is used as a starting material, and a large number of different shRNA expression constructs are produced at the same time through several enzymatic processes. Because its sequence complexity is high enough to cover the entire sequence of target mRNA, this pool of shRNA expression constructs (RNAi library) can be used to find out most optimum sequences (26, 27). In addition, EPRIL is compatible with a complex mixture of cDNAs. When double-stranded cDNA library converted from cellar transcripts is used as starting material, RNAi library comprised of shRNAs targeting the entire genes is generated (23).

We have recently developed an improved version of EPRIL; the main improved point is substitution of a magnetic beads-based system for polyacrylamide gel electrophoresis (PAGE) for purification of intermediate products. The improved EPRIL allows us to produce much more shRNA expression constructs with less starting material compared with the original version. Furthermore, the procedure time is shortened and parallel handling of multiple samples is much easier. In this chapter, we describe a detailed protocol for the improved EPRIL focusing on a case using single cDNA as starting material. However, this protocol also can be used for cDNAs mixture. In such case, a reaction volume should be scaled up correspondingly to the library's complexity you require. We also describe protocols for rapid selection of effective shRNA expression constructs using reporter system.

2. Materials

2.1. Preparation of Vector Plasmid

1. pNAMA2-U6 shRNA-expressing retroviral vector plasmid (Fig. 1a): the plasmid and its sequence data are available on request.
2. Restriction enzymes and buffers: *Bsg*I, *Bbs*I, *Mfe*I, *Eco*RI, 10× NEBuffer 2, 10× NEBuffer 4 (New England Biolabs, NEB).
3. 32 mM *S*-Adenosylmethionine (NEB).
4. 1% Agarose gels.
5. MinElute gel extraction kit (Qiagen).

2.2. Phenol–Chloroform Extraction and Ethanol Precipitation

1. Phenol:Chloroform:Isoamyl alcohol (PCI): 25:24:1 (pH 7.9).
2. 95 and 70% Ethanol.
3. 10 M Ammonium acetate.
4. 20 mg/mL Glycogen.

2.3. Preparation of Adaptors

1. Oligo DNAs for EPRIL Adaptors:

Adaptor 1:	5′-gtcggactcagaccttctccagggatccggagagccacctgtcta-Am-3′ 5′-Btn-tagacaggtggctctccggatccctggagaaggtctgagtc cgac-3′
Adaptor 2:	5′-Phos-ctcgagggcaattggctgcaccccctgcagtgcagggggg agcagccaattgccctcgagnn-3′
Adaptor 3:	5′-Btn-ccagccgcgctacgcgccgcgcggaagacgctttg-3′ 5′-Phos-caaagcgtcttccgcgcggcgcgtagcgcggctgg-Am-3′

(Am, Btn, and Phos indicate amination, biotinylation, and phosphorylation, respectively. Adaptor 2 contains two degenerate bases, denoted as "n," which represents a, c, g, or t.)

2. 12% Native polyacrylamide gels.
3. Diffusion buffer: 0.5 M ammonium acetate, 10 mM $MgCl_2$, 0.1% SDS, and 1 mM EDTA.
4. UltraFree-MC Centrifugal filter units 0.45 μm (Millipore).
5. TE: 10 mM Tirs–HCl (pH 7.5), 1 mM EDTA.

2.4. Enzymatic Production of RNAi Library

1. DNase I: Recombinant DNase I (Takara).
2. BSA: 10 mg/mL BSA (NEB).
3. DNase I dilution buffer: 50 mM Tris–HCl (pH 7.5), 0.1 μg/μL BSA.
4. T4 DNA polymerase, 10× T4 DNA polymerase buffer (Takara).

a

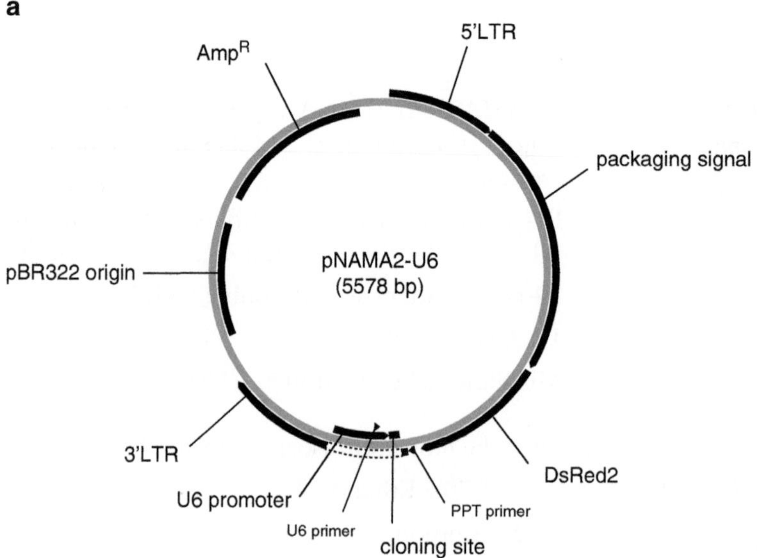

U6 primer 5'-CAGCACAAAAGGAAACTCACCC -3'
PPT primer 5'-AGTCTCCAGAAAAAGGGGGGA -3'

b

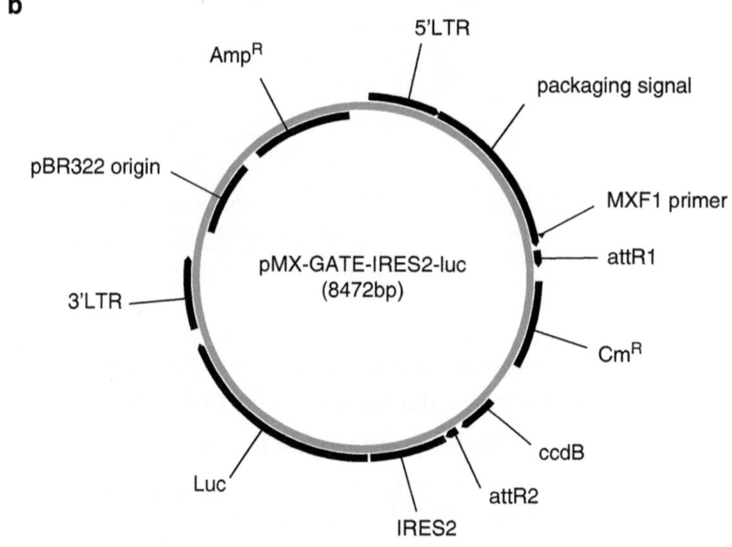

MXF1 primer 5'- AGTAGACGGCATCGCAGCTT -3'

Fig. 1. Plasmid map for pNAMA2-U6 and pMX-GATE-IRES2-Luc. (**a**) pNAMA2-U6 plasmid is a retroviral vector plasmid containing an RNA polymerase III-driven shRNA expression cassette. This plasmid is an improved version of pNAMA-U6, which is used in original EPRIL. The shRNA expression cassette (mouse U6 promoter, cloning site for insert DNA and terminal signal of polymerase III) is placed within *Nhe*I site of 3′ LTR. We use U6 primer or PPT primer for sequencing of insert. (**b**) pMX-GATE-IRES2-Luc is a retroviral vector plasmid containing a Gateway® destination cassette. We use MXF1 primer for sequencing of insert. LTR, long terminal repeat; DsRed2, DsRed2 coding sequence (*Bbs*I site is eliminated); Amp^R, β-lactamase expression cassette; ccdB, ccdB expression cassette; Cm^R, chloramphenicol *O*-acetyltransferase expression cassette; IRES2, internal ribosome entry site of encephalomyocarditis virus; luciferase, firefly luciferase coding sequence.

5. dNTP mix: dNTP mixture, 2.5 mM each (Takara).
6. *Bst* DNA polymerase large fragment, 10× ThermoPol reaction buffer (NEB).
7. Restriction enzymes and buffer: *Mme*I, Nt.BstNBI, *Bpm*I, *Xho*I, 10× NEBuffer 2 (NEB).
8. Ligase solution: DNA ligation kit Ver.2.1, solution I (Takara).
9. Streptavidin beads: Dynabeads MyOne Streptavidin C1 (Invitrogen).
10. 2× B&W buffer: 10 mM Tris–HCl (pH 7.5), 1 mM EDTA, 2 M NaCl.
11. Magnet stand: MPC-S magnet (Invitrogen).
12. PicoGreen assay kit (Invitrogen).
13. Competent cells: ErectroMAX DH5α-E (Invitrogen).
14. Electroporation cuvette: 0.1-cm gap cuvette electrodes (BM Equipment).
15. MicroPulser Electroporator (Bio-Rad).
16. S.O.C. medium (Invitrogen).
17. Luria Broth (LB): For 1 L, 10 g tryptone, 5 g yeast extract, 5 g NaCl, and 2 g glucose.
18. Agar LB plates: For 1 L, 10 g tryptone, 5 g yeast extract, 5 g NaCl, 2 g glucose, and 15 g agar.

2.5. Isolation of shRNA Expression Constructs from RNAi Libraries

1. 96-Deep-well plates (Thermo Fisher Scientific).
2. Gas-permeable seals (Thermo Fisher Scientific).
3. 96-Well PCR plates (Thermo Fisher Scientific).
4. Heat seals (Thermo Fisher Scientific).
5. Circle-Grow (Q-bio).
6. Glycerol.
7. Perfect Prep Plasmid 96 Vac Kit (Eppendorf, now available from 5 Prime).

2.6. Selection of Effective shRNA Expression Constructs

1. pMX-GATE-IRES2-Luc vector plasmid (Fig. 1b): the plasmid and its sequence data are available on request.
2. Gateway® BP Clonase enzyme mix, Gateway® LR Clonase enzyme mix, 5× BP Clonase buffer, proteinase K solution, and 0.75 M NaCl (Invitrogen).
3. pDONR 222 plasmid (Invitrogen).
4. GP293 packaging cells (Clontech).
5. Growth medium: Dulbecco's modified Eagle's medium containing 10% fetal calf serum.
6. 10-cm cell culture dishes (Thermo Fisher Scientific).

7. 96-well cell culture plates (Thermo Fisher Scientific).
8. 96-well v-bottom polypropylene plates (Thermo Fisher Scientific).
9. pVSV-G: VSV-G encoding plasmid (Clontech).
10. Lipofectamine 2000 (Invitrogen).
11. Opti-MEM I (Invitrogen).
12. 293FT cells (Invitrogen).
13. Bright-Glo Luciferase Assay System (Promega).
14. White 96-well plates (Thermo Fisher Scientific).

3. Methods

3.1. Preparation of Vector Plasmid

1. Digest 1 μg of pNAMA2-U6 with 5 U of *Bsg*I at 25°C (see Note 1) for 1 h in 25 μL of NEbuffer 4 containing 80 μM SAM. Purify with phenol–chloroform extraction and ethanol precipitation (see Subheading 3.2).
2. Digest the *Bsg*I-digested pNAMA2-U6 with 5 U of *Bbs*I at 37°C for 1 h in 25 μL of NEBuffer 2. Purify with phenol–chloroform extraction and ethanol precipitation.
3. Digest the *Bsg*I, *Bbs*I-digested pNAMA2-U6 with 2 U of *Mfe*I and 2 U of *Eco*RI at 37°C for 1 h in 10 μL of NEBuffer 4.
4. Run the digested plasmid on 1% agarose gel, cut the 5.6-kb band, and purify using MinElute gel extraction kit according to the manufacturer's instruction.

3.2. Phenol–Chloroform Extraction and Ethanol Precipitation

1. Add an equal volume of PCI, mix gently, centrifuge at 20,000 × *g* for 5 min at 4°C, and transfer the aqueous layer to a new tube.
2. Add 1/4 volume of 10 M ammonium acetate and 1/100 volume of 20 mg/mL Glycogen, and mix completely.
3. Add 4 volumes of 95% ethanol, mix completely, centrifuge at 20,000 × *g* for 20 min at 4°C, and discard the supernatant.
4. Add 700 μL of 70% ethanol, centrifuge at 20,000 × *g* for 5 min at 4°C, and discard the supernatant. Air-dry the pellet at room temperature.

3.3. Preparation of Adaptors

1. For Adaptors 1 and 3, dilute the two oligonucleotides in TE at a concentration of 10 pmol/μL each. For Adaptor 2, dilute one oligonucleotide similarly. Incubate at 95°C for 10 min, and slowly cool down to room temperature.
2. Run the annealed oligonucleotides on 12% native polyacrylamide gel, and cut out the band.

3. Break up the gel-slice with a spatula, incubate in 5 volumes of diffusion buffer with slow rotation at room temperature for 1 h. Centrifuge at 20,000×*g* for 5 min at 24°C, collect the supernatant into a new tube. Likewise, incubate in diffusion buffer once again, then combine two supernatants.
4. To completely remove fine gels, apply the combined supernatant to an UltraFree-MC column, centrifuge at 12,000×*g* for 5 min at 24°C. Purify with phenol–chloroform extraction and ethanol precipitation.
5. Dissolve in TE and determine the concentration of the adaptors.

3.4. Enzymatic Production of RNAi Library

EPRIL comprises following seven steps. These steps refer to those in Fig. 2.

3.4.1. EPRIL-Step 1: Random Fragmentation of Source cDNA

1. Prepare 120 μL of DNase reaction mixture containing 2.4 μg of double-stranded cDNA fragments (see Note 2), 50 mM Tris–HCl (pH 7.5), 1 mM $MnCl_2$, and 0.1 μg/μL BSA (see Note 3).
2. Dilute DNase I at a concentration of 1.4×10^{-3} U/μL (3,500-fold) with DNase I dilution buffer.
3. Preincubate the diluted DNase I and the DNase reaction mixture at 15°C for 5 min. Add 8.4 μL of the diluted DNase I to the mixture, and incubate at 15°C for 10 min.
4. Stop the reaction by adding 3 μL of 0.5 M EDTA, and chill on ice. Purify with phenol–chloroform extraction and ethanol precipitation.
5. Incubate with 16 U of T4 DNA polymerase at 37°C for 5 min in 75 μL of T4 DNA polymerase buffer containing 0.1 mM dNTP mix and 0.1 μg/μL BSA. Purify with MinElute PCR purification kit (see Note 4) according to the manufacturer's instruction. Elute in 20 μL of H_2O (see Note 5).

3.4.2. EPRIL-Step 2: Excising 20 or 21 nt Fragment from the DNase I-Fragmented cDNA

1. Mix the product from EPRIL-Step 1 with 60 pmol of Adaptor 1, adjust the volume to 30 μL with H_2O. Add 30 μL of ligase solution, and incubate at 24°C for 1 h. Purify with phenol–chloroform extraction and ethanol precipitation.
2. Dissolve the precipitate in 199.5 μL of ThermPol buffer containing 0.25 mM dNTP mix. After 5 min preincubation at 65°C, add 0.5 μL (4 U) of *Bst* DNA polymerase, incubate at 65°C for 5 min, and chill on ice. Purify with phenol–chloroform extraction and ethanol precipitation.
3. Digest the product with 20 U of *Mme*I at 37°C for 1 h in 150 μL of NEbuffer 4 containing 50 μM SAM. Purify with phenol–chloroform extraction and ethanol precipitation.

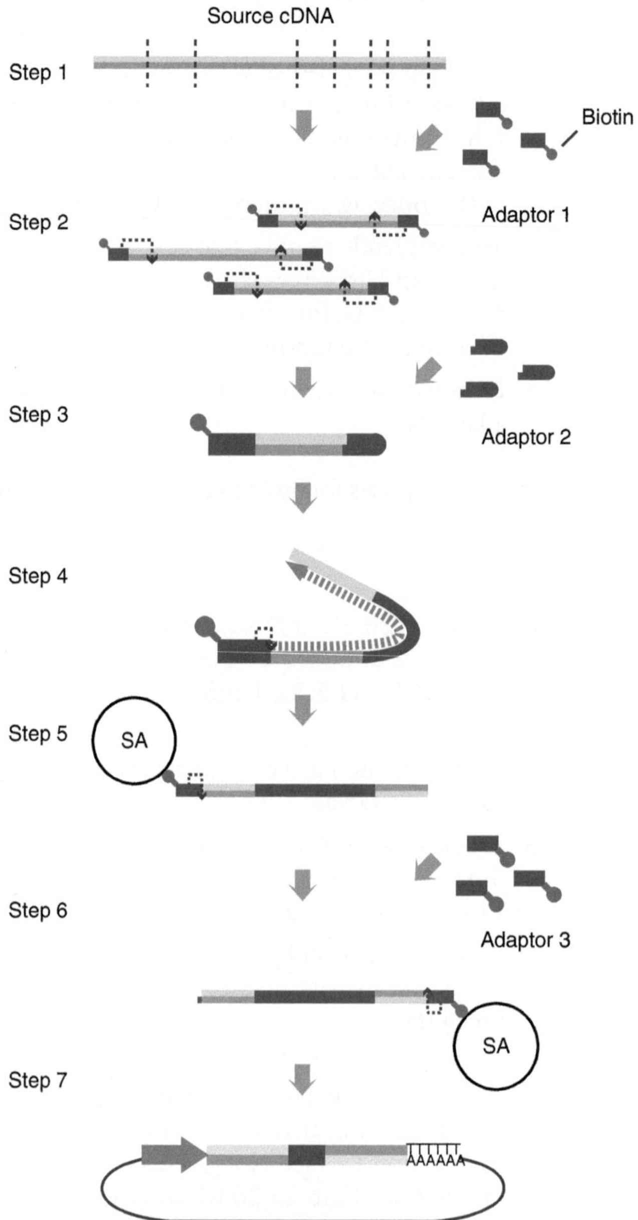

Fig. 2. Schematic drawing of process for EPRIL. EPRIL-Step 1: Source cDNA is randomly fragmented by DNase I and its ends are repaired. EPRIL-Step 2: The DNase I-fragmented cDNA are ligated to Adaptor 1. Adaptor 1 has recognition sites for *Mme*I and *Bpm*I, and the 5′ end of its one strand is biotinylated. Subsequent *Mme*I digestion excises 20 or 21 nt fragment from the DNase I-fragmented cDNA. EPRIL-Step 3: The *Mme*I-digested products are ligated to Adaptor 2. Adaptor 2 is a hairpin-shaped adaptor, hence the ligated product is a single-stranded hairpin DNA. Note that the 3′ protruding end of Adaptor 2 consists of two degenerate bases to make it compatible with all *Mme*I-digested ends. EPRIL-Step 4: The ligated products are digested with nicking enzyme, Nt.BstNBI, to create a nick at immediately after Adaptor 1 sequence, and subsequently polymerase reaction is carried out. As a result, a single-stranded hairpin DNA is converted into a double-stranded DNA bearing an inverted repeat sequence linked by a spacer. EPRIL-Step 5: The products are captured onto streptavidin-coated magnetic beads (SA), then liberated by digestion with *Bpm*I. EPRIL-Step 6: The products are ligated to Adaptor 3. Adaptor 3 has a recognition site for *Bbs*I, and the 5′ end of its one strand is biotinylated. Similar to EPRIL-Step 5, the ligated products are captured onto beads, then liberated by digestion with *Bbs*I. EPRIL-Step 7: The products are ligated into a vector plasmid, and the extra sequence in the spacer is truncated. Use of magnetic beads in EPRIL-Steps 5 and 6 efficiently eliminates by-products and excess adaptors.

3.4.3. EPRIL-Step 3: Addition of Hairpin-Shaped Adaptor

Mix the product from EPRIL-Step 2 with 240 pmol of Adaptor 2, adjust the volume to 50 μL with H_2O, add 50 μL of Ligase Solution, and incubate at 16°C for 1 h. Purify with phenol–chloroform extraction and ethanol precipitation.

3.4.4. EPRIL-Step 4: Conversion of a Single-Stranded Hairpin DNA into a Double-Stranded DNA Encoding an Inverted Repeat Sequence

1. Digest the product from EPRIL-Step 3 with 240 U of Nt.BstNBI at 55°C for 1 h in 400 μL of NEbuffer 3. Purify with phenol–chloroform extraction and ethanol precipitation.
2. Dissolve in 199 μL of ThermPol buffer containing 0.25 mM dNTP mix. After 5 min preincubation at 65°C, add 1.0 μL (8 U) of *Bst* DNA polymerase, incubate at 65°C for 5 min, and chill on ice. Purify with phenol–chloroform extraction and ethanol precipitation. Dissolve in 50 μL of H_2O.

3.4.5. EPRIL-Step 5: Purification Using Streptavidin Beads

1. Transfer 30 μL of streptavidin beads into a new tube, wash two times with 100 μL of 2× B&W buffer using a magnet stand, and resuspend in 50 μL of 2× B&W buffer. Add the product from EPRIL-Step 4, mix gently, and incubate at 24°C for 15 min.
2. Wash two times with 1× B&W buffer and three times with NEBuffer 3, resuspend with 100 μL of NEBuffer 3 containing 20 U of *Bpm*I and 0.1 μg/μL of BSA, and incubate at 37°C for 1 h. Using a magnet stand, transfer the supernatant into a new tube, and purify with phenol–chloroform extraction and ethanol precipitation.

3.4.6. EPRIL-Step 6: Addition of a Sticky End

1. Mix the product from EPRIL-Step 5 with 5 pmol of Adaptor 3, adjust the volume to 12 μL with H_2O, add 12 μL of ligase solution, and incubate at 16°C for 1 h. Purify with phenol–chloroform extraction and ethanol precipitation. Dissolve in 50 μL of H_2O.
2. Prepare streptavidin beads as described in step 1 of Subheading 3.4.5, add the product, mix gently, and incubate at 24°C for 15 min.
3. Wash two times with 1× B&W buffer and three times with NEBuffer 2, resuspend with 600 μL of NEBuffer 2 containing 300 U of *Bbs*I, and incubate at 37°C for 1 h. Using a magnet stand, transfer the supernatant into a new tube, and purify with phenol–chloroform extraction and ethanol precipitation.
4. Run the product on 12% native polyacrylamide gel, and stain with SYBR Green. You should see the 100-bp band (Fig. 3, Lane 2). Cut the band from gel, purify as described in steps 3 and 4 of Subheading 3.3, and dissolve in 10 μL of H_2O. Determine the concentration of the product (insert DNA) by PicoGreen assay kit according to the manufacturer's instruction (see Note 6).

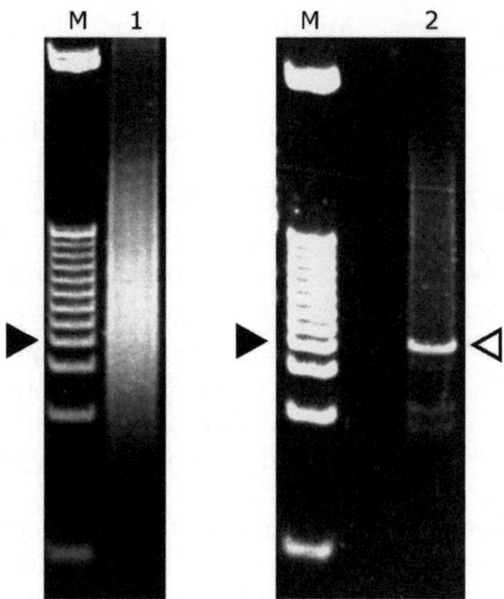

Fig. 3. PAGE analysis of intermediates. *Lane 1*, Products from EPRIL-Step 1; *Lane 2*, Products from EPRIL-Step 6 (*white arrowhead*); *Lane M*, 25 bp ladder marker (*black arrowhead* points 100 bp band).

3.4.7. EPRIL-Step 7: Vector Ligation and Shortening of the Spacer

1. Mix 5.53 fmol (=20 ng) of the digested vector plasmid (see Subheading 3.1) with 16.6 fmol (=1 ng) of the insert DNA (the products from EPRIL-Step 6), adjust the volume to 5 μL with H_2O, add 5 μL of ligase solution, and incubate at 16°C for 1 h. Also prepare a control ligation without insert DNA. Purify with phenol–chloroform extraction and ethanol precipitation.
2. Incubate with 1.2 U of T4 DNA polymerase at 37°C for 5 min in 15 μL of T4 DNA polymerase buffer containing 0.1 μg/μL BSA (see Note 7). Purify with phenol–chloroform extraction and ethanol precipitation.
3. Digest with 1 U of *Xho*I at 37°C for 1 h in 15 μL of NEBuffer 2, then run the product on 1% agarose gel, cut the 5.6-kb band, and purify using MinElute gel extraction kit.
4. Digest with 1 U of *Mfe*I at 37°C for 1 h in 15 μL of NEBuffer 4. Purify with phenol–chloroform extraction and ethanol precipitation. Dissolve in 5 μL of H_2O.
5. Add 5 μL of ligase solution, and incubate at 16°C for 1 h. Purify with phenol–chloroform extraction and ethanol precipitation.
6. Digest the product with 1 U of *Mfe*I and *Eco*RI at 37°C for 1 h in 15 μL of NEBuffer 4. Purify with phenol–chloroform extraction and ethanol precipitation. Dissolve in 2 μL of H_2O.
7. Transform competent cells with the product by electroporation (see Note 8). Chill an electroporation cuvette and the product on ice. Add 40 μL of competent cells to the product,

mix gently, and transfer the mixture into the cuvette. Apply one pulse (see Note 9), put on ice, and immediately add 1 mL S.O.C. medium.

8. Transfer the mixture to a 15-mL polypropylene round tube, and incubate at 37°C for 1 h with shaking at 225 rpm. Transfer the cells to a microtube, centrifuge at 1,000 × *g* for 5 min at 4°C, and wash twice with 1 mL of LB.
9. Plate out appropriate dilution on an agar LB plate containing 100 μg/mL carbenicillin, and incubate overnight at 37°C (see Note 10).
10. As necessary, make pooled library glycerol stocks as follows. Overlay the agar LB plate containing colonies with LB, and harvest the cells with a cell scraper. Incubate at 37°C for 30 min with shaking at 200 rpm. Add 100 μL of the bacterial culture into a microtube containing 100 μL of 50% glycerol/LB, mix gently, and transfer 100 μL of the glycerol mixture into a new microtube (replicate). Immediately freeze and store at –80°C.

3.5. Isolation of shRNA Expression Constructs from RNAi Libraries

Each colony produced by EPRIL (Subheading 3.4.7, step 9) harbors a ready-to-use shRNA expression construct. You can obtain different shRNA constructs from individual colonies. In this section, we describe protocols for isolation of shRNA expression constructs from an RNAi library and preparation of plasmid DNAs in 96-well format.

3.5.1. Making Glycerol Stocks of Individual shRNAs

1. Pre-fill wells in a 96 deep-well plate with 0.5 mL of LB containing 100 μg/mL carbenicillin. Pick single colonies, inoculate into the plate, seal with a gas-permeable seal, and incubate at 37°C for 6–8 h with shaking at 1,200 rpm.
2. Pre-fill wells in a 96-well PCR plate with 50 μL of 50% glycerol/LB. Transfer 50 μL per well of the bacterial culture into the plate, mix gently.
3. Transfer 50 μL per well of the glycerol mixture into a new 96-well PCR plate (replicate copy). Seal with heat seals, immediately freeze, and store at –80°C.

3.5.2. Preparation of shRNA Plasmid DNAs

1. Pre-fill wells in a 96 deep-well plate with 1.2 mL of Circle-Grow containing 100 μg/mL carbenicillin. Add 1 μL of glycerol stocks into the plate, seal with a gas-permeable seal, and incubate at 37°C for 18 h with shaking at 1,200 rpm.
2. Transfer 200 μL per well of the bacterial culture into a new plate and harvest the bacterial cells by centrifugation at 3,000 × *g* for 5 min at 4°C.
3. Prepare plasmids using Perfect Prep Plasmid 96 Vac Kit (see Note 11), and determine the concentration of the plasmid DNAs by PicoGreen assay kit (see Note 12).

3.6. Selection of Effective shRNA Expression Constructs

For rapid selection of effective shRNA constructs, we use a home-made Gateway® destination vector plasmid, pMX-GATE-IRES2-Luc (Fig. 1b). By cloning of the target cDNA fragment into this plasmid, we can create a reporter construct that expresses head-to-tail linked mRNAs of target cDNA and firefly luciferase. With this reporter construct, one can conveniently evaluate the gene silencing efficacy of shRNA constructs by monitoring the reduction in luciferase activity. In this section, we describe a protocol for creation of the reporter constructs and a selection procedure (see Note 13). In Fig. 4, we show an example of one result from our data.

3.6.1. Cloning a Target cDNA into pMX-GATE-IRES2-Luc Vector Using Gateway® One-Step BP/LR Reaction (see Note 14)

1. Prepare attB-flanked cDNA fragments by PCR (Fig. 4a, see Notes 15 and 16).
2. Mix 200 μg of the fragments with 350 ng of pDONR 222, 5 μL of 5× BP Clonase buffer, and adjust the volume to 20 μL with TE. Add 5 μL of BP Clonase enzyme mix, and incubate at 25°C for 4 h.
3. Transfer 5 μL of the reaction mix to a new tube and keep this aliquot at –20°C (see Note 17).
4. Add 1 μL of 0.75 M NaCl, 450 ng of pMX-GATE-IRES2-Luc to the remaining 20 μL reaction mix, and adjust the volume to 24 μL with H_2O. Add 6 μL of LR Clonase enzyme mix, and incubate at 25°C for 2 h.
5. Add 3 μL of proteinase K solution and incubate at 37°C for 10 min.
6. Transform the appropriate competent cells with 10 μL of the reaction mix, plate out on an agar LB plate containing 100 μg/mL carbenicillin, and incubate overnight at 37°C.

3.6.2. Packaging of Reporter Retrovirus

1. Seed GP293 packaging cells in 12 mL of growth medium in 10-cm cell culture dish, and incubate at 37°C in a 5% CO_2 incubator until the cells are ~95% confluent.
2. Just before transfection, carefully change the growth medium with 10 mL of fresh growth medium.
3. Transfect the cells as follows. Dilute 24 μg of the reporter plasmid DNA and 2 μg of pVSV-G in 1 mL of Opti-MEM I. Dilute 60 μL of Lipofectamine 2000 in 1 mL of Opti-MEM I, mix gently, and incubate at room temperature for 5 min. Combine the diluted plasmid and the diluted Lipofectamine 2000, mix gently, and incubate at room temperature for 20 min. Carefully add 2 mL of the mixture to the cells, and incubate at 37°C in a 5% CO_2 incubator for 20–24 h.
4. Carefully change the growth medium with 12 mL of fresh growth medium, and incubate at 32°C in a 5% CO_2 incubator for 24 h.

a

5'-gggg acaagtttgtacaaaaaagcaggct tcctgtcagggtgggtacggct-3'

attB1

5'-gggg accactttgtacaagaaagctggg tcctaaaacttgacttgcttcggaactctgcatac-3'

attB2

b

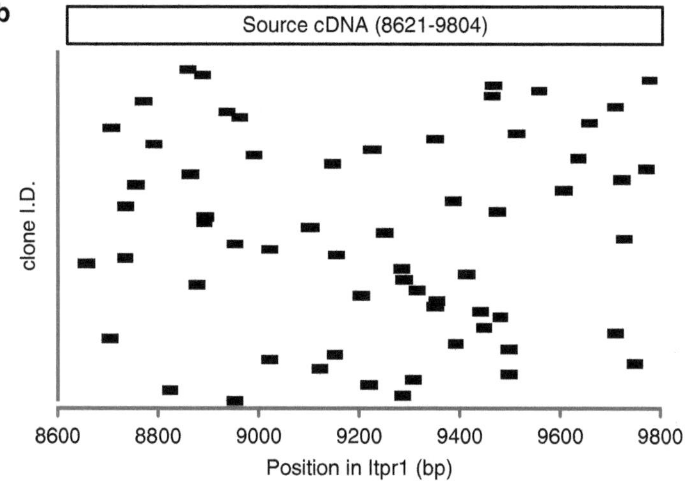

c

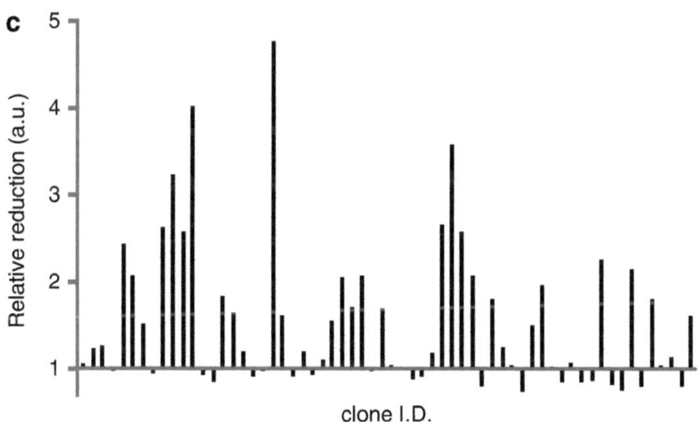

Fig. 4. An example of production of RNAi library and selection of effective shRNA expression constructs. We produced an RNAi library from a cDNA encoding mouse Itpr1 gene and randomly isolated 65 clones, of which 62 clones carried the shRNA constructs. The source cDNA for EPRIL was prepared by PCR using primers containing attB sequence. This PCR product was also used for creation of the reporter construct. (**a**) Features of forward (*top*) and reverse (*bottom*) primer. (**b**) The result of sequencing analysis. The horizontal axis represents the position of Itpr1 cDNA (NCBI refseq, GI: 24475598). Each short bar indicates position of shRNA sequence of individual clones. *White box* represents the source cDNA region. (**c**) The result of the reporter assay. 293FT cells were coinfected with reporter virus and shRNA virus. Two days after infection, cells were analyzed for luciferase activity using the Bright-Glo assay kit. Relative reduction values in luciferase activity are shown.

5. Harvest the medium (reporter virus-containing medium) into a 15-mL tube, and discard the packaging cells by centrifugation at 200×*g* for 5 min at 4°C.

3.6.3. Packaging of shRNA Retrovirus

1. Seed 4.8×10^4 GP293 packaging cells in 80 μL per well of growth medium in a 96-well cell culture plate, and incubate at 37°C in a 5% CO_2 incubator for 4–6 h. The cells should be ~95% confluent at the time of transfection.
2. Transfect the cells as follows. Dispense 12 μL per well of Opti-MEM I containing 30 ng of pVSV-G into a 96-well v-bottom polypropylene plate. Add 6 μL per well of shRNA plasmid DNAs (60 μg/μL). Dilute 113 μL of Lipofectamine 2000 in 1.5 mL of Opti-MEM I in a microtube, mix gently, and incubate at room temperature for 5 min. Dispense 12 μL per well of the diluted Lipofectamine 2000 into the DNA plate, mix gently, and incubate at room temperature for 20 min. Add 20 μL per well of the DNA/Lipofectamine mixture into the cell plate, and incubate at 37°C in a 5% CO_2 incubator for 20–24 h.
3. Carefully change 80 μL per well of the growth medium with fresh growth medium, and incubate at 32°C in a 5% CO_2 incubator for 24 h.
4. Harvest the medium (shRNA virus-containing medium) into a 96-well PCR plate, and discard the packaging cells by centrifugation at $200\times g$ for 5 min at 4°C.

3.6.4. Coinfection with Reporter Virus and shRNA Virus, and Luciferase Assay

1. Seed 1.5×10^4 293FT cells in 100 μL per well of growth medium in a 96-well cell culture plate, and incubate at 37°C in a 5% CO_2 incubator until the cells are adherent (4–6 h).
2. Carefully remove 80 μL per well of the growth medium, add 50 μL of reporter virus-containing medium, and 50 μL of shRNA virus-containing medium. Incubate at 37°C in a 5% CO_2 incubator for 24 h.
3. Carefully change 100 μL per well of the growth medium with fresh growth medium, and incubate at 37°C in a 5% CO_2 incubator for 24 h.
4. Prepare Bright-Glo reagent as follows. Thaw Bright-Glo buffer at room temperature, add to Bright-Glo substrate, and mix by inversion until the substrate is thoroughly dissolved.
5. Add 100 μL of Bright-Glo reagent to each well, transfer the mixture into a white 96-well plate, incubate at room temperature for 5 min, and measure luminescence intensity in an appropriate plate reader (Fig. 4c).

4. Notes

1. Although the manufacturer's instruction recommends 37°C as the optimal incubation temperature of *Bsg*I, we have found that incubation at 25°C is more effective.

2. As source cDNAs for EPRIL, any double-stranded DNAs such as PCR products and restriction-digested fragments can be used. A recommended size range is from 0.5 to1.5 kb.
3. The cleave mode of DNase I is dependent on the presence of ions. In EPRIL, owing to cleave both DNA strands at approximately the same site, we use buffer containing Mn^{2+}.
4. In this step, phenol–chloroform extraction and ethanol precipitation is not recommended. Use of MinElute PCR purification kit is required for elimination of small fragments.
5. We strongly recommend that you check the product from EPRIL-Step 1 on PAGE. Run 2 μL (1/10 volume) of the product on 12% native polyacrylamide gel, and stain with SYBR Green. Appropriately digested products should appear as smear with an average size of 100–200 bp (Fig. 3, Lane 1). In case of over or under digestion is suspected, adjust the concentration of DNase I.
6. Typically, we obtained 2–4 ng of insert DNAs from 2.4 μg of cDNA fragments.
7. Do not add dNTPmix, since this procedure is going to eliminate of liner DNA.
8. Although we use electro competent cells to achieve maximum transformation efficacy, chemical competent cells also can be used.
9. We use "Ec1" program of MicroPulser Electroporator. Typically, the actual voltage and time constant are 1.8 kV and 5.1–5.2 ms, respectively.
10. Typically, we obtained $>10^5$ transformants from 20 ng vector.
11. Although this kit does not ensure endotoxin-free, we successfully use the plasmids for transfection and viral packaging.
12. Typically, we obtained 1 μg of plasmid DNA from 200 μL of bacterial culture.
13. Although we here describe retroviral infection experiment, transient expression experiment also may be used for selection of effective shRNA expression constructs.
14. Refer to manufacturer's instruction for more information about Gateway® Technology.
15. The attB-flanked PCR products also can be used as source cDNAs for EPRIL.
16. Plasmids containing attB-flanked cDNA also can be used.
17. If you cannot obtain any transformant, use this aliquot to assess the efficiency of the BP reaction according to the manufacturer's instruction.

Acknowledgments

This work was supported by fellowship of the Japan Society for the Promotion of Science and grants from The Nakajima Foundation, The Cell Science Research Foundation, and the Suzuken Memorial Foundation.

References

1. Fire, A., Xu, S., Montgomery, M. K., Kostas, S. A., Driver, S. E., and Mello, C. C. (1998) Potent and specific genetic interference by double-stranded RNA in *Caenorhabditis elegans. Nature* **391**, 806–11.
2. Elbashir, S. M., Harborth, J., Lendeckel, W., Yalcin, A., Weber, K., and Tuschl, T. (2001) Duplexes of 21-nucleotide RNAs mediate RNA interference in cultured mammalian cells. *Nature* **411**, 494–8.
3. Caplen, N. J., Parrish, S., Imani, F., Fire, A., and Morgan, R. A. (2001) Specific inhibition of gene expression by small double-stranded RNAs in invertebrate and vertebrate systems. *Proc. Natl. Acad. Sci. USA* **98**, 9742–7.
4. Kim, D., Behlke, M. A., Rose, S. D., Chang, M., Choi, S., and Rossi, J. J. (2005) Synthetic dsRNA Dicer substrates enhance RNAi potency and efficacy. *Nat. Biotechnol.* **23**, 222–6.
5. Siolas, D., Lerner, C., Burchard, J., Ge, W., Linsley, P. S., Paddison, P. J., Hannon, G. J., and Cleary, M. A. (2005) Synthetic shRNAs as potent RNAi triggers. *Nat. Biotechnol.* **23**, 227–31.
6. Brummelkamp, T. R., Bernards, R., and Agami, R. (2002) A system for stable expression of short interfering RNAs in mammalian cells. *Science* **296**, 550–3.
7. Miyagishi, M., and Taira, K. (2002) U6 promoter-driven siRNAs with four uridine 3′ overhangs efficiently suppress targeted gene expression in mammalian cells. *Nat. Biotechnol.* **20**, 497–500.
8. Lee, N. S., Dohjima, T., Bauer, G., Li, H., Li, M., Ehsani, A., Salvaterra, P., and Rossi, J. (2002) Expression of small interfering RNAs targeted against HIV-1 rev transcripts in human cells. *Nat. Biotechnol.* **20**, 500–5.
9. Paul, C. P., Good, P. D., Winer, I., and Engelke, D. R. (2002) Effective expression of small interfering RNA in human cells. *Nat. Biotechnol.* **20**, 505–8.
10. Yu, J., DeRuiter, S. L., and Turner, D. L. (2002) RNA interference by expression of short-interfering RNAs and hairpin RNAs in mammalian cells. *Proc. Natl. Acad. Sci. USA* **99**, 6047–52.
11. Paddison, P. J., Caudy, A. A., Bernstein, E., Hannon, G. J., and Conklin, D. S. (2002) Short hairpin RNAs (shRNAs) induce sequence-specific silencing in mammalian cells. *Genes Dev.* **16**, 948–58.
12. Sui, G., Soohoo, C., Affar, E. B., Gay, F., Shi, Y., Forrester, W. C., and Shi, Y. (2002) A DNA vector-based RNAi technology to suppress gene expression in mammalian cells. *Proc. Natl. Acad. Sci. USA* **99**, 5515–20.
13. de Fougerolles, A., Vornlocher, H., Maraganore, J., and Lieberman, J. (2007) Interfering with disease: a progress report on siRNA-based therapeutics. *Nat. Rev. Drug Discov.* **6**, 443–53.
14. Castanotto, D., and Rossi, J. J. (2009) The promises and pitfalls of RNA-interference-based therapeutics. *Nature* **457**, 426–33.
15. Jackson, A. L., Bartz, S. R., Schelter, J., Kobayashi, S. V., Burchard, J., Mao, M., Li, B., Cavet, G., and Linsley, P. S. (2003) Expression profiling reveals off-target gene regulation by RNAi. *Nat. Biotechnol.* **21**, 635–7.
16. Birmingham, A., Anderson, E. M., Reynolds, A., Ilsley-Tyree, D., Leake, D., Fedorov, Y., Baskerville, S., Maksimova, E., Robinson, K., Karpilow, J., Marshall, W. S., and Khvorova, A. (2006) 3′ UTR seed matches, but not overall identity, are associated with RNAi off-targets. *Nat. Methods* **3**, 199–204.
17. Ui-Tei, K., Naito, Y., Takahashi, F., Haraguchi, T., Ohki-Hamazaki, H., Juni, A., Ueda, R., and Saigo, K. (2004) Guidelines for the selection of highly effective siRNA sequences for mammalian and chick RNA interference. *Nucleic Acids Res.* **32**, 936–48.
18. Reynolds, A., Leake, D., Boese, Q., Scaringe, S., Marshall, W. S., and Khvorova, A. (2004) Rational siRNA design for RNA interference. *Nat. Biotechnol.* **22**, 326–30.
19. Amarzguioui, M., and Prydz, H. (2004) An algorithm for selection of functional siRNA sequences. *Biochem. Biophys. Res. Commun.* **316**, 1050–8.

20. Yang, D., Buchholz, F., Huang, Z., Goga, A., Chen, C., Brodsky, F. M., and Bishop, J. M. (2002) Short RNA duplexes produced by hydrolysis with *Escherichia coli* RNase III mediate effective RNA interference in mammalian cells. *Proc. Natl. Acad. Sci. USA* **99**, 9942–7.
21. Myers, J. W., Jones, J. T., Meyer, T., and Ferrell, J. E. (2003) Recombinant Dicer efficiently converts large dsRNAs into siRNAs suitable for gene silencing. *Nat. Biotechnol.* **21**, 324–8.
22. Sen, G., Wehrman, T. S., Myers, J. W., and Blau, H. M. (2004) Restriction enzyme-generated siRNA (REGS) vectors and libraries. *Nat. Genet.* **36**, 183–9.
23. Shirane, D., Sugao, K., Namiki, S., Tanabe, M., Iino, M., and Hirose, K. (2004) Enzymatic production of RNAi libraries from cDNAs. *Nat. Genet.* **36**, 190–6.
24. Luo, B., Heard, A., and Lodish, H. (2004) Small interfering RNA production by enzymatic engineering of DNA (SPEED). *Proc. Natl. Acad. Sci. USA* **101**, 5494–9.
25. Buchholz, F., Kittler, R., Slabicki, M., and Theis, M. (2006) Enzymatically prepared RNAi libraries. *Nat. Methods* **3**, 696–700.
26. Hashimoto, A., Hirose, K., and Iino, M. (2005) BAD detects coincidence of G2/M phase and growth factor deprivation to regulate apoptosis. *J. Biol. Chem.* **280**, 26225–32.
27. An, D. S., Donahue, R. E., Kamata, M., Poon, B., Metzger, M., Mao, S., Bonifacino, A., Krouse, A. E., Darlix, J., Baltimore, D., Qin, F. X., and Chen, I. S. Y. (2007) Stable reduction of CCR5 by RNAi through hematopoietic stem cell transplant in non-human primates. *Proc. Natl. Acad. Sci. USA* **104**, 13110–5.

Chapter 9

Construction of Small RNA cDNA Libraries for High-Throughput Sequencing

Cheng Lu and Vikas Shedge

Abstract

Small RNAs (smRNAs) play an essential role in virtually every aspect of growth and development, by regulating gene expression at the post-transcriptional and/or transcriptional level. New high-throughput sequencing technology allows for a comprehensive coverage of smRNAs in any given biological sample, and has been widely used for profiling smRNA populations in various developmental stages, tissue and cell types, or normal and disease states. In this article, we describe the method used in our laboratory to construct smRNA cDNA libraries for high-throughput sequencing.

Key words: High-throughput sequencing, Small RNA, miRNA, siRNA, cDNA library

1. Introduction

The 2006 Nobel Prize in medicine for elucidating the RNA interference (RNAi) phenomenon, less than 10 years since its initial discovery, really highlights the growing interest for research in this field. Widely observed in most eukaryotic organisms, RNAi has emerged as an important mechanism of post-transcriptional gene regulation that is involved in wide-ranging developmental processes and adaptation to environmental stress (1, 2). Small RNAs (smRNAs), which typically range in size from 21 to 24 nucleotides, constitute a vital component of the RNAi machinery. Several different types of smRNAs have been identified, including microRNAs (miRNA) that are 21–22 nucleotides long and derived from endogenous hairpin-shaped primary transcripts encoded by genomic loci that are distinct from that of the silenced genes (2, 3). Another important class of smRNAs that are similar in structure and function to miRNAs is the small interfering RNAs (siRNA).

Chaofu Lu et al. (eds.), *cDNA Libraries: Methods and Applications*, Methods in Molecular Biology, vol. 729,
DOI 10.1007/978-1-61779-065-2_9, © Springer Science+Business Media, LLC 2011

siRNAs are generally 22–24 nucleotides long and are usually derived from longer double-stranded RNA molecules (4, 5).

Although numerous smRNAs have been identified using traditional cloning techniques (6–8), this technique may not be effective to identify rare and tissue-specific smRNAs. High-throughput sequencing technologies (also known as deep sequencing) have facilitated not only the identification but also the quantification of the smRNA molecules, thus opening the doors for quantitative expression analysis (9, 10). Since this approach was first reported in 2005 (11), the sequencing technology has undergone a remarkable evolution. Several novel and highly parallel methods, including 454, SBS, supported oligo ligation detection, not only offer a much higher throughput, but also reduce the cost of sequencing dramatically (12, 13). Technological advances have made it possible to study the smRNAs in ways that were unimaginable even a few years ago. More and more tissue and cell-specific smRNA profiling data are becoming available for numerous model organisms. In combination with immunoprecipitation using specific antibody, smRNA populations associated with RNA-binding proteins (such as Ago and Piwi protein families) have been well characterized (14, 15). With modified protocols, several novel classes of smRNAs have been identified which expand the catalog of cellular smRNAs (16, 17). The new technologies also facilitate genome-wide discovery of potential miRNA targets (18, 19). Here, we describe a step-by-step protocol that we use for the construction of smRNA cDNA libraries for deep sequencing.

2. Materials

2.1. RNA Isolation

1. Trizol reagent (Invitrogen; Carlsbad, CA).
2. Chloroform.
3. Ethanol.
4. Nuclease-free water.

2.2. Low-Molecular Weight and High-Molecular Weight RNA Separation

1. 5 M NaCl.
2. 50% Polyethylene glycol (MW = 8,000).
3. 5 mg/mL Glycogen (Ambion; Austin, TX).
4. Ethanol.
5. Nuclease-free water.

2.3. RNA Gel Purification

1. 10× TBE buffer.
2. 40% (w/v) 19:1 Acrylamide:bis (Ambion; Austin, TX).

3. Urea.
4. *N,N,N,N'*-Tetramethyl-ethylenediamine (TEMED) (Bio-Rad; Hercules, CA).
5. 2× Loading buffer (90% formamide, 0.05% xylene cyanol, 0.05% bromophenol blue).
6. 1× TBE. Store at 4°C.
7. miRNA marker (NEB; Ipswich, MA).
8. 10 bp DNA ladder (Invitrogen; Carlsbad, CA).
9. 10% Ammonium persulfate, freshly prepared.
10. 5 M NaCl.
11. 5 mg/mL Glycogen (Ambion; Austin, TX).
12. Ethanol.
13. Nuclease-free water.

2.4. Adaptor Ligation

1. 5 U/μL T4 RNA ligase supplied with 10× RNA ligase buffer (Ambion; Austin, TX).
2. 40 U/μL RNaseOUT (Invitrogen; Carlsbad, CA).
3. 5′ RNA adaptor: 5′ OH-GGU CUU AGU CGC AUC CUG UAG AUG GAUC-OH 3′ (Dharmacon; Lafayette, CO).
4. 3′ RNA adaptor: 5′ pUC GUA UGC CGU CUU CUG CUU GidT 3′ (Dharmacon; Lafayette, CO) (see Note 1).
5. Nuclease-free water.

2.5. Synthesis of cDNAs and PCR Amplification

1. RT-primer: 5′ CAA GCA GAA GAC GGC ATA CGA 3′.
2. 200 U/μL Superscript III Reverse Transcriptase supplied with 5× first-strand buffer and 100 mM DTT (Invitrogen; Carlsbad, CA).
3. 10 mM dNTP mix.
4. 40 U/μL RNaseOUT (Invitrogen; Carlsbad, CA).
5. Forward PCR primer: 5′ CAA GCA GAA GAC GGC ATA CGA 3′.
6. Reverse PCR primer: 5′ GGT CTT AGT CGC ATC CTG TAG ATG 3′.
7. Phusion high-fidelity DNA polymerase (NEB, Ipswich, MA).
8. Nuclease-free water.

2.6. PCR Product Purification and Cloning

1. Chloroform.
2. Phenol (pH 7.9).
3. Ethanol.
4. 10× TBE.
5. 40% (w/v) 19:1 Acrylamide:bis (Ambion; Austin, TX).

6. *N,N,N,N′*-Tetramethyl-ethylenediamine (TEMED) (Bio-Rad; Hercules, CA).
7. 6× Loading buffer (Promega; Madison, WI).
8. 10 bp DNA ladder (Invitrogen; Carlsbad, CA).
9. 10% Ammonium persulfate, freshly prepared.
10. 5 M NaCl.
11. 5 mg/mL Glycogen (Ambion; Austin, TX).
12. Zero Blunt TOPO PCR cloning kit with TOP10 OneShot Competent Cells (Invitrogen; Carlsbad, CA).
13. LB Broth Base (Invitrogen; Carlsbad, CA).
14. 40 μg/mL X-gal (Invitrogen; Carlsbad, CA).
15. 0.1 M IPTG (Invitrogen; Carlsbad, CA).
16. Nuclease-free water.

3. Methods

Ribonucleases (RNases) are very stable and active enzymes that are present in most samples. Since RNases are difficult to terminate and trace amounts are sufficient to destroy RNA, RNA degradation during RNA isolation or purification steps is the most likely reason for failure to obtain a good library. Therefore, it is critical to use fresh materials and maintain an RNase-free environment during the whole process. Trizol reagent (or the original Chomczynski and Sacchi protocol) has been widely used in RNA isolation for good retention of the smRNA population. It should be noted that not all column-based RNA isolation kits are suitable to recover smRNA fraction. The smRNA-enrichment step [low-molecular weight (LMW) isolation] is recommended after total RNA isolation, particularly for samples with low level of smRNAs. In the LMW fraction, RNA species ≤200 nt (including tRNAs) are highly enriched. SmRNAs with different size have been identified in plants, accordingly smRNAs of the desired size ranges need to be further purified using denaturing polyacrylamide gel electrophoresis (PAGE).

Different protocols are used to capture smRNAs. The majority 21–24 nt smRNAs, which are generated by RNase III-like endonuclease, have a 5′ phosphate and a 3′ hydroxyl group. RNA turnover or degradation products generally carry 5′ hydroxyl groups and 2′ or 3′ phosphates. In this protocol, the ligation reactions are designed to specifically capture smRNAs with 5′ phosphate and 3′ hydroxyl termini. The RNA adaptors will provide known priming sites for first-strand cDNA synthesis and PCR amplification. An overview of smRNA cloning methods is schematically depicted in Fig. 1.

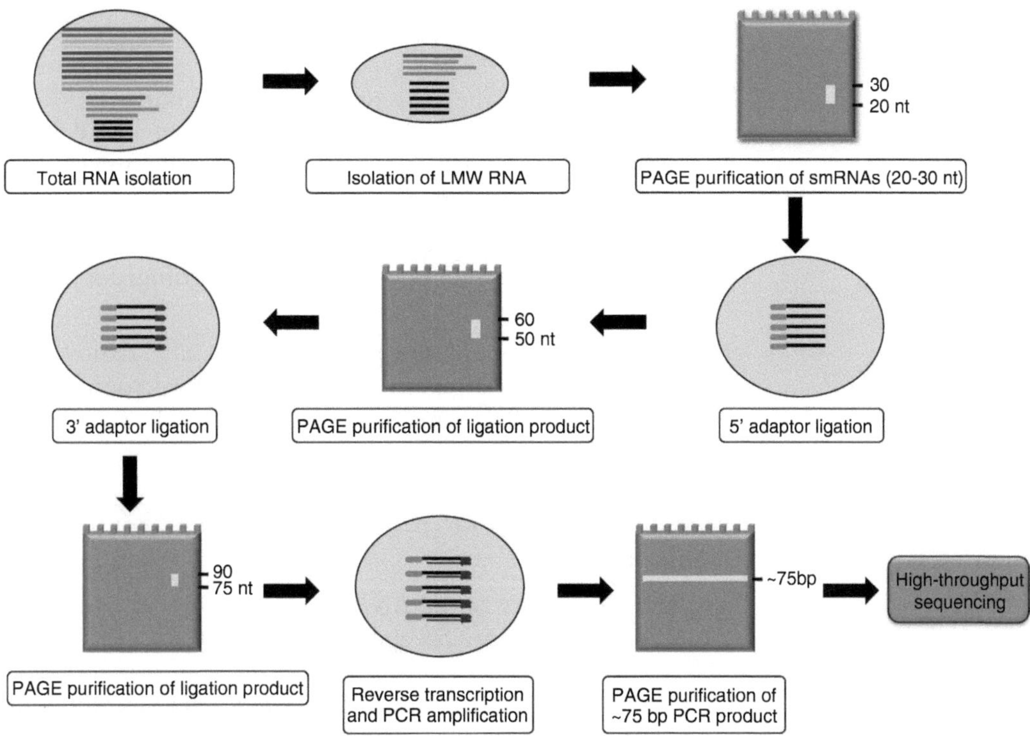

Fig. 1. Outline of small RNA cDNA library construction. The portion of gel within the rectangle was recovered. A ~75 bp band should be easily detected after PCR amplification.

3.1. RNA Isolation and Fractionation of LMW RNA

To avoid the release of ribonucleases, the tissue must remain frozen while grinding. The total RNA isolation procedure is adapted from manufacturer's protocol for TRIzol (Invitrogen), although a wide range of reagents and protocols are available to isolate high-quality total RNA. For plant tissues that contain a high level of polyphenolics, polysaccharides, fibrous materials or secondary metabolites (e.g., seeds), modified reagents, and protocols should be used. A detailed review of various RNA isolation methods can be obtained in a recent article (20).

1. Add 1 g of frozen-ground tissue to a tube containing at least 10 mL of Trizol and vortex vigorously about 30 s (see Note 2).
2. Follow manufacturer's protocol to get the RNA pellet.
3. Dissolve total RNA in 100–200 μL of sterile nuclease-free water.
4. High-molecular weight (HMW) RNAs, mostly rRNA and mRNA are precipitated by adding 50% PEG (MW = 8,000) to a final concentration of 5% and 5 M NaCl to a final concentration of 0.5 M.
5. Incubate the mixture on ice for 30 min.

6. Centrifuge at >12,000×g for 10 min at 4°C to pellet HMW RNAs (see Note 3).
7. Carefully transfer the supernatant to a new microcentrifuge tube without disturbing the pellet. To precipitate LMW RNA, add 2.5 volumes of 100% ethanol, vortex briefly, and place at –20°C for 2 h. You may proceed to the next step or store at –20°C overnight.
8. Centrifuge at maximum speed in a microcentrifuge for at least 30 min.
9. Note the position of the pellet and carefully remove the supernatant using a pipette. Centrifuge the microcentrifuge tube again to collect remaining ethanol. Carefully remove the remaining ethanol using a small pipette tip without disturbing the pellet and allow the pellet to air-dry.
10. Dissolve the pellet in 10–20 μL of nuclease-free water (see Note 4).
11. Check the quality of the RNA by separating 1 μL aliquot of LMW fraction on a 1.5% agarose gel using 0.5× TBE buffer.

3.2. Gel Purification of 20–30 nt smRNA

1. Prepare a 15% polyacrylamide/urea (8 M) gel by combining the following components
   (a) 9.6 g Urea
   (b) 7.5 mL 40% Acrylamide/bis solution
   (c) 10× TBE buffer
   (d) Nuclease-free water to a final volume of 20 mL.
2. Warm the solution to 37°C to dissolve the urea completely. Filter the solution through a nitrocellulose filter and cool to room temperature.
3. Add 120 μL of freshly prepared 10% ammonium persulfate to the solution and mix well.
4. Add 9.2 μL of TEMED. Mix thoroughly by swirling (see Note 5).
5. Pour the solution between glass plates and allow the acrylamide to polymerize at room temperature for at least 30 min.
6. Wash the wells thoroughly with 1× TBE, then prerun the gel for 30 min at 200 V.
7. Add one volume of 2× loading buffer to as much as 10 μL of LMW RNAs. Denature the RNA by incubation for 5 min at 65°C. Then load up to 20 μL of sample per well (1.5 mm thickness, 5 mm width) (see Note 6).
8. Denature 3 μL of miRNA Marker and 10 bp DNA ladder the same way as the LMW RNA and load it into two separate unused lanes.

9. Run the gel at 200 V using 1× TBE buffer until the bromophenol blue dye reaches the lower part of the gel (~60 min).
10. Stain the gel with ethidium bromide for 5 min. After visualization under UV light, cut the gel corresponding to the band size of 20–25 nt, put it into a preweighed 2-mL tube. Determine the weight of the gel slice (see Note 7).
11. Add three volumes (v/w) of nuclease-free 0.3 M NaCl and elute RNA by incubating the tube overnight at room temperature under constant agitation.
12. Collect the supernatant and precipitate the RNA with three volumes of absolute ethanol and 1 μL of glycogen at −80°C for at least 2 h.
13. Pellet smRNA after ethanol precipitation in a microcentrifuge for 30 min at a maximum speed at 4°C.
14. Carefully remove the supernatant and wash the pellet with cold 80% ethanol. Carefully remove the remaining ethanol using a small pipette tip without disturbing the pellet.
15. Allow the RNA pellet to air-dry to evaporate residual ethanol that may inhibit the subsequent ligation reaction. Dissolve the pellet in 10 μL of nuclease-free water.

3.3. 5′ Adaptor Ligation and Purification

Adaptor sequences are needed for reverse transcription and PCR amplification of smRNA cDNA library. Specific adaptors should be used because different sequencing platforms usually require different adaptors. The adaptors and primers listed in Subheading 2 are designed for Illumina SBS technology, so that the sequencing primer-binding site is already present in the 5′ adaptor.

The smRNAs isolated by gel electrophoresis are ligated at their 5′ and 3′ ends to the RNA adaptors with the use of T4 RNA ligase. T4 RNA ligase catalyzes the ligation of a 5′ phosphoryl-terminated nucleic acid donor to a 3′ hydroxyl-terminated nucleic acid acceptor through the formation of a 3′→5′ phosphodiester bond. SmRNAs that are generated by RNase III have 5′ monophosphate and 3′ hydroxyl termini in contrast to most RNA turnover products. To minimize smRNA self-ligation and circularization, excessive amount of adaptors are added in ligation reaction. Alternatively, T4 RNA ligase 2, a truncated T4 RNA ligase, can be used in the ligation reaction. T4 RNA ligase 2 specifically ligates the preadenylated 5′ end of DNA or RNA to the 3′ end of RNA (21, 22). The enzyme does not require ATP for ligation but does need the pre-adenylated substrate, and as a result it cannot ligate the phosphorylated 5′ end of RNA to the 3′ end of RNA (23, 24).

Although majority of 21–24 nt smRNAs have 5′ monophosphate and 3′ hydroxyl termini, some of smRNA species have different 5′ ends that cannot be recovered by the RNA ligation reaction.

For example, to clone smRNAs with 5′ tri-phosphate, dephosphorylation (by alkaline phosphatase) and re-phosphorylation (by T4 polynucleotide kinase) treatments are needed before the ligation reaction.

1. Prepare a ligation reaction mixture by combining the following components: 5 μL smRNAs from the previous step, 2 μL 5′ RNA adaptor (20 μM), 1 μL 10× RNA ligase buffer, and 2 μL T4 RNA ligase (5 U/μL). Incubate the reaction at room temperature for 6 h to overnight (see Note 8).
2. Stop reaction with 10 μL 2× loading buffer. Denature RNA by heating the sample at 65°C for 15 min prior to loading.
3. Prepare a 10% polyacrylamide/urea (8 M) gel as described in steps 1–6 of Subheading 3.2 using the following components:
   (a) 9.6 g Urea
   (b) 5 mL 40% Acrylamide/bis solution
   (c) 10× TBE buffer
   (d) Nuclease-free water to a final volume of 20 mL.
4. Load the entire ligation reaction into one well and run the sample for 1 h at 180 V until the bromophenol blue dye reaches the lower part of the gel.
5. Excise the portion of the sample lane containing RNA molecules of 50–60 nt.
6. Elute, precipitate, and collect the 5′ RNA ligation product as described in steps 10–15 of Subheading 3.2.

3.4. 3′ Adaptor Ligation and Purification

1. Prepare 3′ adaptor ligation reaction mixture by combining the following components: 5 μL 5′ ligated smRNAs from the previous step, 2 μL 3′ RNA adaptor (20 μM), 1 μL 10× RNA ligase buffer, and 2 μL T4 RNA ligase (5 U/μL). Incubate the reaction at room temperature for 6 h to overnight.
2. Stop reaction with 10 μL 2× loading buffer. Denature RNA by heating the sample at 65°C for 15 min prior to loading.
3. Prepare a 7.5% polyacrylamide/urea (8 M) gel as described in steps 1–6 of Subheading 3.2 using the following components:
   (a) 9.6 g Urea
   (b) 3.75 mL 40% Acrylamide/bis solution
   (c) 10× TBE buffer
   (d) Nuclease-free water to a final volume of 20 mL.
4. Load the entire ligation reaction into one well and run the sample for 1 h at 180 V until the bromophenol blue dye reaches the lower part of the gel.
5. Excise the portion of the sample lane containing RNA molecules of 70–90 nt.

6. Elute, precipitate, and collect the 3′ RNA ligation product as described in steps 10–15 of Subheading 3.2.

3.5. Synthesis of cDNAs and PCR Amplification

1. Prepare the following reaction mix: 5 μL of purified ligation product, 3 μL of 100 μM RT-primer, and 3 μL of nuclease-free water. Heat mixture at 65°C for 10 min and spin down to cool.
2. Add the following components on ice in order: 6 μL 5× first-strand buffer, 5.5 μL of 2 mM dNTPs, 3 μL of 100 mM DTT, 1.5 μL RNaseOUT, and 3 μL Superscript II RT (200 U/μL), and incubate for 1 h at 45°C.
3. Mix the following (total 600 μL reaction): 20 μL of cDNA solution, 484 μL water, 60 μL high-fidelity10× PCR buffer (provided by the manufacturer), 12 μL of 10 mM dNTPs, 6 μL of 100 μM 5′ PCR primer, 6 μL of 100 μM 3′ PCR primer, and 12 μL of Phusion high-fidelity DNA polymerase.
4. Transfer 50 μL of the reaction mixture to each of 12 PCR tubes.
5. Perform PCR with an initial incubation at 98°C for 1 min; 15–20 cycles of 98°C for 10 s, 58°C for 30 s, and 72°C for 30 s; and a final incubation at 72°C for 3 min (see Note 9).

3.6. PCR Product Purification and Cloning

For most deep sequencing methods (such as SBS and 454), purified PCR product can be used directly. The cloning and subsequent ABI sequencing steps are for the purpose of quality control only. For a typical *Arabidopsis* sample, about one-third of plasmid inserts should match known miRNAs.

1. Subject PCR to standard chloroform–phenol extraction followed by ethanol precipitation: add 1/10th volume of 5 M NaCl and 2.5 volumes of cold absolute ethanol, mix, and chill at −70°C for at least 15 min. Centrifuge the tubes and dissolve the final precipitates in 100 μL nuclease-free water.
2. Prepare 10% polyacrylamide nondenaturing gel as described in steps 1–6 of Subheading 3.2, except no urea added.
3. Add 100 μL of 6× loading dye in PCR product and load the entire sample into six wells, along with 10 bp DNA ladder as marker.
4. Stain the gel with ethidium bromide, and excise the portion of the gel containing DNA molecules of 70–75 bp.
5. Transfer the gel portions to a 2-mL pre-weighed microcentrifuge tube, crush the gel.
6. Elute, precipitate, and collect the cDNAs as described in steps 10–15 of Subheading 3.2. Dissolve DNA pellet in 15 μL nuclease-free water (see Note 10).

7. Prepare a TOPO Cloning reaction by mixing the following components: 0.2 μL of fresh purified PCR product, 1 μL of 1.2 M NaCl solution, and 1 μL of PCR 4Blunt-TOPO vector. Incubate 10 min at room temperature (see Note 11).
8. Transform One Shot cells with 2 μL of TOPO reaction following manufacturer instruction.
9. Spread 10–50 μL of transformation mix onto LB plates containing 50 μg/mL kanamycin and X-gal/IPTG, and incubate overnight at 37°C.
10. Randomly transfer white or light blue colonies to a 96-well plate and culture them overnight at 37°C.
11. Submit the plate for automated sequencing using M13 forward or M13 reverse primer as sequencing primer.

4. Notes

1. The exact sequences of the adapters can be changed depending on specific needs. In theory, based on RNA adaptor design, no adaptor–adaptor ligation should occur since the RNA adaptor either has no 5′-phosphate or it has a blocked structure that cannot undergo ligation with T4 RNA ligase. p, phosphate; idT, inverted deoxythymidine. All the adaptors should be further purified by HPLC or PAGE.
2. It is essential to use the correct ratio of tissue and TRIzol (Invitrogen) in order to obtain optimal RNA quality. A maximum amount of 1 g plant material can be processed for 10 mL TRIzol (Invitrogen). Don't over-process the starting material, as this will significantly increase RNA degradation. If isolating RNA from tissues rich in RNases, we recommend using only 500 mg of starting material for 10 mL TRIzol (Invitrogen).
3. Usually, the pellet is visible in the collection tube. Because most of the regular sized mRNAs are retained in the pellet, it could be used for other RNA applications.
4. 10 μL Nuclease-free water is typically used to resuspend LMW RNA from 100 μg of total RNA.
5. TEMED is best stored at 4°C in a desiccator. Buy small bottles as it may decline in quality after opening.
6. For best results, always use freshly prepared denaturing polyacrylamide gels. No more than 20 μL of sample should be applied to a well of 1.5 mm thickness and 5 mm width.
7. For most plant samples, two bands can be easily detected after ethidium bromide staining in the range between 20 and 30 nt if enough amount of LMW RNA is used.

8. To minimize degradation and protect RNA integrity, RNase inhibitor (such as RNase OUT) can be added in all the ligation reactions.
9. To get enough cDNA for high-throughput sequencing and to maintain quantitative information at the same time, 15–20 PCR cycles are usually used for cDNA amplification. If small amounts of starting material are used, the PCR cycles can be increased to 25.
10. SBS sequencing primer sites are present in RNA adaptors and PCR primers. Therefore, the purified PCR products are ready for entering SBS.
11. Several PCR polymerases have been tested in our lab. We found that NEB's high-fidelity DNA polymerase produces most reliable results. Because this enzyme generates blunt-end PCR products, the blunt-end TOPO vector should be used in subsequent cloning steps.

Acknowledgments

The authors would like to thank Dr. Pamela Green for encouragement and mentoring on the smRNA study. The work was done in her laboratory. The research was supported by USDA grant #2007-1-0199 to P.J. Green.

References

1. Carthew, R. W., and Sontheimer, E. J. (2009) Origins and mechanisms of miRNAs and siRNAs. *Cell* **136**, 642–655.
2. Voinnet, O. (2009) Origin, biogenesis, and activity of plant microRNAs. *Cell* **136**, 669–687.
3. Kim, V. N., Han, J., and Siomi, M. C. (2009) Biogenesis of small RNAs in animals. *Nat. Rev. Mol. Cell. Biol.* **10**, 126–139.
4. Okamura, K., Chung, W. J., and Lai, E. C. (2008) The long and short of inverted repeat genes in animals: microRNAs, mirtrons and hairpin RNAs. *Cell Cycle* **7**, 2840–2845.
5. Ramachandran, V., and Chen, X. (2008) Small RNA metabolism in *Arabidopsis*. *Trends Plant Sci.* **13**, 368–374.
6. Lagos-Quintana, M., Rauhut, R., Lendeckel, W., and Tuschl, T. (2001) Identification of novel genes coding for small expressed RNAs. *Science* **294**, 853–858.
7. Lau, N. C., Lim, L. P., Weinstein, E. G., and Bartel, D. P. (2001) An abundant class of tiny RNAs with probable regulatory roles in *Caenorhabditis elegans*. *Science* **294**, 858–862.
8. Lee, R. C., and Ambros, V. (2001) An extensive class of small RNAs in *Caenorhabditis elegans*. *Science* **294**, 862–864.
9. Fahlgren, N., Sullivan, C. M., Kasschau, K. D., Chapman, E. J., Cumbie, J. S., Montgomery, T. A., Gilbert, S. D., Dasenko, M., Backman, T. W., Givan, S. A., and Carrington, J. C. (2009) Computational and analytical framework for small RNA profiling by high-throughput sequencing. *RNA* **15**, 992–1002.
10. Sharma, C. M., and Vogel, J. (2009) Experimental approaches for the discovery and characterization of regulatory small RNA. *Curr. Opin. Microbiol.* **12**, 536–546.
11. Lu, C., Tej, S. S., Luo, S., Haudenschild, C. D., Meyers, B. C., and Green, P. J. (2005) Elucidation of the small RNA component of the transcriptome. *Science* **309**, 1567–1569.
12. Lister, R., Gregory, B. D., and Ecker, J. R. (2009) Next is now: new technologies for

sequencing of genomes, transcriptomes, and beyond. *Curr. Opin. Plant Biol.* **12**, 107–118.

13. Shendure, J., Mitra, R. D., Varma, C., and Church, G. M. (2004) Advanced sequencing technologies: methods and goals. *Nat. Rev. Genet.* **5**, 335–344.
14. Girard, A., Sachidanandam, R., Hannon, G. J., and Carmell, M. A. (2006) A germline-specific class of small RNAs binds mammalian Piwi proteins. *Nature* **442**, 199–202.
15. Qi, Y., He, X., Wang, X. J., Kohany, O., Jurka, J., and Hannon, G. J. (2006) Distinct catalytic and non-catalytic roles of ARGONAUTE4 in RNA-directed DNA methylation. *Nature* **443**, 1008–1012.
16. Katiyar-Agarwal, S., Gao, S., Vivian-Smith, A., and Jin, H. (2007) A novel class of bacteria-induced small RNAs in *Arabidopsis*. *Genes Dev.* **21**, 3123–3134.
17. Lee, Y. S., Shibata, Y., Malhotra, A., and Dutta, A. (2009) A novel class of small RNAs: tRNA-derived RNA fragments (tRFs). *Genes Dev.* **23**, 2639–2649.
18. Addo-Quaye, C., Eshoo, T. W., Bartel, D. P., and Axtell, M. J. (2008) Endogenous siRNA and miRNA targets identified by sequencing of the *Arabidopsis* degradome. *Curr. Biol.* **18**, 758–762.
19. German, M. A., Pillay, M., Jeong, D. H., Hetawal, A., Luo, S., Janardhanan, P., Kannan, V., Rymarquis, L. A., Nobuta, K., German, R., De Paoli, E., Lu, C., Schroth, G., Meyers, B. C., and Green, P. J. (2008) Global identification of microRNA-target RNA pairs by parallel analysis of RNA ends. *Nat. Biotechnol.* **26**, 941–946.
20. Accerbi, M., Schmidt, S. A., De Paoli, E., Park, S., Jeong, D. H., and Green, P. J. (2010) Methods for isolation of total RNA to recover miRNAs and other small RNAs from diverse species. *Methods Mol. Biol.* **592**, 31–50.
21. Ho, C. K., and Shuman, S. (2002) Bacteriophage T4 RNA ligase 2 (gp24.1) exemplifies a family of RNA ligases found in all phylogenetic domains. *Proc. Natl. Acad. Sci. USA* **99**, 12709–12714.
22. Ho, C. K., Wang, L. K., Lima, C. D., and Shuman, S. (2004) Structure and mechanism of RNA ligase. *Structure* **12**, 327–339.
23. Aravin, A., and Tuschl, T. (2005) Identification and characterization of small RNAs involved in RNA silencing. *FEBS Lett.* **579**, 5830–5840.
24. Hafner, M., Landgraf, P., Ludwig, J., Rice, A., Ojo, T., Lin, C., Holoch, D., Lim, C., and Tuschl, T. (2008) Identification of microRNAs and other small regulatory RNAs using cDNA library sequencing. *Methods* **44**, 3–12.

Chapter 10

Focusing Mutations Within Random Libraries to Distinct Areas: Protein Domain Library Generation by Overlap Extension

Andreas Gratz and Joachim Jose

Abstract

Directed evolution is an often used approach toward new proteins with tailor-made properties. It consists of random variation of the coding sequence of a protein followed by an appropriate selection procedure or a suitable type of property read out. In many, if not all cases, it is of significant advantage to constrain the randomly mutagenized DNA sequence to that encoding a particular part of the protein or a distinct domain, and not to mutagenize the entire gene of the target protein. For this purpose, a three-step, polymerase-based method was developed, which is independent of two flanking restriction sites adjacent to the nucleotide sequence supposed to be mutagenized, and named protein library generation by overlap extension (PDLGO).

Key words: Error-prone PCR, Random mutagenesis, SOE-PCR, Directed evolution, Esterase, EstE, Library

1. Introduction

During the last decade, directed evolution experiments have shifted from a more conceptual approach to applied studies (1, 2). Directed evolution is derived from the eponymous natural evolution consisting of variation and selection and represents a tool for the design of proteins and enzymes with adapted problem-oriented properties. Directed evolution consists of multiple consecutive rounds of random DNA sequence variation and selection of best-suited phenotype from a library of mutants. The phenotype can be modified gradually until a tailor-made property is obtained. A commonly used strategy to generate subtle genetic diversity in directed evolution is error-prone PCR (epPCR) (3), where a small number of mutations is randomly distributed throughout a DNA sequence.

Chaofu Lu et al. (eds.), *cDNA Libraries: Methods and Applications*, Methods in Molecular Biology, vol. 729,
DOI 10.1007/978-1-61779-065-2_10, © Springer Science+Business Media, LLC 2011

Oftentimes, it is not desirable to diversify the whole protein but rather focus random mutations to regions that are known or supposed to play a major role in the property that should be modified. For instance, if the aim is to alter activity or selectivity of an enzyme, mutations within the active site could be a good starting point, while variation of other parts, e.g., responsible for subcellular localization or signaling, should be avoided. In many cases, this can be achieved by using a classical cassette mutagenesis approach. Here, the activity-determining part of the gene is altered via epPCR and can be combined with the unaltered structure-determining part using an intermediate restriction endonuclease site and T4 DNA ligase.

In a mutagenesis study of EstE, an outer membrane anchored GDSL-type esterase; however, such an endonuclease restriction site could not be found. The sequence of EstE consists of three major parts (Fig. 1): (1) an N-terminal signal peptide, which is processed by a signal peptidase and facilitates EstE's translocation across the inner membrane; (2) the activity-determining sequence harboring the active-site forming catalytic domains of EstE; and (3) the β-barrel, a structurally important domain that anchors the enzyme in the outer membrane of *Escherichia coli*. For a whole-cell EstE activity screening system, anchoring of the enzyme within the outer membrane is crucial. As a consequence, the

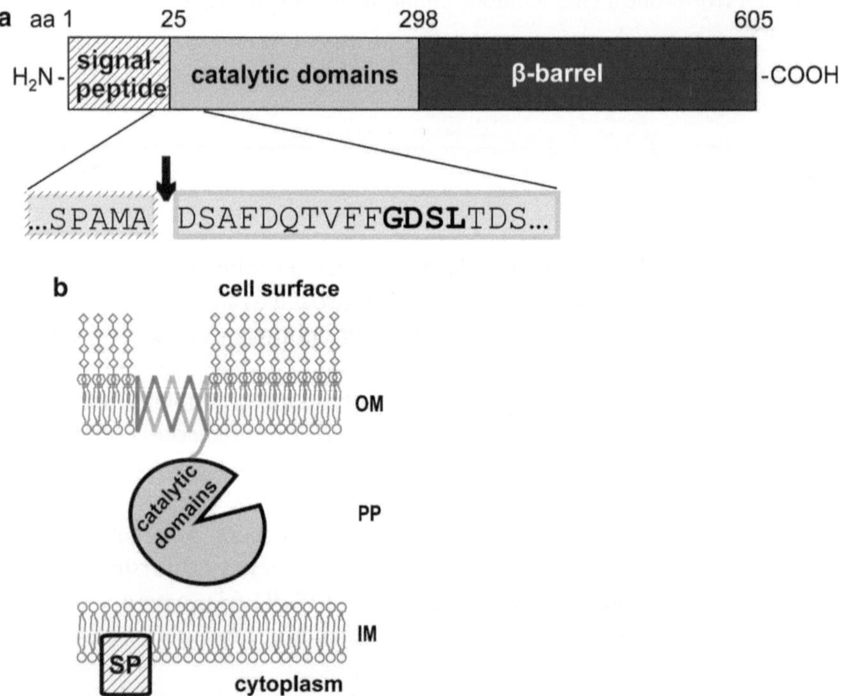

Fig. 1. Structure (**a**) and subcellular localization (**b**) of esterase EstE, when expressed in *E. coli* BL21(DE3). The signal peptidase cleavage site, separating sequences 1 and 2, is indicated by an *arrow*.

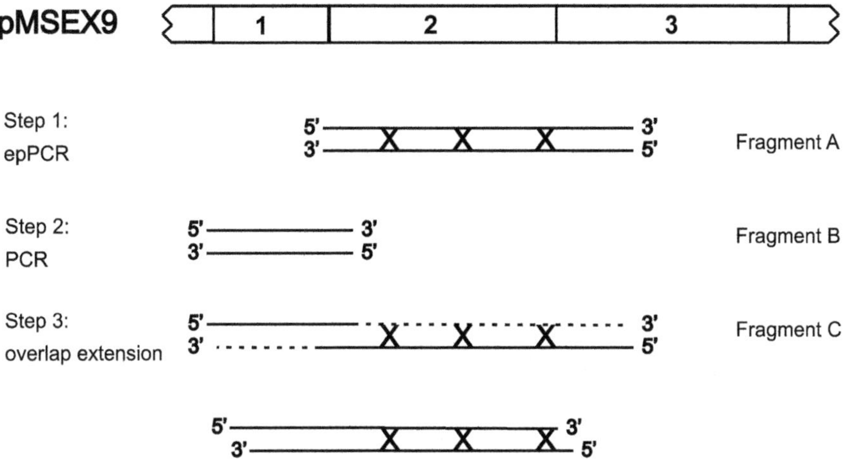

Fig. 2. General strategy of PDLGO. pMSEX9 is the plasmid encoding original EstE, x schematically illustrates mutations introduced by epPCR (PCR A).

sequence of neither part 1 nor part 3 must be altered in a random mutagenesis experiment. Sequence part 2 consists of five conserved blocks of amino acids, which presumably form EstE's active site. The first of these blocks includes the eponymous GDSL motif and is separated from the signal peptidase cleavage site by only eight amino acids (Fig. 1, arrow) (4). No restriction site can be found in between and none could be generated by silent mutation.

However, for our approach, it was inevitable to include the GDSL motif in random, epPCR-based mutagenesis. Some methods are known to create chimerical DNA sequences by using "splicing by overlap extension" PCRs (SOE-PCRs) (5, 6), but so far, none of these sequences to be spliced contained a library of random mutations. Our aim was to combine the classical epPCR strategy with standard high-fidelity (HF) PCR and use an overlap extension step to merge unmodified sequence (1) and randomly mutagenized sequence (2). This strategy comprises of three steps, as illustrated in Fig. 2, and is named protein domain library generation by overlap extension (PDLGO) (7). Besides its application for random protein mutation, PDLGO meanwhile proved to be suitable for functional promoter studies on DNA level as well (8).

2. Materials

2.1. Biological and Chemical Materials

For most purposes, e.g., gel extraction, plasmid isolation restriction digestions, polymerase reaction, etc., kits from commercial suppliers were applied and the ingredients were used according to the manufacturer's instructions. The kits we used are listed, but most probably, any similar kit will do as well.

1. Electrocompetent cells of *E. coli.*
2. Pure water. This is a tender point. We used deionized water which was subsequently purified by filtration using a Milli-Q Biocel (Millipore, Billerica, MA, USA).
3. Proofreading polymerase (*Pfu* polymerase from Stratagene, LaJolla, CA, USA. Vent polymerase from New England Biolabs, Beverly, MA, USA).
4. *Taq* polymerase (e.g., from Eppendorf, Hamburg, Germany).
5. Agarose, NEEO, ultra-quality (Roth, Karlsruhe, Germany) in 1× TAE buffer.
6. Gel purification kit (e.g., Eppendorf Perfect Prep Gel cleanup kit, Eppendorf, Hamburg, Germany).
7. Plasmid mini-prep kit (Qiagen, Hilden, Germany).
8. Plasmid containing the starting sequence to be mutagenized.
9. dNTP mixture (10 mM each, New England Biolabs, Beverly, MA, USA).
10. Manganese chloride (Merck, Darmstadt, Germany).
11. PCR Mastermix (1×) (Eppendorf, Hamburg, Germany):

    1.5 mM magnesium

    200 μM dATPs

    200 μM dGTP

    200 μM dCTP

    200 μM dTTP

    1.25 U Taq DNA polymerase
12. NotI and NdeI restriction endonucleases (New England Biolabs, Beverly, MA, USA).
13. BSA (20% bovine serum albumin in pure water, filter sterilized).
14. T4 DNA ligase and ligase buffer (New England Biolabs, Beverly, MA, USA).
15. SOC medium (pH 7.0):

Tryptone	20 g/L
Yeast extract	5 g/L
NaCl	0.5 g/L
KCl	2.5 mM
Autoclave (20 min, 121°C, 14 atmospheres), then add:	
$MgCl_2$(autoclaved)	10 mM
Glucose (sterile filtrated)	20 mM

16. Tributyrin agar (pH 7.0): supplied with suitable selection agent (e.g., ampicillin: 50 mg/L).

Tryptone	10 g/L
Yeast extract	5 g/L
NaCl	20 g/L
Agar	16 g/L
Glyceryl tributyrate	1%

17. DNA oligonucleotides (used as PCR primers) can be purchased at Sigma–Aldrich (or other providers) and stored as 100 μM stock solution in 10 mM TE buffer at –20°C (see Notes 1 and 2).

2.2. Equipment

1. Microcentrifuge (Mikro 120, Hettich, Tuttlingen, Germany).
2. Electroporation (EC100 Gene Transformer, EC Apparatus, St. Petersburg, NT, USA) and suitable electroporation cuvettes with 1-mm gap width.
3. Thermocycler (Mastercycler Gradient. Eppendorf, Hamburg, Germany).
4. Incubator for temperature range 16–65°C (AccuBlock Digital Dry Bath, Labnet, Edison, NJ, USA).
5. Agarose gel electrophoresis supplies and equipment (Biometra, Göttingen, Germany).

3. Methods

The aim of PDLGO is to generate a random library of a protein, where the random mutations within this library are constrained to a distinct domain of the protein, e.g., the catalytic domain of an enzyme. For this purpose, a three-step polymerase-based method is used, as depicted in Fig. 2. In comparison to standard epPCR or cassette mutagenesis, PDLGO has the advantage of being independent of two flanking restriction sites within the nucleotide sequence which is supposed to be mutagenized. Therefore, the area within a coding sequence, which should contain random mutations, can be picked more or less freely. This area can be separated exactly from adjacent sequences, which should be left unmodified. In particular, this is of imminent importance in case a random library of a multidomain protein is generated, wherein one domain is the exclusive target of variation and the other(s) need to be maintained in an unaltered state.

A typical example would be the membrane-anchoring domain of a receptor, which remains unaltered, whereas the ligand-binding site is randomly mutagenized. Another example would be a secreted protein, wherein the domain needed for secretion remains unaltered and the functional domain is variegated. In principle, this approach is suited to investigate the function of a distinct protein domain by random variation, in the context of other domains, which are kept constant.

As mentioned above, PDLGO is composed of three polymerase-based steps. In the first step, epPCR is used to generate a pool of DNA fragments, containing the random mutations. This fragment is referred to as fragment A. In the second step, HF PCR fragment B is produced, which has two functions. First, it partially overlaps with fragment A, namely, its reverse primer (Brev), with the forward primer of fragment A (Afor) in order to separate the randomly mutagenized DNA region exactly from the DNA region which remains unaltered. Second, it adjusts the length of the final PCR fragment C, sufficient to cover a suitable restriction site upstream of fragment A, which can be used for cloning purposes. In case such a suitable restriction site can be found in close vicinity to the randomized sequence, B is small fragment. In any other case, fragment B can be extended as required. This means, PDLGO is not completely independent of flanking restriction sites, but the location of suitable restriction sites with reference to the DNA sequence to be mutagenized is not critical. In the third polymerase reaction, fragments A and B are hybridized by their overlap and the single strands of this hybrid are filled by a DNA polymerase, resulting in fragment C. Fragment C consists of an unaltered 5′ area, derived from fragment B, and a downstream area containing the random mutations, derived from fragment A. After cleavage with restriction enzymes, it can be cloned into a plasmid that can subsequently be used to transform an adequate host, e.g., *E. coli.* In case a downstream restriction site cannot be found, please precede as described in Note 11.

3.1. Creation of Fragment A by epPCR

The critical step in the creation of fragment A is adjusting the mutation rate. It should be as high as possible in order to obtain a large number of different variants, but should not lead to frame shifts in the DNA or functional loss of the encoded protein (see Notes 3–5). In the present study, this was achieved by the addition of 0.5 mM $MnCl_2$, according to the directed evolution of pyranose oxidase, described previously (9). The addition of 0.5 mM $MnCl_2$ resulted in a small reduction of PCR yield. Therefore, PCR was performed in quadruplicates with 50 μL volume each in order to obtain sufficient amounts of fragment A.

1. Add the following in a vial:

   8 μL DNA (20–50 ng of plasmid encoding the starting sequence)

   48 μL PCR Mastermix (2.5×)

   6 μL Afor (forward primer) (20 mM)

   6 μL Arev (reverse primer) (20 mM)

   20 μL $MnCl_2$ (5 mM)

   112 μL Pure water

   = 200 μL Total volume (divide to four PCR vials, 50 μL each)

2. Run PCR A:

95°C	5 min
29 cycles of	
95°C	30 s
51°C	15 s
72°C	60 s
Terminal elongation:	
72°C	2 min

3. Gel purification:

   Check the result of the PCR by 0.8% agarose gel electrophoresis (a typical result with the 915-bp band of fragment A derived from EstE is shown in Fig. 3). Gel-purify complete PCR products using the Eppendorf Perfect Prep Gel cleanup kit (or any other), following the manufacturer's protocol.

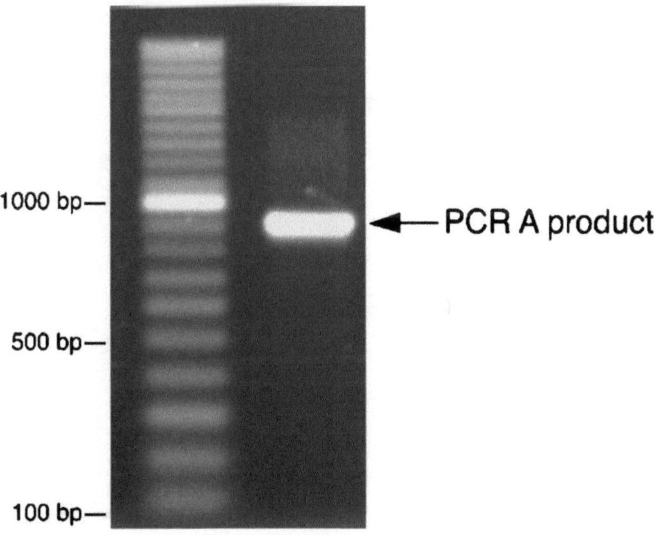

Fig. 3. Typical result of error-prone PCR A.

Dissolve the purified DNA in a final volume of 30 μL and store at −20°C. Run an adequate volume on an agarose gel to check concentration and purity of fragment A.

3.2. Creation of Fragment B by High-Fidelity PCR

1. Add the following in a vial:
   2 μL DNA (20–50 ng of plasmid DNA)
   2.5 μL Bfor (forward primer) (20 mM)
   2.5 μL Brev (reverse primer) (20 mM)
   7.5 μL dNTP mix (containing 10 mM of each dNTP)
   1 U Proofreading polymerase (see Note 6)
   5 μL Polymerase buffer (10×)
   28.5 μLPure water (in case 1 μL polymerase = 1 U)
   = Total volume of 50 μL
2. HF PCR:
   Run PCR B according to the following cycling protocol: 95°C for 5 min, 29 cycles of (95°C for 30 s, 48°C for 15 s, 72°C for 60 s), 72°C for 2 min. Check the result on 1.5% agarose gel (result for fragment B derived from EstE is shown in Fig. 4) and gel-purify complete PCR products using the Eppendorf Perfect Prep Gel cleanup kit (or any other), following the manufacturer's protocol. Dissolve the purified DNA in a final volume of 25 μL and store at −20°C. Run an adequate volume on an agarose gel to check the concentration and purity of fragment B.

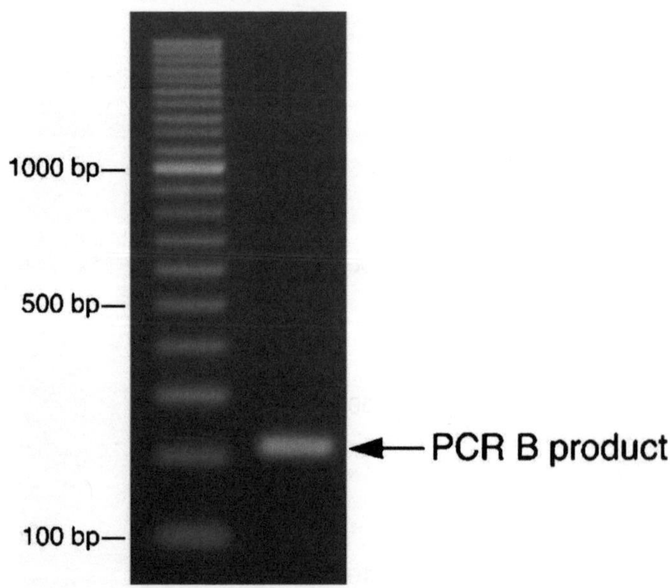

Fig. 4. Typical result of high-fidelity PCR B.

3.3. Creation of Fragment C by SOE-PCR

PCR C is a so-called SOE-PCR, used to combine the DNA sequence(s) of fragment A obtained in step 3 with the DNA sequence of fragment B from step 4 (see Note 7). The tender point in library creation is the amount of fragment A used in PCR C. Therefore, it is important to gather as much fragment A as possible and adjust the concentration of fragment B to equimolar amounts with fragment A.

1. Add the following in a PCR vial:

   25 μL Fragment A (from step 3)

   13 μL Fragment B (from step 4)

   5 μL Polymerase buffer (10×)

   2 μL dNTP Mix (10 mM of each dNTP)

   1 μL *Vent* polymerase (1 U)

   4 μL Pure water

   = Final volume of 50 μL

2. SOE-PCR:
   Run PCR C according to the following cycling protocol: 95°C for 5 min, 46 cycles of (95°C for 1 min, 65°C for 2 min, 72°C for 2 min), 72°C for 3 min. Cool to 4°C (see Note 8).

   Check the result on 1.5% agarose gel (Fig. 5) and gel-purify complete PCR products using the Eppendorf Perfect Prep Gel cleanup kit (or any other), following the manufacturer's protocol. Dissolve the purified DNA in a final volume of 20 μL and store at –20°C as fragment C.

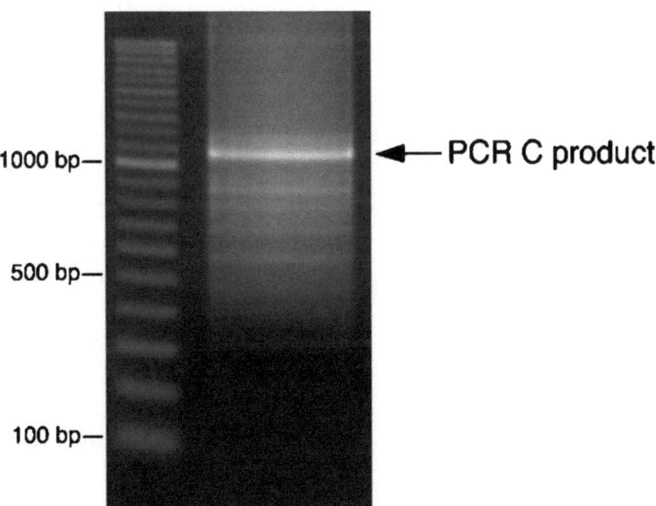

Fig. 5. Typical result of the "splicing by overlap extension" (SOE) reaction (PCR C).

3.4. Library Generation

Digest fragment C by restriction enzymes in order to ligate it into the plasmid digested with the same enzymes. This results in the replacement of the starting sequence by the mutagenized sequence, where the random mutations are strictly focused in the DNA sequence covered by fragment A.

1. Digestion:

   16.2 μL Fragment C

   2.1 μL Restriction buffer (10×)

   2.1 μL BSA

   0.3 μL NotI

   0.3 μL NdeI

   = Total volume of 21 μL

   Incubate at 37°C for 3.5 h and transfer subsequently to 65°C for 20 min in order to inactivate the restriction enzymes

2. Ligation:

   10 μL Digested fragment C

   3 μL Digested plasmid backbone (fragment C: plasmid = 3:1)

   1.5 μL Ligase buffer (10×)

   1 μL Ligase (5 U)

   = Final volume of 15 μL

   Incubate at 16°C for 16 h. Inactivate for 20 min at 65°C and desalt the ligation mixture (see Note 9)

3. Transformation:

   Be sure to chill electroporation cuvettes, competent cells, and the ligation mixture on ice and warm SOC medium at 37°C before electroporation. Transform the complete ligation sample into a 50-μL aliquot of competent *E. coli* cells (1,800 V); in the present example, we used electrocompetent cells of *E. coli* BL21(DE3). Immediately add 1 mL of prewarmed SOC medium and incubate the cells at 37°C for 1 h with smooth agitation (50 rpm) (see Note 10). Plate the complete volume obtained after transformation on tributyrin agar plates in adequate dilutions to obtain around 80 single-cell clones per plate, containing the suitable selection antibiotic (ampicillin). Because esterase activity will result in the formation of so-called halos around the colonies on the tributyrin agar plates, it is important to choose a dilution that leaves sufficient space between the colonies. An excerpt of such a tributyrin agar plate with single colonies, with and without "halos," is shown in Fig. 6.

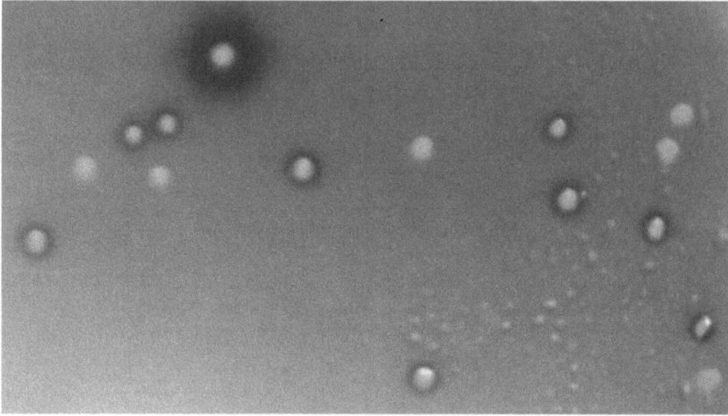

Fig. 6. Detail of a tributyrin LB agar plate showing colonies derived from library transformants with different variants of EstE. The diameter of the "halos" around the colonies correlates with the enzymatic activity of the variants.

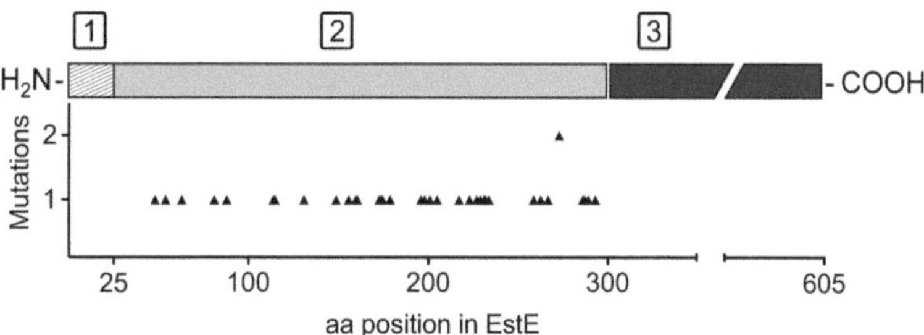

Fig. 7. Distribution of random mutations within the EstE library, introduced by PDLGO. Mutations were exclusively found in the catalytic domain (sequence part 2).

3.5. Library Characterization

Following strictly the strategy described within Subheadings 3.1–3.4, a library of EstE from *Xanthomonas vesicatoria* was obtained with a library size of 1,600 single-cell clones. The single-cell clones were picked from the agar plates and transferred to 96-well microplates for storage and for activity determination. An adequate number of single-cell clones were chosen for DNA sequence analysis in order to determine the degree of random variation and to preclude any sort of unexpected bias. As shown in Fig. 7, the mutations were exclusively located in the catalytic domain of EstE, covered by fragment A. No mutations were observed in the transport domains, represented by the signal peptide and the β-barrel. The mutations within the catalytic domain were randomly distributed. Almost all of the amino acid exchanges appeared only once, only one amino acid position appeared to be mutated twice within the single-cell clones analyzed. This indicated that the strategy for the creation of random libraries, as described here, does not lead to any bias, neither in the area nor in the amino acids that are varied. This is also exemplified by Table 1, in which the single nucleotide exchanges observed in the

Table 1
Summary of PDLGO mutation characteristics in EstE library construction

Dimension of library	1,600 cfu
Mutation frequency [1/1,000 bp]	5.4
Resulting aa exchanges [1/100 aa]	1.14
Ratio transition:transversion	4:1
Mutations in signal peptide sequence	0

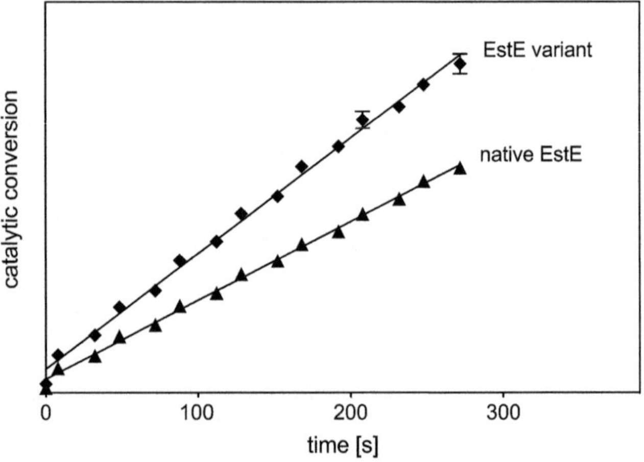

Fig. 8. Increased activity of an EstE variant as obtained by PDLGO toward *p*-nitrophenyl acetate, compared to original EstE.

analyzed mutants are summarized. Within this random library of EstE, enzyme variants with increased activity could be identified (e.g., in Fig. 8).

4. Notes

1. Primer design is crucial in the PDLGO strategy. Primer length and sequence should be determined carefully to enable successful amplification. The amplicon's length should be greater than 200 bp.
2. Forward primer of PCR A and reverse primer of PCR B should overlap in approximately 30 bp to enable easy hybridization and extension (a typical example for EstE-coding sequence is given in Fig. 9).

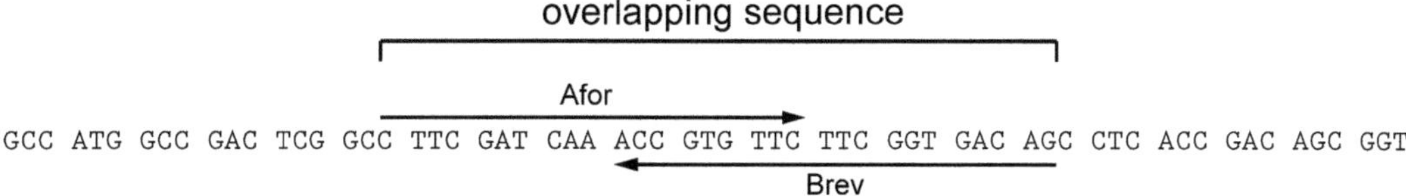

Fig. 9. Binding sites of PCR primers at the border sequence between signal peptide and catalytic domain of EstE (only coding sequence is shown), as well as overlapping sequence between fragment A and fragment B, employed for hybridization in PCR C.

3. The amount of the PCR A product is vital for the size of library variability. Try to get a large amount of PCR product by adjusting (1) the annealing temperature of the PCR primers, (2) the $MgCl_2$ concentration of the PCR buffer, and (3) by using a hot-start PCR if side products are obtained.
4. It is recommendable to test conditions for an effective PCR A first and later apply suitable amounts of $MnCl_2$ to produce a domain library.
5. In PCR A, carefully adjust the $MnCl_2$ concentration. The authors reported that using a concentration of 0.15 mM $MnCl_2$ resulted in no mutations at all, whereas the application of 1.5 mM $MnCl_2$ produced a large fraction of frameshift mutations. Moreover, increasing $MnCl_2$ concentration will decrease PCR yield.
6. If possible, use a proofreading DNA polymerase, e.g., Vent or *Pfu* polymerase, in PCR B to avoid (1) mutations and (2) the attachment of a random nucleotide by *Taq* polymerase, which would decrease the efficiency of hybridization.
7. PCR C is not really a chain reaction, because it just fuses fragment A to fragment C and does not amplify a given sequence. However, the details are similar to that of a PCR, and for convenience, we denoted this step "PCR C".
8. In order to improve overlap extension reaction (PCR C), the authors found it extremely helpful to use Vent DNA polymerase and apply multiple cycles (here 46 cycles) of denaturation (95°C, 1 min), hybridization (65°C, 2 min), and elongation (72°C, 3 min) to increase the efficiency of this step. The hybridization temperature has to be adjusted eventually depending on the length of the overlapping sequence. However, if the overlap becomes too short, using a Klenow filling strategy with reduced temperatures might be considered.
9. Desalting of the ligation mixture can be done using Millipore's 0.025-µm dialysis membranes. Place a corresponding disk in a Petri dish filled with autoclaved ultrapure water (shiny side up!). Transfer the complete ligation mixture onto the disk and leave it for 10–30 min. Transfer the remaining drop into

a new vial. Desalting will most probably increase the efficiency of electroporation.

10. Depending on the circumstances, chemically competent cells with high transformation efficiency can be used as well. In such cases, use the instructions provided by the supplier.
11. In cases where a suitable downstream restriction site cannot be found directly adjacent to fragment A, PDLGO turns from a three-step polymerase-based strategy to a five-step polymerase-based strategy. A third PCR is required, which needs to be a HF PCR as well in order to produce fragment B′. This fragment B′ has the same function downstream of fragment A, as fragment B has upstream of fragment A. Fragment B′ also partially overlaps with fragment A, but at its downstream end, namely, forward primer of fragment B′ (B′for) partially overlaps with the reverse primer of fragment A (Arev). Finally, after obtaining fragment C, fragment B′ is used for a Klenow fragment polymerase reaction as described for fragment B, and the resulting fragment C′ is cleaved and inserted into the desired plasmid.

References

1. Arnold F.H. (1998) Design by directed evolution. *Acc. Chem. Res.* **31**, 125–131.
2. Fasan R., Chen, M.M., Crook, N.C., Arnold, F.H. (2007) Engineered alkane-hydroxylating cytochrome P450BM3 exhibiting nativelike catalytic properties. *Angew. Chem. Int. Ed. Engl.* **46**, 8414–8418.
3. Leung D., Chen, E., Goeddel, D. (1989) A method for random mutagenesis of a defined DNA segment using a modified polymerase chain reaction. *Technique* **1**, 11–15.
4. Talker-Huiber D., Jose, J., Glieder, A., Pressnig, M., Stubenrauch, G., Schwab, H. (2003) Esterase EstE from *Xanthomonas vesicatoria* (Xv_EstE) is an outer membrane protein capable of hydrolyzing long-chain polar esters. *Appl. Microbiol. Biotechnol.* **61**, 479–487.
5. Horton R.M., Hunt, H.D., Ho, S.N., Pullen, J.K., Pease, L.R. (1989) Engineering hybrid genes without the use of restriction enzymes: gene splicing by overlap extension. *Gene* 77, 61–68.
6. Heckman K.L., Pease, L.R. (2007) Gene splicing and mutagenesis by PCR-driven overlap extension. *Nat. Protoc.* **2**, 924–932.
7. Gratz A., Jose, J. (2008) Protein domain library generation by overlap extension (PDLGO): a tool for enzyme engineering. *Anal. Biochem.* **378**, 171–176.
8. Vega J., Puebla, C., Vasquez, R., Farias, M., Alarcon, J., Pastor-Anglada, M., Krause, B., Casanello, P., Sobrevia, L. (2009) TGF-{beta}1 inhibits expression and activity of hENT1 in a nitric oxide-dependent manner in human umbilical vein endothelium. *Cardiovasc. Res.* **82**, 458–467.
9. Bastian S., Rekowski, M.J., Witte, K., Heckmann-Pohl, D.M., Giffhorn, F. (2005) Engineering of pyranose 2-oxidase from Peniophora gigantea towards improved thermostability and catalytic efficiency. *Appl. Microbiol. Biotechnol.* **67**, 654–663.

Chapter 11

Generation of Families of Construct Variants Using Golden Gate Shuffling

Carola Engler and Sylvestre Marillonnet

Abstract

Current standard cloning methods based on the use of restriction enzymes and ligase are very versatile, but are not well suited for high-throughput cloning projects or for assembly of many DNA fragments from several parental plasmids in a single step. We have previously reported the development of an efficient cloning method based on the use of type IIs restriction enzymes and restriction–ligation. Such method allows seamless assembly of multiple fragments from several parental plasmids with high efficiency, and also allows performing DNA shuffling if fragments prepared from several homologous genes are assembled together in a single restriction–ligation. Such protocol, called Golden Gate shuffling, requires performing the following steps: (1) sequences from several homologous genes are aligned, and recombination sites defined on conserved sequences; (2) modules defined by the position of these recombination sites are amplified by PCR with primers designed to equip them with flanking *Bsa*I sites; (3) the amplified fragments are cloned as intermediate constructs and sequenced; and (4) finally, the intermediate modules are assembled together in a compatible recipient vector in a one-pot restriction–ligation. Depending on the needs of the user, and because of the high cloning efficiency, the resulting constructs can either be screened and analyzed individually, or, if required in larger numbers, directly used in functional screens to detect improved protein variants.

Key words: DNA shuffling, High-throughput cloning, Restriction–ligation, Type IIs restriction enzymes, Seamless cloning, Modular cloning, Directed evolution

1. Introduction

The discovery of restriction enzymes almost 40 years ago provided biologists with the ability to create recombinant DNA molecules at will, in practice opening the field of modern molecular biology (1). Since then, cloning with restriction enzymes and ligase has become the base for work in molecular biology and has been hugely successful. However, despite such success, generating constructs is still a relatively slow and tedious process that has several limitations.

Chaofu Lu et al. (eds.), *cDNA Libraries: Methods and Applications*, Methods in Molecular Biology, vol. 729,
DOI 10.1007/978-1-61779-065-2_11, © Springer Science+Business Media, LLC 2011

In particular, each new construct requires the design of a specific cloning strategy, and only one or a few DNA fragments can be ligated together into a recipient vector in one step. Among some of the significant progresses made in cloning in the past few years, recombination-based cloning has been developed to facilitate the construction of plasmids for applications where a gene of interest needs to be transferred into several different acceptor expression vectors (2). Recombination-based cloning allows making constructs in a one-pot reaction with great efficiency. However, such approach still suffers several drawbacks, including the fact that unwanted recombination sites are left at the cloning junctions and that only a limited number of fragments (up to 4) can be cloned in a vector in one step.

We have recently developed a cloning method called Golden Gate cloning that overcomes many of the limitations of current cloning methods (3). This cloning strategy allows nine separate fragments to be cloned in a defined linear order into an acceptor vector in one step, with more than 90% of the colonies obtained containing the desired construct (4). Interestingly, this cloning method is not based on new exotic enzymes or genetic elements, rather it relies on the use of several previously known elements that, when used in concert, provide extremely high efficiency. The most important of these elements consists of the use of type IIs restriction enzymes combined with restriction–ligation, ensuring a very high cloning efficiency, and that only the correct desired construct remains at the end of the restriction–ligation. This cloning method not only has applications for cloning of any construct in general, but is also useful for DNA shuffling. Here, we provide a detailed protocol on how to perform Golden Gate shuffling, but the information provided is by default also useful for making any other type of construct that a user would want to make.

The principle of Golden Gate cloning/shuffling is based on the ability of type IIs enzymes to cleave outside of their recognition site sequence, allowing two DNA fragments flanked by compatible restriction sites to be digested and ligated seamlessly (5–8) (Fig. 1a). Since the ligated product of interest does not contain the original type IIs restriction site, it will not be subject to redigestion in a restriction–ligation. However, all other products that reconstitute the original site will be redigested, allowing their components to be made available for further ligation, leading to the formation of an increasing amount of desired product with increasing time of incubation. Since the sequence of the overhangs at the ends of the digested fragments can be chosen to be any four nucleotide sequence of choice, more than one fragment of interest can be assembled together in a defined linear order in a single restriction–ligation. Moreover, if fragments of several homologous genes are available in the same restriction–ligation mix, DNA shuffling is possible.

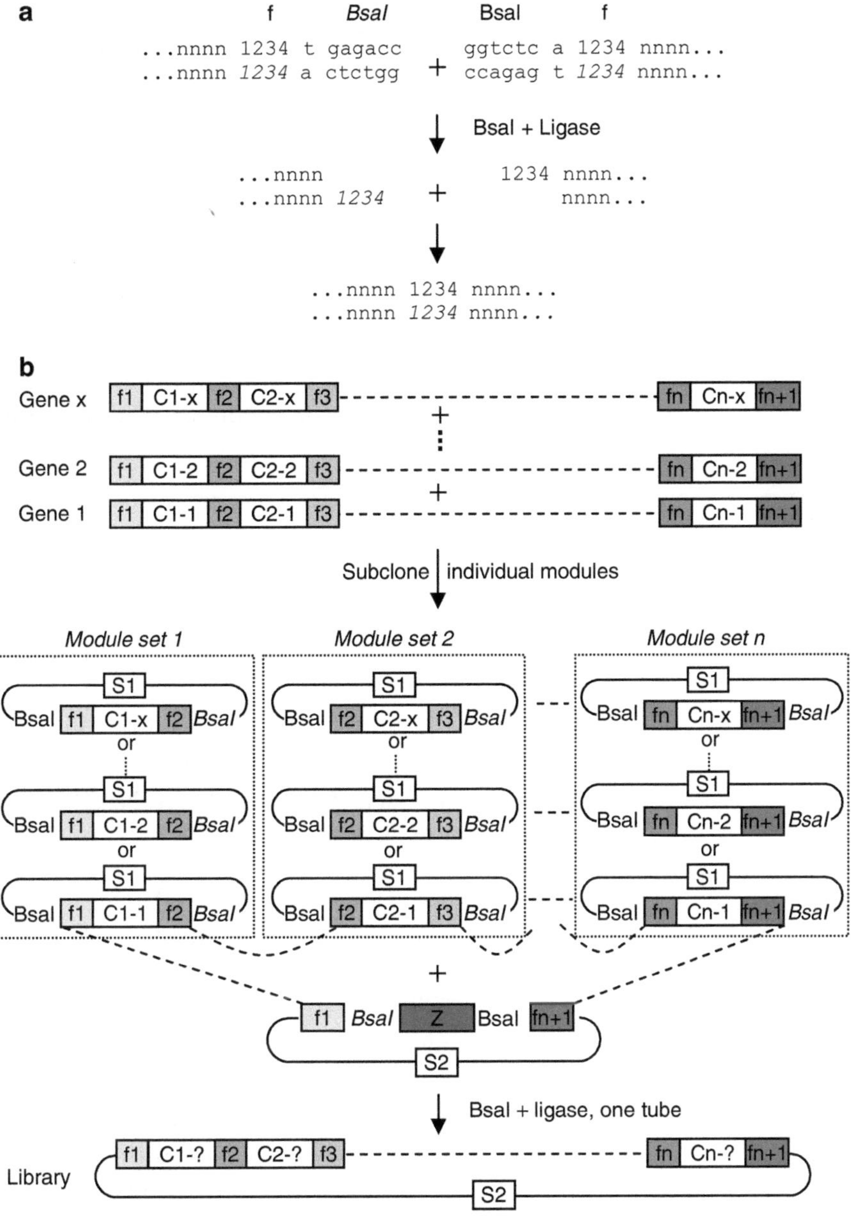

Fig. 1. DNA shuffling strategy. (**a**) Two DNA ends terminated by the same four nucleotides (sequence f, composed of nucleotides 1234, complementary nucleotides noted in *italics*) flanked by a *Bsa*I recognition sequence form two complementary DNA overhangs after digestion with *Bsa*I. (**b**) For shuffling, genes of interest are aligned, and recombination points consisting of four nucleotide sequences (f1 to fn+1) are defined on conserved sequences. Module fragments (core sequence, C1 to Cn, plus flanking four nucleotide sequences) are amplified by PCR and cloned in an intermediate cloning vector. Module fragment plasmids and the acceptor vector are assembled in one restriction–ligation with *Bsa*I and ligase. S1 and S2, two different selectable markers. *Z* lacZ alpha gene fragment.

Basically, using Golden Gate shuffling requires performing the following steps (Fig. 1b): (1) an alignment between the sequences to be shuffled is made and recombination sites are defined on conserved sequences, (2) modules defined by the position of these recombination sites are amplified by PCR with primers designed to equip them with flanking *Bsa*I sites, (3) the amplified fragments are cloned as intermediate constructs and sequenced, and (4) finally, the intermediate modules are assembled together in a compatible recipient vector in a one-pot restriction–ligation. Depending on the needs of the user, and because of the high cloning efficiency, the resulting constructs can either be screened and analyzed individually, or, if required in larger numbers, directly used in functional screens to detect improved protein variants.

2. Materials

2.1. Polymerase Chain Reaction

1. Novagen KOD Hot Start DNA polymerase (Merck KGaA, Darmstadt), supplied with 10× buffer, 25 mM $MgSO_4$, and 2 mM dNTPs.
2. Custom-made primers can be ordered from any of the many commercial vendors (for example, Invitrogen, Karlsruhe).
3. NucleoSpin® Extract II kit (Macherey Nagel, Dueren), for purification of PCR products.

2.2. Cloning

1. Restriction endonuclease *Sma*I (10 U/μL) (NEB, New England Biolabs Inc., Ipswich, MA, USA), supplied with 10× NEBuffer 4 (200 mM Tris–acetate pH 7.5, 100 mM magnesium acetate, 500 mM potassium acetate, and 10 mM dithiothreitol).
2. Restriction endonuclease *Bsa*I (10 U/μL) (NEB), supplied with 10× NEBuffer 4.
3. T4 DNA ligase 3 U/μL or T4 DNA ligase (HC) 20 U/μL (Promega, Mannheim), both supplied with 10× ligation buffer (300 mM Tris–HCl pH 7.8, 100 mM $MgCl_2$, 100 mM DTT, and 10 mM ATP).
4. For measuring of DNA concentration, we use the NanoDrop ND1000 (Peqlab, Erlangen).
5. Luria-Bertani (LB) medium: 1% bacto-tryptone, 0.5% yeast extract, 1% NaCl in deionized water, and adjusted to pH 7.0 with 5 N NaOH. For plates, 1.5% agar is added.
6. Antibiotics carbenicillin (used instead of ampicillin) and kanamycin: filter-sterilized stocks of 50 mg/mL in H_2O (stored in aliquots at –20°C) are diluted 1:1,000 (final concentration: 50 μg/mL) in an appropriate amount of medium after the medium has been autoclaved and cooled down.

For spectinomycin, a stock of 40 mg/mL is made and is used at a final concentration of 100 μg/mL (dilution 1:400).

7. 5-Bromo-4-chloro-3-indolyl-β-d-galactopyranoside (X-gal): stock solution of 20 mg/mL in dimethylformamide (DMF). For preparation of plates, the stock is diluted 1:500 (final concentration: 40 μg/mL) in an appropriate amount of LB agar after autoclaving/melting and cooling down.

2.3. Preparation of Chemically Competent Cells

1. Solution TFB1: 30 mM potassium acetate, 10 mM $CaCl_2$, 50 mM $MnCl_2$, 100 mM RbCl, and 15% glycerol; adjust to pH 5.8 (with 1 M acetic acid), filter-sterilize, and store at 4°C (ready to use) or at room temperature (cool down before use).
2. Solution TFB2: 100 mM MOPS (or PIPES), 75 mM $CaCl_2$, 10 mM RbCl, and 15% glycerol; adjust to pH 6.5 (with 1 M KOH), filter-sterilize, and store at 4°C (ready to use) or at room temperature (cool down before use).
3. The OD_{600} of bacterial cultures is measured in a SmartSpec™3000 spectrophotometer (Biorad, Muenchen).

2.4. Screening of Colonies

1. DNA minipreps: NucleoSpin® Plasmid Quick Pure (Macherey Nagel, Dueren).
2. Restriction endonucleases (NEB or Fermentas, St. Leon-Rot), all supplied with 10× buffer and if necessary also with 100× BSA (dilute 1:10 and store in aliquots at –20°C).
3. DNA ladder: GeneRuler™ 1-kb DNA Ladder Plus (Fermentas) is used as marker for gel electrophoresis.
4. 50× TAE buffer: 242.0 g Tris, 57.1 mL acetic acid, and 100 mL 0.5 M EDTA, pH 8.0, in 1 L of deionized water.
5. Gels: agarose (0.7–1.5%) in 1× TAE is melted in a microwave oven and one drop of 0.025% ethidium bromide solution (Carl Roth GmbH, Karlsruhe) is added per 100 mL of melted agarose solution.
6. Running buffer of agarose gels is 1× TAE.
7. Gels are checked visually using a Syngene GelVue transilluminator (VWR, Darmstadt), and pictures are taken by using a Quantity one® gel analysis software (Biorad).
8. DNA maps of plasmids are made by using the Vector NTI software (Invitrogen).

2.5. Sequencing

1. DNA/constructs to sequence are sent to an external contractor (GATC Biotech, Konstanz). Sequence data are analyzed using the DNASTAR's Lasergene software.
2. Primers M13RP (CAGGAAACAGCTATGACC) and/or M13FP (TGTAAAACGACGGCCAGT) are used for sequencing of inserts cloned in pUC19-derived vectors.

3. Methods

3.1. Selection of Recombination Points

The first step for performing DNA shuffling consists of selecting "recombination sites" within several parental sequences. We use the term "recombination sites" in functional analogy to the recombination sites used in recombination systems such as the phage P1 Cre-loxP recombination system. However, for the purpose described here, "recombination sites" can be any three or four nucleotide sequences of choice defined as such (with the limitations described below), and will serve as the sequence where restriction enzyme digestion and ligation will take place; no real recombinase will in fact be used. We call the sequence between recombination sites as "modules" (more generally, plasmids containing such sequences are also referred to as modules). One "set" of modules consists of all equivalent modules prepared from homologous sequences and flanked by the same two recombination sites (Fig. 1b).

1. A first and obvious requirement for choosing recombination sites is to select them within sequences conserved between all parental sequences. This requirement is easy to satisfy since only four nucleotides need to be conserved for each recombination site (one amino acid) if an enzyme such as *Bsa*I is used for shuffling (digestion with *Bsa*I results in a four-nucleotide overhang), and only three nucleotides if an enzyme such as *Sap*I is used (see Note 1). Therefore, the first step in gene shuffling consists of performing an alignment of the amino acid sequence, and then of the nucleotide sequence (we use the vector NTI program, but any other program will also be suitable). Performing a sequence alignment can, however, be omitted if nonhomologous parental sequences are used (see Note 2).
2. A second requirement for recombination sites is to avoid selecting the same sequence twice, as this would lead to illegitimate recombination and deletion of the sequences between the two sites. It is also important to make sure that the sequence of any site does not match the sequence of any of the other chosen sites, both on the same and on the complement strand. For example, choice of the sequence ATTC will preclude the choice of the sequence GAAT for any of the other recombination sites used for this shuffling experiment. Use of two such sites would sometimes lead to ligation of two inappropriate fragments, one in the opposite orientation. This would lead to the formation of molecules that will not be able to form circular plasmids, but that would continue to ligate to further modules and form long linear multimeric concatemers.

3. A third requirement is to avoid the 16 palindromic sequences, since any palindromic DNA end can be ligated to another copy of the same DNA fragment in opposite orientation, and lead to the same problem as described above. For enzymes leading to a four-nucleotide extension, 240 different sequences are, therefore, available.
4. Finally, a fourth but optional requirement can be defined to maximize the efficiency of DNA shuffling. We have observed that inappropriate ligation of fragments can occur between ends with four-nucleotide overhangs that match for three of four consecutive nucleotides, for example, as in sequences GGTG and AGTG, or GGTG and CACT. Therefore, combination of two such sites should be avoided if possible (see Note 3).

Other than the minimal requirements defined above, the number of recombination sites, as well as their position within the gene to shuffle, is chosen depending on the needs and the goal of the user for each specific protein. Therefore, the size (see Note 4) and number of fragments to shuffle will vary for each gene and each experiment. We have tested up to eight recombination points within a gene (nine module sets), but a higher number should be possible as well, although probably with reduced efficiency.

The following steps consist of amplifying the defined modules by PCR, cloning them in intermediate vectors, and sequencing them. Alternatively, these steps can be replaced by simply ordering these modules from a gene synthesis company (see Note 5).

3.2. PCR Amplification of the Modules

Modules defined by the position of the recombination sites need to be amplified by PCR using primers designed to add two *Bsa*I sites flanking each module. Primers are designed such that the overhangs created by digestion of the amplified products with *Bsa*I (or any other type IIs enzyme chosen) correspond to the sequence of the chosen recombination sites. Therefore, the sequence ttg gtctca is added to each primer sequence, for example, ttggtctca CAGG nnnnn (CAGG being the recombination site, followed by 16–20 nucleotides of target sequence). For nine modules prepared from three homologs, 54 primers need to be made.

Moreover, a requirement for Golden Gate shuffling is to not have any internal *Bsa*I sites present within any of the DNA fragments used for shuffling. Indeed, the presence of a *Bsa*I site within one of the modules would lead to redigestion of the shuffled DNA molecules containing such fragment at the end of the assembly step. These linear molecules will not transform *Escherichia coli*. Therefore, any such site needs to be removed before this step, and doing so at the time of generation of the entry clones is appropriate. We have previously described a method to remove internal *Bsa*I sites upon cloning in the entry vector, which requires using a specific cloning vector corresponding to the combination of recombination sites flanking the given fragment (3). However, this method is not useful for DNA shuffling since new and different recombination sites are chosen for each new shuffling experiment, and therefore, specific cloning vectors are not available. Instead, removal of internal *Bsa*I sites from PCR fragments can be done easily using gene SOEing (9) (see Note 6).

PCR is usually performed using plasmid DNA, but genomic DNA or cDNAs can also be used as starting material, depending on the source available to the user; for small modules, template DNA is not even necessary (see Note 7). For amplification, we use the enzyme KOD Hot Start DNA polymerase since it has a very low error rate and, unlike Taq polymerase or many enzyme mixes, produces DNA products with blunt ends (other thermostable polymerases that a user would prefer to use are of course perfectly acceptable as well). Blunt ends are advantageous, as the products can be easily cloned in any standard vector such as pUC19 by blunt-end cloning (see below), and therefore, cloning them does not require the purchase of a kit.

1. The PCR mix is set up following the manufacturer's instructions, for example, using KOD polymerase, with the following conditions: 1 μL of plasmid DNA (5–20 ng/μL), 5 μL of 10× buffer, 3 μL of 25 mM $MgSO_4$, 5 μL of 2 mM dNTPs, 1.5 μL each of 10 μM sense and antisense primers, and 1 μL of KOD Hot Start DNA polymerase (10 U/μL, final concentration 0.02 U/μL) in a total reaction volume of 50 μL.
2. PCR is performed using the following cycling conditions: (1) incubation at 95°C for 2 min for polymerase activation, (2) denaturation at 95°C for 20 s, (3) annealing at 58°C for 10 s – the temperature for the annealing step can be adjusted for specific primers, but the temperature of 58°C usually works well for primers designed as described above, (4) extension at 70°C, the duration depends on the length of the expected fragment (from 10 s/kb for fragments smaller than 500 bp up to 25 s/kb for fragments larger than 3 kb, see manufacturer's instructions); steps 2–4 are repeated 35 times and are followed by a final extension step at 70°C for 20 s to 2 min (depending on fragment length). The reaction is then incubated at 12°C until taken out of the thermocycler.
3. Of the PCR product obtained, 2 μL is then analyzed by gel electrophoresis to make sure that a product of the correct size has been amplified.
4. The amplified fragment is purified from remaining primers, potential primer dimers, and remaining polymerase enzyme by using the NucleoSpin® Extract II kit and following the kit protocol. DNA is eluted from the column with 30–50 μL of

elution buffer (5 mM Tris–HCl, pH 8.5). In case several bands were amplified rather than only the expected fragment, the same kit can also be used to cut and extract the appropriate DNA fragment from an agarose gel.

3.3. Blunt-End Cloning of the Modules

Many commercial kits are available for cloning PCR products, including the pGEM-T kit (Promega), pJET (Fermentas), and TOPO® TA kit (Invitrogen). However, PCR products can also be cloned very efficiently without purchasing a kit, by using blunt-end cloning performed with a restriction–ligation (10, 11). This method is very efficient and has the advantage that the DNA fragment of interest to be cloned does not need to be flanked by any specific sequence, and, therefore, restrictions on primer design are minimal (see Note 8). Another advantage of this cloning method is that any plasmid of choice by a user can be used as long as it contains a unique blunt site, preferably in a reporter gene such as *LacZ*. This is useful since cloning vectors for generating entry clones for shuffling need to fulfill preferentially two requirements: (1) they should preferably not contain any restriction site for the type IIs enzyme chosen for shuffling (see Note 9) and (2) the antibiotic resistance gene of the entry vector should preferably be different from the one in the destination vector. Since several commercial cloning vectors have a *Bsa*I restriction site in the ampicillin resistance gene (for example, pGEM-T or pJET), we have made our own entry cloning vectors that simply consist of pUC19 lacking a *Bsa*I restriction site (see Note 10).

1. Add 0.5 μL of vector (50 ng), 1 μL of PCR product (50–100 ng), 2 μL of 10× ligation buffer (Promega), 1 μL of *Sma*I enzyme (10 U; NEB), 1 μL of ligase (3 U; Promega), and 14.5 μL of water (total volume of 20 μL) into a tube. The reaction mix is incubated for 1–2 h at room temperature or in a 25°C incubator, if one is available.
2. The entire ligation mix is transformed to DH10B chemically competent cells and plated on LB plates with X-gal and the appropriate antibiotic (the transformation protocol is described below in paragraph Subheading 3.6).
3. White colonies (or sometimes pale blue when small inserts are cloned) are picked and inoculated in 5 mL of LB medium containing the appropriate antibiotic.
4. Plasmid DNA is extracted using the NucleoSpin® Plasmid Quick Pure kit from Macherey Nagel following the manufacturer's instructions.
5. Plasmid DNA can be checked by restriction enzyme digestion using *Bsa*I and analysis of the digested DNA by agarose gel electrophoresis. A fragment of the size of the expected module should be visible.

6. Two minipreps are sent for sequencing using primers M13RP and/or M13FP.
7. When the correct sequence has been verified, DNA concentration of the plasmid prep is measured using the NanoDrop ND1000 (Peqlab).

3.4. Construction of the Destination Vector

A destination vector compatible with the entry modules needs to be made. There are many ways of making a destination vector, but what is important is that the final vector should respect the following criteria. First, it should contain two *Bsa*I sites (or any other type IIs enzyme chosen) with cleavage sites compatible with the beginning of the first and the end of the last entry module sets. The vector backbone should not contain any other *Bsa*I restriction site. And finally, it should have an antibiotic selectable marker different from the one used in the entry clones. No description is provided here for making such a vector, since each vector will require a specific construction strategy.

3.5. DNA Shuffling

Once entry constructs and the recipient vector are made and sequenced, performing DNA shuffling only requires pipetting all components into a reaction mix, incubating the mix in a thermocycler, and transforming it into competent cells. An important factor is to add an equimolar amount of DNA for each of the module sets and the destination vector. Since a module set usually contains several modules, the amount of DNA for each individual module of a set containing x different modules (x alternative homologous sequences) should contain only $1/x$ the amount of DNA compared to the vector; for example, each module from a set containing three modules should have a third of the amount of DNA compared with the recipient vector.

1. A restriction–ligation is set up by pipetting 40 fmol (or 100 ng, see Note 11) of each module set and of the vector, 2 μL of 10× ligation buffer, 10 U (1 μL) of *Bsa*I, and either 3 U (1 μL) of ligase for assembly of 2–4 module sets or 20 U (1 μL) of HC ligase for assembly of more than four module sets, in a total volume of 20 μL into a tube.
2. The restriction–ligation mix is incubated in a thermocycler. For assembly of 2–4 module sets, incubation for 60–120 min at 37°C is sufficient. If more module sets are ligated together, the incubation time is increased to 6 h, or cycling is used as following: 2 min for 37°C followed by 3 min for 16°C, both repeated 50 times (see Note 12).
3. Restriction–ligation is followed by a digestion step (5 min at 50°C) and then by heat inactivation for 5 min at 80°C. The final incubation step at 80°C is very important and should not be omitted. Its purpose is to inactivate the ligase at the end of the restriction–ligation. Omitting this step would lead

to religation of some of the insert and plasmid backbone fragments still present in the mix, when it is taken out of the thermocycler before transformation. Such unwanted products might be ligated more efficiently than they are redigested by the type IIs enzyme at room temperature. Therefore, a larger percentage of colonies would contain such type of undesired ligation products.

3.6. Transformation of the Library in Competent Cells

The entire ligation is transformed into chemically competent DH10B cells (see Note 13).

1. Frozen chemically competent cells (100 μL per tube) are thawed on ice.
2. The entire ligation is added to the cells, and the mix incubated on ice for 30 min.
3. The cells and DNA mix is heat shocked for 90 s at 42°C in a water bath.
4. The cells are allowed to recover on ice for 5 min.
5. To the cells, add 1 mL of LB medium, and incubate the tube at 37°C in a shaker-incubator (150 rpm) for 45 min to 1 h.
6. After incubation, 25–100 μL of the transformation are plated on LB agar plates containing antibiotic and X-gal. Plating of an aliquot of the transformation is necessary to estimate the number of independent constructs that will be obtained. The remainder of the transformation can be inoculated into 5 mL of liquid LB with the appropriate antibiotic if users want to grow the entire library.
7. The plates and liquid culture are incubated overnight at 37°C.
8. Many white and very few blue colonies should be obtained on the plate. A few white colonies from the plate can be picked for preparation of miniprep DNA. Plasmid DNA can be analyzed by restriction digestion and sequencing to estimate the number of correct clones.
9. Miniprep DNA is also prepared from the liquid culture. This DNA prep should represent a library of constructs containing shuffled DNA. Depending on the specific goal of the shuffling experiment, clones can either be functionally screened individually or as a library. The shuffled plasmid library may be transformed in any target organism of choice for functional screening.

3.7. Preparation of Chemically Competent DH10B Cells

Chemically competent or electrocompetent *E. coli* cells can either be purchased from a commercial vendor or made in the laboratory. The protocol that we use is as follows:

1. *E. coli* strain DH10B is inoculated from a glycerol stock onto an LB plate; the inoculum is streaked on the plate using a loop

so as to obtain individual colonies. The plate is incubated overnight at 37°C.

2. Inoculate 5 mL of LB from a single colony and incubate the flask overnight in a shaker-incubator (37°C, shaking 150 rpm).
3. The following day, transfer 2 mL of this culture to a flask containing 200 mL of LB and incubate for around 2 h until OD_{600} reaches 0.6.
4. Cool down the cells on ice for 10 min. The cells are pelleted in a centrifuge for 5 min at 4,500 rpm (4,000 × *g*) at 4°C. The cells are resuspended in 0.4 volume of ice-cold TFB1.
5. Repeat the centrifugation. Resuspend the pellet in 1/25 volume of ice-cold TFB2.
6. The cells are aliquoted 100 μL per tube and shock-frozen in liquid nitrogen. The aliquots are stored at −80°C.

4. Notes

1. Several different type IIs enzymes can be used for construct assembly. We have, for example, tested the enzymes *Bsa*I, *Bpi*I, and *Esp*3I. For all three, restriction–ligation can be performed efficiently in ligase buffer from Promega. All three have a 6-bp recognition sequence and have a four-nucleotide cleavage site located one (*Bsa*I and *Esp*3I) or two nucleotides (*Bpi*I) away from the recognition sequence. Enzymes of the type of *Sap*I such as *Lgu*I can also be used efficiently in a restriction–ligation (12). These enzymes have a 7-bp recognition sequence, meaning that it occurs more rarely than 6-bp cutters, and therefore, fewer sites will have to be removed from sequences of interest to clone (discussed in paragraph Subheading 3.2). However, these enzymes have a three-nucleotide cleavage site, meaning that only 64 different sequences are available to choose from for use as recombination sites.
2. Shuffling does not necessarily require making an alignment of several sequences. A user might want to shuffle sequences with no homology at all; for example, test a range of different promoters and terminators for optimal expression of a coding sequence. In such a case, two recombination sites would be chosen, one between the promoter and the coding sequence, and one between the coding sequence and the terminator. DNA shuffling would require the following three sets of entry modules: (1) the first set containing as many promoter modules as desired, (2) the second set would contain only

one module consisting of the coding sequence, and (3) the third set would contain again several terminator modules.

3. Such requirement can easily be fulfilled if only few recombination sites are necessary within a gene. However, it becomes more difficult to fulfill when a large number of sites are used in one cloning experiment. However, if the choice of such sites cannot be avoided, shuffling will still be possible with nevertheless reasonable efficiency.
4. The minimal size of modules that we have tested is 38 bp (including the recombination sites, but excluding the flanking *Bsa*I recognition sites). In theory, a module needs to be long enough for the two strands to remain annealed under restriction–ligation conditions, in practice 37°C. This means that even smaller modules could be made, which would be useful if a user wants to focus his efforts on a very small region of a protein of interest.
5. For generation of entry clones, the steps consisting of PCR amplification, cloning, and sequencing can be avoided if the entry modules are simply ordered from a gene synthesis company. This can be a useful option if the user does not have access to a DNA sequence of interest, for example, a sequence from sequence databases obtained from a metagenomics project. If a sequence is ordered from a gene synthesis company, the fragment to be synthesized should be ordered directly with the appropriate type IIs restriction sites flanking the sequence of interest. Also, it is useful to make sure upon ordering that the synthesized fragment does not contain any internal *Bsa*I sites, and that the cloning vector in which the ordered DNA fragment will be cloned does not also contain additional sites for the type IIs enzyme chosen. Finally, it is also useful to make sure that the antibiotic resistance gene from the vector in which the synthesized fragment is cloned is different from the one from the vector that will be used for assembly of the shuffled library.
6. Basically, two primers overlapping an internal *Bsa*I site are made, one in each orientation, with a mismatch designed to introduce a silent mutation in the type IIs restriction site. Two separate PCRs are performed with primers designed to amplify the two halfs of the module. The PCR products are purified on a column, and a mix of both is used as a template for a second PCR performed using both flanking primers only (the two primers flanking the given module). This PCR is purified on a column, cloned, and sequenced.
7. For small modules of up to 80 nucleotides, PCR amplification does not necessarily require a DNA template. For example, two complementary primers can be ordered covering the entire sequence of the module (including the flanking type

IIs restriction sites). Both primers are annealed in water and directly used for blunt-end cloning in the cloning vector. For larger but still small modules, two overlapping primers can be ordered that are complementary at their 3′ end on a length of 20–25 nucleotides. A double-stranded DNA fragment can be obtained by performing a PCR with both primers without a template. In theory, one single PCR cycle should be sufficient, but using 35 cycles as for normal PCR also works well.

8. One restriction for blunt-end cloning is that the ends of the primer should not recreate a *Sma*I site (or any other blunt-end restriction site used for cloning) after ligation of the PCR product (i.e., the DNA fragment to subclone should not start with the sequence GGG or finish with CCC). A second restriction is that the fragment to be cloned should not contain an internal restriction site for the enzyme used for cloning. If this is the case, another enzyme should be chosen for cloning, for example, *Eco*RV (and a cloning vector containing a unique *Eco*RV site in the polylinker should be used).

9. The presence of a *Bsa*I site in the vector backbone of the entry modules does not prevent from using them for performing Golden Gate shuffling, since plasmids containing the final shuffled sequences should not contain this vector backbone. However, the presence of such a site in all entry constructs would lead to continuous ligation and redigestion at this site, which would unnecessarily consume some ATP from the ligation mix, at the expense of the desired ligation events.

10. The widely used pUC19 vector also contains a *Bsa*I site in the ampicillin-resistance gene. A simple strategy, enzymatic inverse PCR (13), can be used to eliminate the internal *Bsa*I site in pUC19. The entire plasmid can be amplified with two primers overlapping with the *Bsa*I site: primers bsarem1 (ttt ggtctc a ggtt ctcgcggtatcattgcagc) and bsarem2 (ttt ggtctc a aacc acgctcaccggctccag). These primers are designed to introduce a single silent nucleotide mutation in the *Bsa*I recognition site in the vector. The primers are themselves flanked by two *Bsa*I restriction sites that form two compatible overhangs after *Bsa*I enzyme digestion. After amplification of the entire plasmid with both primers, the PCR is purified with a column (to remove remaining polymerase and nucleotides). The linear fragment is subjected to restriction–ligation using *Bsa*I and ligase, and transformed in *E. coli*.

11. In practice, if all module plasmids and the vector have approximately the same size (4–5 kb), simply adding 100 ng of DNA of each module set and of the vector will work relatively well. However, if plasmids with widely different sizes are used, calculating an equimolar amount should provide a higher cloning efficiency. The following formula (from the NEB catalog) can

be used: 1 μg of a 1,000-bp DNA fragment corresponds to 1.52 pmol.

12. We have found that both types of programs work well when high concentration ligase is used, but both programs can be tested in parallel by the users to optimize ligation efficiency.
13. Any other *E. coli* strain can also be used. If higher transformation efficiency is required, the restriction–ligation mix can be transformed in electrocompetent *E. coli* cells. In this case, DNA from the restriction–ligation mix should first be ethanol-precipitated and resuspended in 10 μL of water.

Acknowledgments

The authors would like to thank Dr. Stefan Werner for critical reading of this manuscript.

References

1. Roberts, R. J. (2005) How restriction enzymes became the workhorses of molecular biology. *Proc. Natl. Acad. Sci. USA* **102**, 5905–5908.
2. Katzen, F. (2007) Gateway® recombinational cloning: a biological operating system. *Expert Opin. Drug Discov.* **2**, 571–589.
3. Engler, C., Kandzia, R., and Marillonnet, S. (2008) A one pot, one step, precision cloning method with high throughput capability. *PLoS One* **3**, e347.
4. Engler, C., Gruetzner, R., Kandzia, R., and Marillonnet, S. (2009) Golden gate shuffling: a one-pot DNA shuffling method based on type IIs restriction enzymes. *PLoS One* **4**, e5553.
5. Lebedenko, E. N., Birikh, K. R., Plutalov, O. V., and Berlin, Y. A. (1991) Method of artificial DNA splicing by directed ligation (SDL). *Nucleic Acids Res.* **19**, 6757–6761.
6. Szybalski, W., Kim, S. C., Hasan, N., and Podhajska, A. J. (1991) Class-IIS restriction enzymes – a review. *Gene* **100**, 13–26.
7. Berlin, Y. A. (1999) DNA splicing by directed ligation (SDL). *Curr. Issues Mol. Biol.* **1**, 21–30.
8. Lu, Q. (2005) Seamless cloning and gene fusion. *Trends Biotechnol.* **23**, 199–207.
9. Horton, R. M., Ho, S. N., Pullen, J. K., Hunt, H. D., Cai, Z., and Pease, L. R. (1990) Gene splicing by overlap extension. *Biotechniques* **8**, 528–535.
10. Bolchi, A., Ottonello, S., and Petrucco, S. (2005) A general one-step method for the cloning of PCR products. *Biotechnol. Appl. Biochem.* **42**, 205–209.
11. Liu, Z. G., and Schwartz, L. M. (1992) An efficient method for blunt-end ligation of PCR products. *Biotechniques* **12**, 28–30.
12. Kotera, I., and Nagai, T. (2008) A high-throughput and single-tube recombination of crude PCR products using a DNA polymerase inhibitor and type IIS restriction enzyme. *J. Biotechnol.* **137**, 1–7.
13. Stemmer, W. P., and Morris, S. K. (1992) Enzymatic inverse PCR: a restriction site independent, single-fragment method for high-efficiency, site-directed mutagenesis. *Biotechniques* **13**, 214–220.

Chapter 12

Application of Full-Length cDNA Resources to Gain-of-Function Technology for Characterization of Plant Gene Function

Youichi Kondou, Mieko Higuchi, Takanari Ichikawa, and Minami Matsui

Abstract

Generation and characterization of mutants are important for the investigation of gene function. Gain-of-function technology is one of the most useful approaches for the systematic production of mutant resources. Full-length cDNAs have been collected from various plant species and have become important resources for functional genomics. We have developed a novel gain-of-function technology for the identification of gene function using a full-length cDNA library, and this system has been named as FOX hunting system (Full-length cDNA Over-eXpressing gene hunting system). In this system, full-length cDNAs are randomly expressed in *Arabidopsis*. We also generated rice FOX *Arabidopsis* lines in which full-length cDNAs from rice were expressed in *Arabidopsis*, and we demonstrated that gene function derived from heterologous organisms can be analyzed systematically using the FOX hunting approach. In this protocol, we describe the process of generating *Arabidopsis* mutants expressing rice full-length cDNA libraries and the methods of identifying genes from the isolated mutants.

Key words: Full-length cDNA resources, *Arabidopsis*, FOX hunting system, Gain-of-function, Heterologous gene expression, Transgenic plants

1. Introduction

Whole genome sequences of several plants have been determined since the elucidation of that of *Arabidopsis thaliana* (1). The locations of genes on chromosomes and characterizations of structures in transcriptional units like introns and exons have been established by gene prediction, sequence homology, sequence motif analysis, and other computational analysis. Although these programs for the annotation of genes using genome sequence have been developed, the accuracy of the predictions cannot be certified without experimental analysis. Collection of full-length

Chaofu Lu et al. (eds.), *cDNA Libraries: Methods and Applications*, Methods in Molecular Biology, vol. 729,
DOI 10.1007/978-1-61779-065-2_12, © Springer Science+Business Media, LLC 2011

cDNAs (fl-cDNAs) is one way to gather such experimental evidence, because they contain all the information for functional RNAs and proteins. Indeed, sequencing of approximately 240,000 *Arabidopsis* fl-cDNAs allowed the correct annotation of thousands of gene structures (2). In addition to *Arabidopsis*, large sets of fl-cDNAs have been collected from many other plants; rice (3), *Physcomitrella patens* (4), wheat (5), poplar (6), cassava (7), soybean (8), Sitka spruce (9), *Thellungiella halophia* (10), and barley (11). These fl-cDNA resources can also be used for various analyzes such as cDNA microarray, two-hybrid screening, and X-ray crystallography (12).

There are two main genetic approaches towards dissecting plant gene function; loss-of-function and gain-of-function technology. The primary way is loss-of-function technology, in which genetic mutants generated by loss-of-function mutations are analyzed. The mutations can be caused by point mutations or small deletions derived from chemical or fast neutron mutagenesis, respectively (13). Insertional mutagenesis using transferred DNA (T-DNA) tagging or transposon tagging methods have also been developed as useful ways of generating loss-of-function mutants, because genes inserted with these tags can be determined using sequence information of the T-DNA or transposon (14, 15). The alternative approach is gain-of-function technology, in which genetic mutants generated by random activation of endogenous genes or by transformation of extrinsic genes are analyzed. Gain-of-function technology allows analysis of functionally redundant members of gene families, because many loss-of-function mutants, in which one redundant gene is disrupted, do not show clear phenotypes (16, 17). It has been reported that most predicted proteins in *Arabidopsis*, rice, and other plant species are members of gene families (1, 18). In addition, the problem of phenotypic lethality by loss-of-function can be overcome (19, 20).

Recently, the application of fl-cDNA resources to gain-of-function technology has been developed for the characterization of plant gene function (21). We named this method as the FOX hunting system (Full-length cDNA Over-eXpressing gene hunting system). An outline of this system is shown in Fig. 1. Fl-cDNAs are cloned downstream of the cauliflower mosaic virus (*CaMV*) *35S* promoter in a T-DNA vector using a mixture of each fl-cDNA at approximately the same molar ratio. This fl-cDNA expression library is integrated into the plant genome via transformation by *Agrobacterium*. Mutants showing interesting phenotypes are isolated and the introduced fl-cDNAs can be easily determined using vector-specific primers. Therefore, linkage between fl-cDNAs and gene function can be directly characterized in this system. We generated 15,000 FOX *Arabidopsis* lines using the RIKEN *Arabidopsis* fl-cDNA collection including about 10,000 nonredundant fl-cDNAs and characterized the functions of some genes

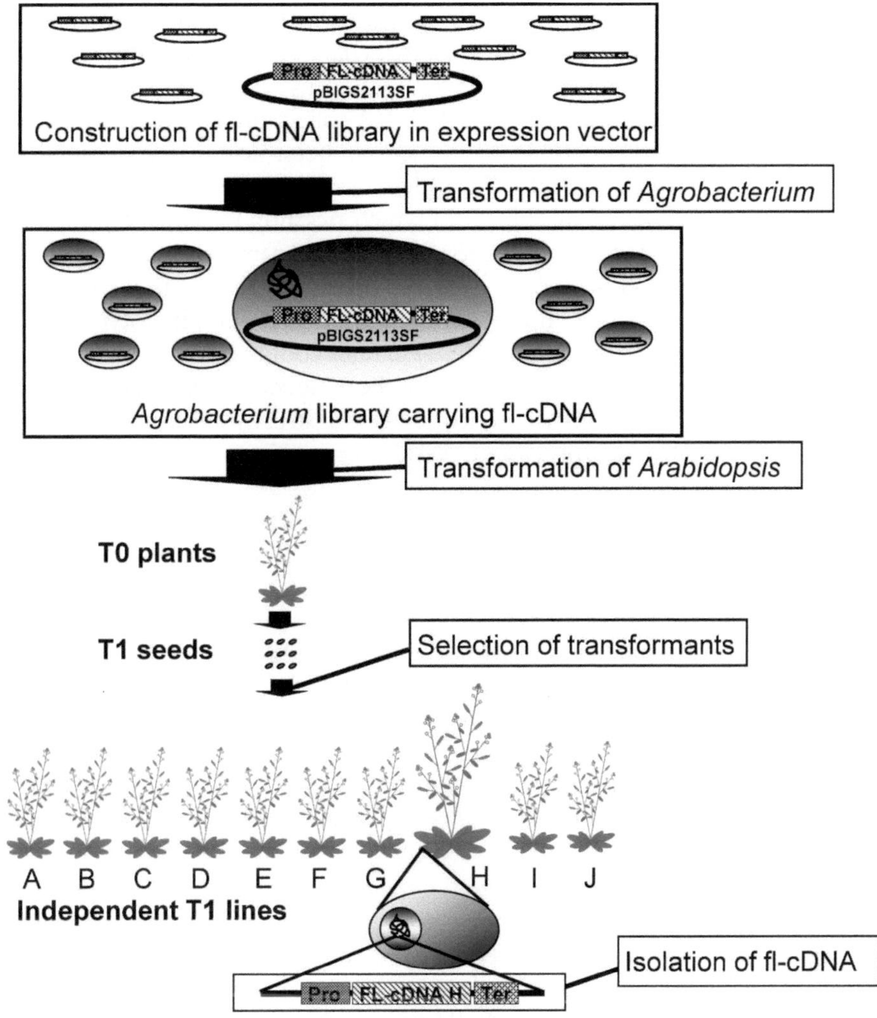

Fig. 1. Outline of the FOX hunting system. The fl-cDNA library is cloned into pBIGS2113SF and then the expression vector carrying the fl-cDNAs is introduced into *Agrobacterium*. *Arabidopsis* plants are transformed with the *Agrobacterium* library. The T_0 plants generated are self-pollinated, and then many independent T_1 FOX seeds are obtained. From the T_1 FOX lines, a mutant line (in this case, the "H" line) is identified during the screening process. The fl-cDNA introduced in FOX line "H" can be identified easily by PCR using T-DNA-specific primers and sequencing.

as a first trial (21). Characterizations of several important genes have been reported using these FOX *Arabidopsis* lines (22, 23).

The FOX hunting system can be applied to almost all plant species, because information about the genome is not necessary, only fl-cDNAs are required. This feature highlights the great advantage of the FOX hunting system in analyzing gene function in different plants, because determination of the whole genome sequence in many species is still difficult. *Arabidopsis* is an attractive organism to act as the host plant for the FOX hunting system, because *in planta Agrobacterium*-mediated transformation using vacuum infiltration has been developed and it has brought high-throughput production of *Arabidopsis* transformants (24).

Moreover, *Arabidopsis* has a short generation time and its size is compact, and these features have enabled high-throughput analysis of heterologous gene function using over-expression. As a model case, we used a rice fl-cDNA collection composed of about 13,000 rice fl-cDNAs to investigate heterologous gene function. More than 33,000 independent *Arabidopsis* transgenic lines (rice FOX *Arabidopsis* lines), in which rice fl-cDNAs were expressed under the control of the *CaMV 35S* promoter, were produced (25). After screening of these lines, the fl-cDNAs were reintroduced into rice and it was confirmed that the phenotypes observed in the rice FOX *Arabidopsis* lines could be recapitulated in rice (25). Two heat tolerant rice FOX *Arabidopsis* lines, one expressing *OsHsfA2e* that encodes a heat stress transcription factor and the other expressing *ONAC063* encoding a NAC transcription factor, were isolated from the lines (26, 27). These results demonstrate that it is possible to characterize gene function in many plant species using collections of their fl-cDNAs in the heterologous expression system in *Arabidopsis.* Several types of resources and databases for the FOX hunting system have been generated and are available as previously reported (28).

This chapter provides the procedure for the production of rice FOX *Arabidopsis* lines using rice fl-cDNAs and the determination of those that cause interesting phenotypes. It provides an example of the application of fl-cDNA resources to gain-of-function technology for analyzing gene function in various plant species.

1.1. Screening Process for FOX Lines

The screening strategy is shown in Fig. 2. T_1 plants selected by antibiotic resistance are self-pollinated and the T_2 seeds can then be used for screening. T_1 seeds can also be screened in parallel with antibiotic selection. Isolated candidates are self-pollinated to generate T_2 seeds. The T_2 seeds undergo a secondary screening to confirm the phenotype observed in the T_1 plants. After confirmation of the phenotype in the T_2 generation, the introduced fl-cDNA is isolated from the candidate FOX line. The isolated fl-cDNA is reintroduced into *Arabidopsis* to confirm that its expression is responsible for the observed phenotype.

2. Materials

2.1. Construction of Expression Vector Harboring Rice Fl-cDNA Library

1. pBIGS2113SF is used as the expression vector for ectopic expression in *A. thaliana*.
2. 3 M sodium acetate (pH 4.8): dissolve 40.8 g $NaOAc \cdot 3H_2O$ in 100 mL of water. Adjust the pH to 4.8 with glacial acetic acid before autoclaving.

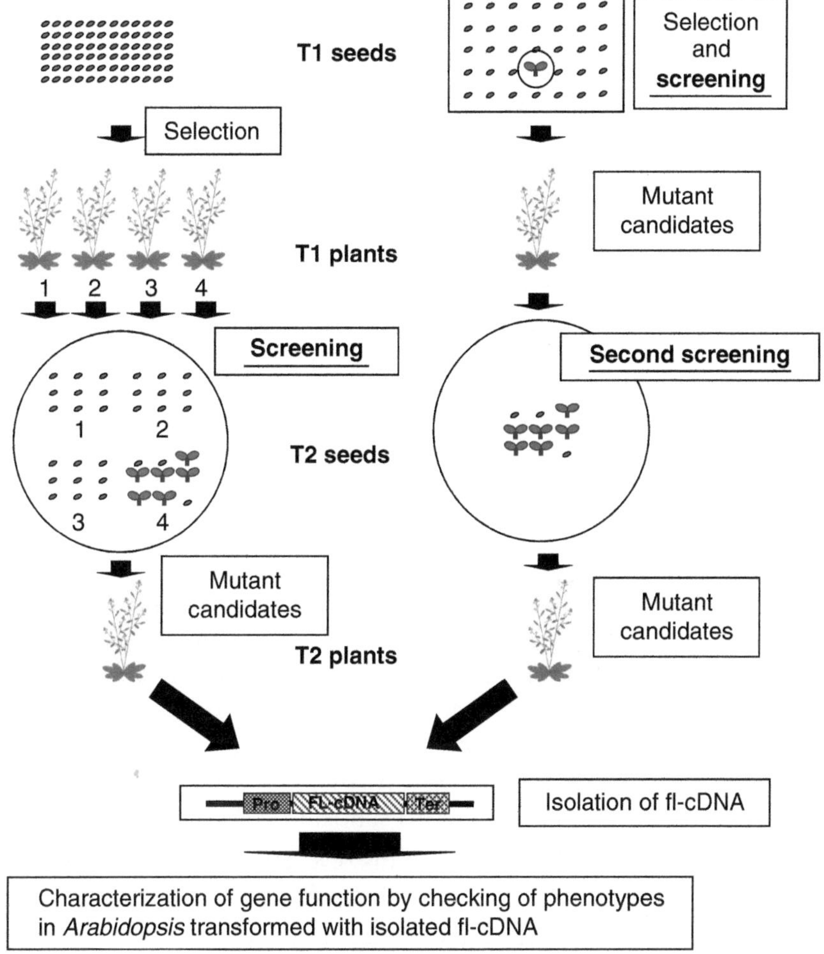

Fig. 2. Screening process of the FOX hunting system. Two types of screening can be carried out using T_2 plants (*left side*) or T_1 plants (*right side*). Screening can be done using T_2 plants. When T_1 plants are subjected to screening, antibiotic selection should be applied simultaneously. Self-pollinated T_2 plants can be used to confirm the observed phenotype. The fl-cDNA that caused the observed phenotype is determined through observation of plants retransformed with the isolated fl-cDNA.

3. Ligation solution: mix 1 μL of 10× ligation buffer in 8 μL of distilled water.

2.2. Plant Material and Growth Conditions

1. *Arabidopsis thaliana* plants (Columbia-0) are grown in a controlled growth chamber at 22°C under long-day conditions (light: 16 h, dark: 8 h) (see Note 1).
2. Mix 1.5 kg of PRO-MIX (Premier Tech Ltd.) and 0.9 kg of vermiculite and add liquid fertilizer appropriate for the cultivation of *Arabidopsis* plants. The soil is used after autoclaving at 120°C for 30 min (see Note 2).

2.3. Preparation of Agrobacterium and Escherichia coli Cultures

1. LB medium: dissolve 10 g of tryptone peptone, 5 g of yeast extract, and 5 g of NaCl in 1,000 mL of water and autoclave at 120°C for 20 min. For agar medium, add 16 g of agar powder to 1,000 mL of LB medium.
2. SOC medium: dissolve 20 g of tryptone peptone, 5 g of yeast extract, 0.19 g of KCl, 2.03 g of $MgCl_2 \cdot 6H_2O$, 2.46 g of $MgSO_4 \cdot 7H_2O$, 0.58 g of NaCl, and 3.6 g of glucose in 1,000 mL water and autoclave.
3. Kanamycin and gentamycin stock solution: dissolve 50 and 10 mg/mL of kanamycin and gentamycin in water, respectively. Sterilize by filtering through a 0.22 μm membrane and store at −20°C.
4. *Escherichia coli* strain: DH10B.
5. *Agrobacterium* strain: GV3101 pMP90.

2.4. Transformation of Arabidopsis Plants

1. Infiltration medium: dissolve the appropriate amount of Murashige and Skoog (MS) inorganic salts to make a 0.5× solution and 50 g of sucrose in 1,000 mL of water. Add 112 μL of Gamborg's 1,000× vitamin solution, 10 μL of benzylaminopurine stock solution, and 200 μL of Silwet L-77 (Agri-Turf Supplies, Inc.) (see Note 3).
2. Benzylaminopurine stock solution: dissolve 1 mg of benzylaminopurine (Wako Pure Chemical Industries, Ltd.) in 1 mL of dimethyl sulfoxide (DMSO).
3. BAM plate (29): dissolve 101 mg of KNO_3 in 1,000 mL of water, add 8 g of Bacto Agar, and autoclave.
4. 0.2% agar: add 2 g of Bacto Agar to 1,000 mL of water and autoclave.
5. Bleach solution: mix 10 mL of sodium hypochlorite solution and 100 μL of Triton X-100 in 90.9 mL of water.
6. Hygromycin B and cefotaxime stock solution: dissolve 20 and 100 mg/mL of hygromycin B and cefotaxime in water, respectively. Sterilize by filtering through a 0.22 μm membrane and store at −20°C.

2.5. Recloning of Rice Fl-cDNA into Expression Vector

1. Enzyme for the polymerase chain reaction (PCR): PrimeSTAR HS DNA polymerase is purchased from Takara Bio Inc. 2× PrimeSTAR GC buffer and a solution of 2.5 mM of each dNTP are supplied with the enzyme.
2. Reaction solution for PCR and colony PCR: mix 50 μL/100 μL of 2× PrimeSTAR GC buffer I, 8 μL/100 μL of the dNTP mixture, 20 pmol/100 μL of each primer, an appropriate volume of DNA template, and 2.5 U/100 μL of PrimeSTAR HS DNA polymerase in water.

3. Methods

3.1. Production of Agrobacterium Library with Rice Fl-cDNA Expression Library

3.1.1. Cloning of Rice Fl-cDNA Library into Expression Vector

1. Make the reaction solution by mixing 20 μL of the expression vector (20 ng/μL), 40 μL of the normalized rice fl-cDNA library (30 ng/μL), 10 μL of 10× *Sfi*I buffer, 1 μL of the BSA solution, 24 μL of distilled water, and 5 μL of *Sfi*I, and incubate overnight at 37°C (see Note 4).
2. Mix 5 μL of *Sfi*I into the reaction solution and incubate for at least 3 h at 50°C. Spin down the tubes every hour to reduce condensation.
3. Add 1/10 volume of 3 M sodium acetate (pH 4.8) and 1 volume of isopropanol.
4. Pellet the DNA by centrifugation in a microcentrifuge.
5. Remove the supernatant and wash the DNA pellets twice with 70% ethanol and dry.
6. Dissolve the DNA pellets in 9 μL of ligation solution and mix in 1 μL of T4 DNA ligase (400 U/microl).
7. Incubate the ligation reaction mixture overnight at 16°C.
8. Thaw 40 μL of electrocompetent *E. coli* cells, strain DH10B, on ice and place a cuvette with a 1 mm electrode gap on ice (see Note 5).
9. Add 2 μL of ligated DNA to the competent cells and mix by pipetting.
10. Transfer the cells to the ice-cold cuvette.
11. Electroporate the cells at 4 ms, 1.5 kV, 200 Ω, and 25 μF.
12. Immediately add 500 μL of SOC medium prewarmed at 37°C to the cuvette and mix the cells by pipetting.
13. Move the cells into a 1.5 mL tube and incubate for 1 h at 37°C.
14. Add SOC medium to the cells to obtain approximately 5,000–10,000 colonies per 10 cm Petri dish. Spread the diluted cells on LB agar medium containing 50 μg/mL kanamycin. Approximately 150,000 colonies are needed to produce the *Agrobacterium* library, so 15–30 plates are required.
15. Incubate the Petri dishes overnight at 37°C.
16. Add 1 mL of LB medium to each Petri dish and scrape the colonies using a spreader.
17. Collect the LB medium from all the Petri dishes and isolate plasmid DNA from the cells.

3.1.2. Transformation of Expression Vector into Agrobacterium Cells

1. Thaw electrocompetent *Agrobacterium* cells, strain GV3101 pMP90, on ice and also place a cuvette with a 2 mm electrode gap on ice.
2. Add 2 μL of the plasmid DNA mixture to 40 μL of *Agrobacterium* competent cells and mix by vortexing.
3. Transfer the cells to the ice-cold cuvette.
4. Electroporate the cells at 4 ms, 2.5 kV, 200 Ω, and 25 μF.
5. Immediately add 500 μL of SOC medium prewarmed at 28°C to the cuvette and mix by pipetting.
6. Move the cells into a 1.5 mL tube and incubate for 1–3 h at 28°C.
7. Dilute the cells with SOC medium to obtain approximately 5,000–10,000 colonies per 10 cm Petri dish. About 150,000 colonies are required to produce the *Agrobacterium* library. Spread the diluted cells on LB medium containing 50 μg/mL of kanamycin and 10 μg/mL of gentamycin.
8. Incubate the Petri dishes for 2 days at 28°C.
9. Put 1 mL of LB medium on to each Petri dish and scrape the colonies using a spreader. Collect the LB medium containing the cells from all the Petri dishes. This should be equivalent to 150,000 colonies. These cell cultures are used as the *Agrobacterium* library for the production of the rice FOX *Arabidopsis* lines.

3.2. Transformation of Arabidopsis Plants with Expression Vector Library Harboring Rice Fl-cDNAs via Agrobacterium

1. Put 2 μL of *Agrobacterium* cells carrying the rice fl-cDNA expression library into 200 mL of LB medium containing 50 μg/mL of kanamycin and 10 μg/mL of gentamycin, and grow until the OD_{600} reaches 1.2–1.5 at 28°C (see Note 6).
2. Decant the *Agrobacterium* culture into a tube and centrifuge at 6,000 × *g* for 13 min.
3. Remove the supernatant and resuspend the pellet to an OD_{600} of 0.8 with infiltration medium.
4. Invert the *Arabidopsis* plant pots over the infiltration medium containing the *Agrobacterium* cells and dip the inflorescences into the medium for 30 s.
5. Put the plant pots into a plastic bag and seal.
6. Return to the growth chamber overnight. Open the plastic bag and leave overnight again.
7. Take the pots out of the plastic bag the following day.
8. Grow the *Arabidopsis* plants until they are ready to harvest and collect the T_1 seeds.

3.3. Selection of Transgenic Arabidopsis Plants

1. Put 0.25 g of T_1 seeds in a tube and add 70% ethanol. Keep the tube for 1 min.
2. Discard the 70% ethanol and then add the bleach solution. Keep the tube for 10 min.
3. Remove the bleach completely and wash the seeds with sterile water three times.
4. Decant off the water and resuspend the seeds in 0.2% water agar.
5. Spread the seeds in the agar on solid BAM medium containing 20 μg/mL of hygromycin B and 100 μg/mL of cefotaxime sodium salt (on 10×14 cm square Petri dishes) (see Note 7).
6. Place the plates at 4°C for least 2 days under dark conditions to induce germination.
7. Incubate the seedlings under long-day conditions for 5–10 days. Alternatively, if the T_1 plants are to be used for screening they can be grown under various conditions (e.g., high salt, high temperature, high light, etc.) (see Notes 8–10).
8. Select the *Arabidopsis* seedlings that have true leaves and healthy roots, and transfer them to soil. Grow these T_1 plants, the rice FOX *Arabidopsis* lines, to harvest and collect the T_2 seeds.

3.4. Screening and Determination of Rice Fl-cDNAs from Rice FOX Arabidopsis Lines

1. Screen the rice FOX *Arabidopsis* lines in the T_1 or T_2 generation under appropriate conditions (see Note 11).
2. Extract chromosomal DNA from candidate *Arabidopsis* plants (see Note 12).
3. Add less than 200 ng of DNA template to a 50 μL reaction volume for PCR. Use the primer set F1 and R1 given in Table 1 (see Notes 13 and 14).
4. For amplification of rice fl-cDNAs, perform an initial denaturation at 94°C and then continue with 30 cycles of 94°C, 30 s; 52°C, 30 s; 72°C, 3 min, finishing with a 10 min 72°C extension cycle.
5. Separate the PCR products in a 0.8% agarose gel and extract the amplified PCR fragments from each band (see Notes 15–17).

3.5. Retransformation of Arabidopsis Plants with Expression Vectors Harboring Isolated Rice Fl-cDNAs via Agrobacterium

3.5.1. Recloning of Rice Fl-cDNA into Expression Vector

1. Make the reaction solution by mixing 0.9 μL of expression vector (5–20 ng/μL), 7 μL of the amplified PCR products, 1 μL of 10× H buffer, 0.1 μL of the BSA solution, and 0.5 μL of *Sfi*I, and incubate overnight at 37°C (see Note 18).
2. Mix 0.5 μL of *Sfi*I into the reaction solution and incubate at 50°C for a minimum of 3 h with a spin down every hour.
3. Add 1/10 volume of 3 M sodium acetate (pH 4.8) and 1 volume of isopropanol to the reaction solution to precipitate the DNA.
4. Wash the DNA pellets twice with 70% ethanol and dry.

Table 1
Primers used for amplification of fl-cDNAs from genomic DNA of FOX lines

Forward primer		Reverse primer	
Name	Sequence	Name	Sequence
F1	GGAAGTTCATTTATTCGGAGAG	R1	GGCAACAGGATTCAATCTTAAG
F2	CATTTATTCGGAGAGGTACGTAT	R2	GGATTCAATCTTAAGAAACTTTATTGCCAA
F3	GTACGTATTTTTACAACAATTACCAACAAC	R3	CAAATGTTTGAACGATCGGGGAAAT
F4	ATTACATTTTACATTCTACAACTACATCT	R4	GATCCTCTAGAGGCCCTTAT
F5	CCCCCCCCCCCCD (A or G or T)	R5	AAAAAAAAAAAAB (C or G or T)

The primers (F1–F4, R1–R4) are designed outside of the *Sfi*I sites in pBIGS2113SF. These primers can be used for amplification of *Arabidopsis* and rice fl-cDNAs. The F5 and R5 are adaptor sequences used in the *Arabidopsis* fl-cDNA cloning (31) and can also be used for sequencing the *Arabidopsis* fl-cDNAs. We usually use the F1 and R1 primer sets to amplify the fl-cDNAs.

5. Dissolve the DNA in 4.5 μL of the ligation solution and add 0.5 μL of T4 DNA ligase.
6. Incubate the ligation reaction mixture at 16°C overnight.
7. Add 1/10 volume of 3 M sodium acetate (pH 4.8) and 2 volumes of ethanol for precipitation of the ligated DNA, wash the DNA pellets with 70% ethanol, dry and resuspend in 5 μL of distilled water (see Note 19).

3.5.2. Transformation of Expression Vector into E. coli and Agrobacterium Cells

1. Transform *E. coli* cells with 5 μL of the ligated DNA.
2. Directly amplify the DNA fragment integrated into the expression vector from each antibiotic-resistant colony using colony PCR analysis and electrophorese to check the size of the DNA fragment.
3. If the size of the fragment is correct, repick the single colony into liquid LB medium containing 50 μg/mL of kanamycin. Shake the cell culture overnight at 37°C.
4. Extract plasmid DNA from the culture.
5. Check the whole sequence of the integrated rice fl-cDNA using the primer sets given in Table 1 (see Note 20).
6. Use 1 μL of plasmid DNA for transformation of *Agrobacterium* cells.
7. Pick antibiotic-resistant colonies into 2 mL of liquid LB medium containing 50 μg/mL of kanamycin and 10 μg/mL of gentamycin, and shake the culture overnight at 28°C for subculturing.

3.5.3. Observation of Phenotypes of Retransformants

1. Retransform *Arabidopsis* plants using the *Agrobacterium* cultures above.
2. Grow T_1 plants of the retransformants and collect the T_2 seeds.
3. Check the phenotypes in the T_2 generation (see Note 21).

4. Notes

1. Healthy *Arabidopsis* plants should be used for floral dipping. They should have numerous unopened floral buds and few siliques at the time of transformation.
2. The soil has to be completely wet with water before autoclaving for effective sterilization.
3. The benzylaminopurine and Gamborg's vitamin solution can be left out of the infiltration medium, because they have little effect on the transformation efficiency.

4. It is important in the FOX hunting system to use a mixture of nonredundant fl-cDNAs at approximately the same molar ratio as a normalized fl-cDNA library. The fl-cDNA species may be biased in accordance with the abundance of fl-cDNAs in the library, if nonnormalized fl-cDNA libraries are used.
5. Electrocompetent *E. coli* cells with a competency of at least 1×10^8 transformants/μg of pUC19 plasmid should be used to make the expression library. Improved results can be acquired by using cells with high transformation competency.
6. The concentration of the cells in the culture is not vital. *Arabidopsis* plants can be transformed irrespective of the density of the *Agrobacterium* cells.
7. When hygromycin B-resistant plants were not obtained, transformants could be picked up using 1/10× Murashige and Skoog (MS) medium plates containing hygromycin B and cefotaxime.
8. In some cases, transformants might show phenotypes related to growth defects. Therefore, it may take longer time for the transformants to have grown sufficiently on the selection medium to discriminate between these and hygromycin B-sensitive individuals.
9. 250 mg of T_1 seeds (approximately 17,000 seeds) can produce about 30–40 transformants.
10. T_1 seeds can be used for screening under various conditions in parallel with antibiotic selection. For example, plants resistant to salt stress were isolated on half-strength MS medium containing 200 mM NaCl (30).
11. The phenotypes of candidates are dominant, since the FOX hunting system is a gain-of-function technology.
12. Many protocols for the extraction of plant genomic DNA can be used. The methods for isolation of pure and high molecular weight DNA are recommended for amplifying the introduced fl-cDNAs by PCR.
13. PrimeSTAR HS DNA polymerase is usually used for fl-cDNA amplification. It may be necessary to determine which PCR enzyme is best for different plant species. For example, we used PrimeSTAR HS DNA polymerase with GC buffer for amplification of rice fl-cDNAs because of the high GC content in rice. LA-Taq (Takara BioInc.) or KOD-FX (Toyobo Co., Ltd) is recommended when it is difficult to amplify the introduced fl-cDNAs.
14. The F1 and R1 primer set is usually effective for amplification of the fl-cDNAs. Other primers shown in Table 1 can also be used if a PCR fragment cannot be obtained using the F1 and R1 primers. In some cases, the fl-cDNAs could be amplified by nested PCR using the first PCR product and an internal primer sets.

15. Repeat the amplification of fl-cDNAs using different conditions if the size of the PCR fragment is less than 200 bp, since a fragment of this size is obtained when empty vector is used as the template DNA.
16. Sometimes we obtained several PCR fragments because there were multiple T-DNA insertions in the *Arabidopsis* genome. We found on average 2.6 *Arabidopsis* fl-cDNA insertions in the FOX *Arabidopsis* lines (21). In such cases, all fragments should be recovered and then each one used to generate retransformed plants to confirm which fl-cDNA caused the phenotype.
17. For determination of the introduced rice fl-cDNA, direct sequencing of PCR fragments can be carried out using different internal primers than those used for PCR amplification. It is better to perform sequencing from both ends. We found that tandem fl-cDNAs were inserted into the expression vector in some FOX *Arabidopsis* lines.
18. If the PCR fragment cannot be integrated into the expression vector, it is recommended that higher concentrations of the PCR fragment and expression vector be added to the reaction mixture for *Sfi*I digestion.
19. This step can be omitted, because ethanol precipitation of the ligated DNA is used to eliminate salts in the ligation reaction mixture to allow for high transformation efficiency by electroporation. Mix 1 μL of the ligation reaction mixture with 30 μL of *E. coli* competent cells if the ligation reaction mixture is directly added to the competent cells.
20. The rice fl-cDNA sequences inserted into the expression vector have to be checked for PCR errors at this step.
21. Correlation between the expression level of the rice fl-cDNA and the phenotype should be checked if the phenotype of the rice FOX *Arabidopsis* line is recaptured in the retransformed plants.

Acknowledgments

This work is supported by a Special Coordination Fund for Promoting Science and Technology awarded to Minami Matsui, Kenji Oda, and Hirohiko Hirochika. This study is also supported by a Grant-in-Aid for Young Scientists (B) from the Ministry of Education, Culture, Sports and Technology of Japan (19710055) to Youichi Kondou and (21780315) to Mieko Higuchi. We thank Dr. Hirofumi Kuroda, Ms. Yoko Horii, and Dr. Yuko Tsumoto (RIKEN Plant Science Center) for technical support. We appreciate the helpful discussions with Dr. Takeshi Yoshizumi (RIKEN Plant Science Center).

References

1. AGI (2000) Analysis of the genome sequence of the flowering plant *Arabidopsis thaliana*. *Nature* **408**, 796–815.
2. Seki, M., Narusaka, M., Kamiya, A., Ishida, J., Satou, M., Sakurai, T., Nakajima, M., Enju, A., Akiyama, K., Oono, Y., *et al.* (2002) Functional annotation of a full-length *Arabidopsis* cDNA collection. *Science* **296**, 141–145.
3. Kikuchi, S., Satoh, K., Nagata, T., Kawagashira, N., Doi, K., Kishimoto, N., Yazaki, J., Ishikawa, M., Yamada, H., Ooka, H., *et al.* (2003) Collection, mapping, and annotation of over 28,000 cDNA clones from *japonica* rice. *Science* **301**, 376–379.
4. Nishiyama, T., Fujita, T., Shin, I. T., Seki, M., Nishide, H., Uchiyama, I., Kamiya, A., Carninci, P., Hayashizaki, Y., Shinozaki, K., *et al.* (2003) Comparative genomics of *Physcomitrella patens* gametophytic transcriptome and *Arabidopsis thaliana*: implication for land plant evolution. *Proc. Natl Acad. Sci. U S A* **100**, 8007–8012.
5. Ogihara, Y., Mochida, K., Kawaura, K., Murai, K., Seki, M., Kamiya, A., Shinozaki, K., Carninci, P., Hayashizaki, Y., Shin, I. T., *et al.* (2004) Construction of a full-length cDNA library from young spikelets of hexaploid wheat and its characterization by large-scale sequencing of expressed sequence tags. *Genes Genet. Syst.* **79**, 227–232.
6. Nanjo, T., Sakurai, T., Totoki, Y., Toyoda, A., Nishiguchi, M., Kado, T., Igasaki, T., Futamura, N., Seki, M., Sakaki, Y., *et al.* (2007) Functional annotation of 19,841 *Populus nigra* full-length enriched cDNA clones. *BMC Genomics* **8**, 448.
7. Sakurai, T., Plata, G., Rodriguez-Zapata, F., Seki, M., Salcedo, A., Toyoda, A., Ishiwata, A., Tohme, J., Sakaki, Y., Shinozaki, K., *et al.* (2007) Sequencing analysis of 20,000 full-length cDNA clones from cassava reveals lineage specific expansions in gene families related to stress response. *BMC Plant Biol.* **7**, 66.
8. Umezawa, T., Sakurai, T., Totoki, Y., Toyoda, A., Seki, M., Ishiwata, A., Akiyama, K., Kurotani, A., Yoshida, T., Mochida, K., *et al.* (2008) Sequencing and analysis of approximately 40,000 soybean cDNA clones from a full-length-enriched cDNA library. *DNA Res.* **15**, 333–346.
9. Ralph, S. G., Chun, H. J., Kolosova, N., Cooper, D., Oddy, C., Ritland, C. E., Kirkpatrick, R., Moore, R., Barber, S., Holt, R. A., *et al.* (2008) A conifer genomics resource of 200,000 spruce (*Picea spp.*) ESTs and 6,464 high-quality, sequence-finished full-length cDNAs for Sitka spruce (*Picea sitchensis*). *BMC Genomics* **9**, 484.
10. Taji, T., Sakurai, T., Mochida, K., Ishiwata, A., Kurotani, A., Totoki, Y., Toyoda, A., Sakaki, Y., Seki, M., Ono, H., *et al.* (2008) Large-scale collection and annotation of full-length enriched cDNAs from a model halophyte, *Thellungiella halophila*. *BMC Plant Biol.* **8**, 115.
11. Sato, K., Shin, I. T., Seki, M., Shinozaki, K., Yoshida, H., Takeda, K., Yamazaki, Y., Conte, M., and Kohara, Y. (2009) Development of 5006 full-length CDNAs in barley: a tool for accessing cereal genomics resources. *DNA Res.* **16**, 81–89.
12. Seki, M., and Shinozaki, K. (2009) Functional genomics using RIKEN *Arabidopsis thaliana* full-length cDNAs. *J. Plant Res.* **122**, 355–366.
13. Ostergaard, L., and Yanofsky, M. F. (2004) Establishing gene function by mutagenesis in *Arabidopsis thaliana*. *Plant J.* **39**, 682–696.
14. Krysan, P. J., Young, J. C., and Sussman, M. R. (1999) T-DNA as an insertional mutagen in *Arabidopsis*. *Plant Cell* **11**, 2283–2290.
15. Ramachandran, S., and Sundaresan, S. (2001) Transposons as tools for functional genomics. *Plant Physiol. Biochem.* **39**, 243–252.
16. Ito, T., and Meyerowitz, E. M. (2000) Overexpression of a gene encoding a cytochrome P450, *CYP78A9*, induces large and seedless fruit in *Arabidopsis*. *Plant Cell* **12**, 1541–1550.
17. Nakazawa, M., Yabe, N., Ichikawa, T., Yamamoto, Y. Y., Yoshizumi, T., Hasunuma, K., and Matsui, M. (2001) *DFL1*, an auxin-responsive *GH3* gene homologue, negatively regulates shoot cell elongation and lateral root formation, and positively regulates the light response of hypocotyl length. *Plant J.* **25**, 213–221.
18. Goff, S. A., Ricke, D., Lan, T. H., Presting, G., Wang, R., Dunn, M., Glazebrook, J., Sessions, A., Oeller, P., Varma, H., *et al.* (2002) A draft sequence of the rice genome (*Oryza sativa L. ssp. japonica*). *Science* **296**, 92–100.
19. Tani, H., Chen, X., Nurmberg, P., Grant, J. J., SantaMaria, M., Chini, A., Gilroy, E., Birch, P. R., and Loake, G. J. (2004) Activation tagging in plants: a tool for gene discovery. *Funct. Integr. Genomics* **4**, 258–266.
20. Weigel, D., Ahn, J. H., Blazquez, M. A., Borevitz, J. O., Christensen, S. K., Fankhauser, C.,

Ferrandiz, C., Kardailsky, I., Malancharuvil, E. J., Neff, M. M., *et al.* (2000) Activation tagging in *Arabidopsis*. *Plant Physiol.* **122**, 1003–1013.

21. Ichikawa, T., Nakazawa, M., Kawashima, M., Iizumi, H., Kuroda, H., Kondou, Y., Tsuhara, Y., Suzuki, K., Ishikawa, A., Seki, M., *et al.* (2006) The FOX hunting system: an alternative gain-of-function gene hunting technique. *Plant J.* **48**, 974–985.
22. Breuer, C., Kawamura, A., Ichikawa, T., Tominaga-Wada, R., Wada, T., Kondou, Y., Muto, S., Matsui, M., and Sugimoto, K. (2009) The trihelix transcription factor GTL1 regulates ploidy-dependent cell growth in the *Arabidopsis* trichome. *Plant Cell* **21**, 2307–2322.
23. Okazaki, K., Kabeya, Y., Suzuki, K., Mori, T., Ichikawa, T., Matsui, M., Nakanishi, H., and Miyagishima, S. Y. (2009) The PLASTID DIVISION1 and 2 components of the chloroplast division machinery determine the rate of chloroplast division in land plant cell differentiation. *Plant Cell* **21**, 1769–1780.
24. Bechtold, N., and Pelletier, G. (1998) In planta *Agrobacterium*-mediated transformation of adult *Arabidopsis thaliana* plants by vacuum infiltration. *Methods Mol. Biol.* **82**, 259–266.
25. Kondou, Y., Higuchi, M., Takahashi, S., Sakurai, T., Ichikawa, T., Kuroda, H., Yoshizumi, T., Tsumoto, Y., Horii, Y., Kawashima, M., *et al.* (2009) Systematic approaches to using the FOX hunting system to identify useful rice genes. *Plant J.* **57**, 883–894.
26. Yokotani, N., Ichikawa, T., Kondou, Y., Matsui, M., Hirochika, H., Iwabuchi, M., and Oda, K. (2008) Expression of rice heat stress transcription factor OsHsfA2e enhances tolerance to environmental stresses in transgenic *Arabidopsis*. *Planta* **227**, 957–967.
27. Yokotani, N., Ichikawa, T., Kondou, Y., Matsui, M., Hirochika, H., Iwabuchi, M., and Oda, K. (2009) Tolerance to various environmental stresses conferred by the salt-responsive rice gene *ONAC063* in transgenic *Arabidopsis*. *Planta* **229**, 1065–1075.
28. Kuromori, T., Takahashi, S., Kondou, Y., Shinozaki, K., and Matsui, M. (2009) Phenome analysis in plant species using loss-of-function and gain-of-function mutants. *Plant Cell Physiol.* **50**, 1215–1231.
29. Nakazawa, M., and Matsui, M. (2003) Selection of hygromycin-resistant *Arabidopsis* seedlings. *BioTechniques* **34**, 28–30.
30. Du, J., Huang, Y. P., Xi, J., Cao, M. J., Ni, W. S., Chen, X., Zhu, J. K., Oliver, D. J., and Xiang, C. B. (2008) Functional gene-mining for salt-tolerance genes with the power of *Arabidopsis*. *Plant J.* **56**, 653–664.
31. Seki, M., Carninci, P., Nishiyama, Y., Hayashizaki, Y., and Shinozaki, K. (1998) High-efficiency cloning of *Arabidopsis* full-length cDNA by biotinylated CAP trapper. *Plant J.* **15**, 707–720.

Chapter 13

Construction of Yeast Surface-Displayed cDNA Libraries

Scott Bidlingmaier and Bin Liu

Abstract

Using yeast display, heterologous protein fragments can be efficiently displayed at high copy levels on the *Saccharomyces cerevisiae* cell wall. Yeast display can be used to screen large expressed protein libraries for proteins or protein fragments with specific binding properties. Recently, yeast surface-displayed cDNA libraries have been constructed and used to identify proteins that bind to various target molecules such as peptides, small molecules, and antibodies. Because yeast protein expression pathways are similar to those found in mammalian cells, human protein fragments displayed on the yeast cell wall are likely to be properly folded and functional. Coupled with fluorescence-activated cell sorting, yeast surface-displayed cDNA libraries potentially allow the selection of protein fragments or domains with affinity for any soluble molecule that can be fluorescently detected. In this report, we describe protocols for the construction and validation of yeast surface-displayed cDNA libraries using preexisting yeast two-hybrid cDNA libraries as a starting point.

Key words: Yeast, Surface cDNA display, Library

1. Introduction

Various techniques have been developed to screen large expressed protein libraries for proteins or protein fragments with specific binding properties, including the yeast two-hybrid system and phage display (1–9). However, these systems have limitations. Because it relies on the internal coexpression of "bait" and "prey" fusion proteins, the two-hybrid system cannot be used to identify proteins that bind to externally synthesized or modified proteins or compounds. Phage display is limited by potential expression bias against eukaryotic proteins expressed in a prokaryotic host and the low number of fusion proteins displayed on each phage particle (2, 6). To address these limitations, yeast surface-displayed libraries of human cDNA

Chaofu Lu et al. (eds.), *cDNA Libraries: Methods and Applications*, Methods in Molecular Biology, vol. 729,
DOI 10.1007/978-1-61779-065-2_13, © Springer Science+Business Media, LLC 2011

fragments can be utilized. Heterologous protein fragments can be efficiently displayed at high copy levels on the *Saccharomyces cerevisiae* cell wall (10), and yeast surface-display technology has been successfully used to affinity mature human antibody fragments and map antibody-binding epitopes (11, 12). Because yeast protein expression pathways are similar to those found in mammalian cells, human protein fragments displayed on the yeast cell wall are likely to be properly folded and functional. Yeast surface-displayed human cDNA libraries have been successfully used to screen for proteins that bind to various target molecules (13–16). Coupled with fluorescence-activated cell sorting (FACS), yeast surface-displayed cDNA libraries potentially allow the selection of protein fragments or domains with affinity for any soluble molecule that can be fluorescently detected. In this report, we describe protocols for the construction and validation of yeast surface-displayed cDNA libraries using preexisting yeast two-hybrid cDNA libraries as a starting point.

2. Materials

2.1. Generation of Frameshift Variants of pYD1

1. pYD1 (see Note 1).
2. *Bam*HI restriction enzyme.
3. 10× React 3 (supplied with restriction enzymes, New England Biolabs).
4. Klenow fragment.
5. 10× React 2 (supplied with klenow fragment, New England Biolabs).
6. Mung bean nuclease.
7. 10× Mung bean nuclease buffer (supplied with mung bean nuclease).
8. Spin prep PCR isolation kit (Qiagen).
9. Elution buffer EB (supplied with Qiagen spin prep kit).
10. T4 DNA ligase.
11. 10× Ligation buffer (supplied with T4 DNA ligase).
12. 1,000× Ampicillin: 1 g ampicillin, add ddH_2O to 10 mL, sterilize by filtration and aliquot, and store aliquots at –20°C.
13. 150 mm 2× YT + ampicillin plates: 16 g tryptone, 10 g yeast extract, 5 g NaCl, 17 g agar, bring volume to 1 L with ddH_2O, adjust pH to 7.0, sterilize by autoclaving, let cool until not too hot to touch and add 1 mL 1,000× ampicillin,

pour plates and allow to cool at RT, and store plates at 4°C until ready to use.

14. Gap5 primer: 5′-TTAAGCTTCTGCAGGCTAGTG-3′.

2.2. Library Construction

1. Random-primed, size-selected cDNA libraries in pYesTrp (Invitrogen).
2. *Eco*RI restriction enzyme (New England BioLabs).
3. Shrimp alkaline phosphatase (SAP) (Fermentas).
4. Spin prep gel isolation kits (Qiagen).
5. Ultracel YM-30 microconcentrator (Millipore).
6. 10G Supreme electrocompetent cells (Lucigen).
7. Transformation recovery medium (supplied with 10G Supreme cells).
8. 1 mm Electroporation cuvettes (Molecular BioProducts).
9. Electroporators (Eppendorf 2510, Bio-Rad Gene Pulser II).
10. 2× YT: 16 g tryptone, 10 g yeast extract, 5 g NaCl, bring volume to 1 L with ddH_2O, adjust pH to 7.0, and sterilize by autoclaving.
11. Maxiprep kit (Qiagen).

2.3. Transformation of Library into Yeast

1. EBY100 *S. cerevisiae* strain (see Note 1).
2. YPD: 10 g of yeast extract, 20 g of bacteriological peptone, 20 g dextrose, bring volume to 1 L with ddH_2O, and autoclave to sterilize.
3. 10× SD-CAA for making plates: 70 g yeast nitrogen base without amino acids, 50 g bacto casamino acids, 100 g dextrose, bring volume to 500 mL with ddH_2O, and filter sterilize.
4. 50% PEG-3350: 250 g PEG-3350, bring volume to 500 mL with ddH_2O, filter sterilize.
5. 1 M Lithium acetate: 33 g lithium acetate, bring volume to 500 mL with ddH_2O, filter sterilize.
6. SD-CAA plates: 5.4 g Na_2HPO_2, 7.4 g NaH_2PO_4, 17 g agar, bring volume to 900 mL with ddH_2O and autoclave, let cool until not too hot to touch and add 100 mL 10× SD-CAA, pour plates of desired size (100 or 150 mm) and allow to cool at RT, and store plates at 4°C until ready to use.
7. 2× SR-CAA yeast growth media: 20 g raffinose, 14 g yeast nitrogen base, 10 g bacto casamino acids, 5.4 g Na_2HPO_4, 7.4 g NaH_2PO_4, bring volume to 1 L with ddH_2O, and filter sterilize.

2.4. Test Library Induction

1. 20% Galactose: 100 g galactose, bring volume to 500 mL with ddH_2O, filter sterilize.
2. Mouse anti-Xpress (Invitrogen).
3. Mouse anti-V5 (Invitrogen).
4. Goat antimouse-CY5 (Jackson ImmunoResearch).

3. Methods

The methods below are divided into four categories: generation of frameshift variants of pYD1 vector (Subheading 3.1); library construction (Subheading 3.2); transformation of library into yeast (Subheading 3.3); and test library induction (Subheading 3.4). The construction of yeast display libraries and the generation of frameshifted variants of the original pYD1 yeast display vector have been described previously (13, 14). The pYD1 frameshift variants theoretically allow for the in-frame expression of a larger number of cDNA inserts. The cDNA in the protocol we describe comes from premade Invitrogen cDNA libraries designed for yeast two-hybrid experiments. We chose libraries designed for two-hybrid experiments because they have been random-primed and size-selected (0.3–1.5 kB), and will thus display domain-sized protein fragments that may be more efficiently expressed and folded than full-length proteins. The choice of cDNA source material will depend on the final application for the library. A wide variety of premade cDNA libraries are available, and it should be possible to adapt the presented cloning scheme to use other cDNA sources. Other protocols for the de novo generation of cDNA libraries have been described and can also be adapted for the generation of yeast surface-displayed cDNA libraries based on the protocols we describe.

3.1. Generation of pYD1(+1) and pYD1(−1) Frameshift Variants of the pYD1 Vector

1. Set up the following digestion:

3 μL 10× React 3
10 μg pYD1 plasmid
2 μL *Bam*HI
Bring volume to 30 μL with ddH_2O
Incubate at 37°C for 3 h

2. Run digestion on 0.75% agarose gel, cut out linearized band with a clean razor blade, and isolate DNA using Qiagen gel isolation kit following manufacturer's protocols. Elute in 50 μL EB buffer.

3. Set up the following reactions:

Klenow fill in for pYD(±1)	***Mung bean chew back for pYD(−1)***
3 μL 10× React 2	3 μL 10× Mung bean reaction buffer
2 μL 0.5 mM dNTP mix	20 μL Linearized pYD1 (≈2 μg)
20 μL Linearized pYD1 (≈2 μg)	1 μL Mung bean nuclease (10 U/μL)
2 μL Klenow fragment (5 U/μL)	6 μL ddH_2O
4 μL ddH_2O	Incubate at 30°C for 30 min

Incubate at RT for 30 min.

4. Purify Klenow and mung bean reactions using the Qiagen PCR purification kit according to manufacturer's directions. Elute in 50 μL buffer EB.
5. Set up ligations to recircularize:

2 μL 10× ligation buffer
10 μL Klenow filled in pYD1 *or* mung bean chewed back pYD1
7 μL ddH_2O
1 μL T4 DNA ligase (400 U/μL)
Incubate ligations at RT for 2 h

6. Transform ligations into bacteria using any standard method, plate on 2× YT-ampicillin plates, and incubate overnight at 37°C.
7. Prepare minipreps from colonies using any standard method compatible with DNA sequencing and analyze by sequencing with the Gap5 primer to identify the correct clones.
8. Prepare large scale DNA preps of pYD1(+1) and pYD1(−1) for future library cloning.

3.2. DNA Digestion, Ligation, and Transformation

1. Set up digests of the pYD1, pYD1(+1), and pYD1(−1) yeast display vectors, and the cDNA library (see Note 2) as follows:

3 μL 10× React 3	5 μL 10× React 3
10 μg pYD1 plasmid	20 μg cDNA library plasmid
2 μL *Eco*RI	2 μL *Eco*RI
Bring volume to 30 μL with ddH_2O	Bring volume to 50 μL with ddH_2O

Incubate at 37°C for 3 h.

For simplification, the pYD1, pYD1(+1), and pYD1(−1) digests can be pooled.

2. To dephosphorylate the ends of the linearized pYD1 vectors, add 6 μL SAP (1 U/μL) to the pooled vector digestions and incubate at 37°C for 1 h. Heat inactivate the SAP at 65°C for 15 min.
3. Run the vector and library digests on a 1% agarose gel. Using clean razor blades, cut the linearized vector and the desired size range of cDNA inserts (0.3–1.5 kB). Isolate the DNA from the gel slices using the Qiagen gel isolation kit following the manufacturer's protocol. Verify the quality of the isolated DNA fragments by running a fraction on an agarose gel. Determine the concentration of the purified fragments using a spectrophotometer.
4. Set up ligations as follows (see Note 3):

Control ligation	Library ligation
2 μL 10× ligation buffer	2 μL 10× ligation buffer
700 ng linearized vector	700 ng linearized vector
1 μL T4 DNA ligase	300 ng cDNA insert
Bring volume to 20 μL with ddH_2O	1 μL T4 DNA ligase
	Bring volume to 20 μL with ddH_2O

Incubate ligations for 16 h at 16°C.

5. Desalt ligations with an YM-30 microconcentrator according to manufacturer's instructions. Spin at least 20 volumes (400 μL) of ddH_2O through the filter unit. The retentate volume should be approximately 50 μL.
6. Prewarm transformation recovery medium at 37°C. Place electroporation cuvettes on ice. Thaw 10G supreme cells completely on ice (10–20 min) and aliquot 25 μL of cells to prechilled 1.5-mL Eppendorf tubes on ice. Add 1–3 μL of each of the ligations to the tubes with cells and stir gently without pipetting up and down.
7. Transfer 25 μL of the cell/ligation mixture to the chilled electroporation cuvettes, and tap the cuvette a few times to distribute the cells across the bottom of the chamber. Electroporate using the following conditions (see Note 4):

   10 μF

   600 Ω

   1,800 V

8. Immediately add approximately 1 mL of the prewarmed recovery medium to the cuvette, resuspend the cells, and transfer to 17 mm culture tubes.
9. Incubate tubes at 37°C with shaking at 250 rpm for 1 h. Dry 2× YT-ampicillin plates at 37°C while transformations are recovering.
10. Plate 1/10,000 of each transformation on a 2× YT-ampicillin plate for tittering, and plate the remaining transformation cultures on three more plates at approximately 350 μL/plate. Incubate the plates overnight at 37°C.
11. Calculate the number of total colonies for the control and library transformations using the titer plates. The library transformation should have at least 100 times as many colonies as the control (see Note 5).
12. Prepare minipreps from at least 20 colonies from the library titer plate and sequence using the Gap5 primer to verify the quality and diversity of the library. The colonies can also be analyzed using colony PCR (Fig. 1) (see Note 6).
13. Repeat the transformation of the library ligation until the desired library size is achieved (see Note 7).
14. Add 3 mL 2× YT + ampicillin to each of the library transformation plates and resuspend cells with a flame-sterilized spreader. Collect cells from all the plates in one 50 mL conical flask using a pipetman.

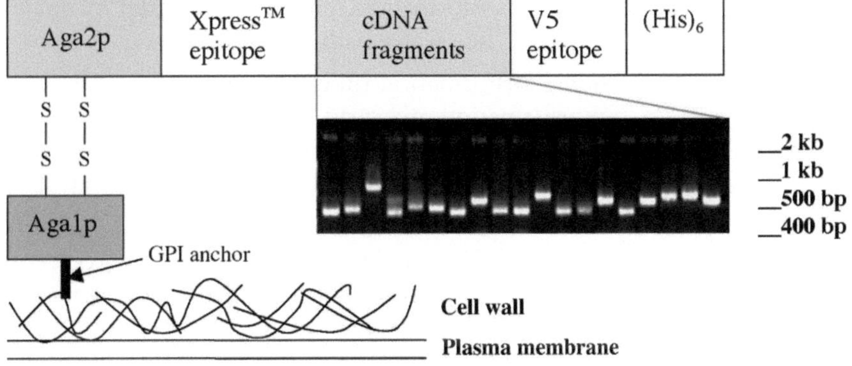

Fig. 1. Construction of yeast surface-displayed human cDNA library. Size-selected (0.3–1.5 kB) human testis cDNAs were cloned into the yeast display vector pYD1 to create a yeast surface-display library containing 5×10^7 members. Human protein fragments are expressed as fusions with Aga2p. The cDNA inserts are flanked by epitope-tags for monitoring the level of expression (Xpress, V5). PCR analysis shows that the cDNA inserts range in size from 0.3 to 1.5 kB (This research was originally published in Mol Cell Proteomics (13). © the American Society for Biochemistry and Molecular Biology).

15. Prepare several bacterial freezer stocks of the library by mixing 0.5 mL of 50% glycerol and 0.5 mL of the collected transformants in a cryotube and store at –80°C.
16. Prepare a Qiagen maxiprep of the remaining library transformants according to manufacturer's protocols. Determine the concentration of the library DNA prep using a spectrophotometer.

3.3. Transformation of Library into Yeast

1. Inoculate a 5-mL YPD culture with EBY100 and grow overnight with shaking at 30°C and 200 rpm.
2. Determine concentration of the overnight yeast culture by measuring the OD_{600} of a 1:20 dilution using a spectrophotometer (1 $OD_{600} = 2 \times 10^7$/mL).
3. Use the overnight culture to inoculate a 50-mL YPD culture at 0.5 OD_{600} and incubate at 30°C, 200 rpm for 4–5 h (at least two cell divisions).
4. Harvest cells by centrifugation at $300 \times g$ for 5 min, wash in 25 mL sterile ddH_2O, and resuspend in 1 mL sterile ddH_2O. Transfer to a 1.5-mL Eppendorf tube, wash again with 1 mL sterile ddH_2O, and resuspend in 1 mL sterile ddH_2O.
5. Add 100 μL EBY100 cell suspension to ten 1.5-mL Eppendorf tubes, spin at top speed for 30 s, and remove supernatant.
6. To each tube of cells, add the following in this order:

   240 μL 50% PEG-3350

   36 μL 1.0 M lithium acetate

   5 μg plasmid DNA

   Bring volume to 360 μL with sterile ddH_2O

   Vortex transformation mix vigorously and incubate tubes at 42°C in a water bath for 40 min.
7. Spin transformation tubes at top speed for 30 s and remove transformation mix. Add 200 μL sterile ddH_2O to each tube and resuspend by vortexing. Pool transformants by transferring to a 15-mL conical flask (approximately 2 mL total volume).
8. Prewarm 150 mm SD-CAA plates at 30°C. To titer the transformation, add 10 μL of pooled transformants to 1 mL of sterile ddH_20, mix and plate 20, or 2 μL (1/10,000 or 1/100,000) by pipetting into a 100-μL puddle of sterile water on the SD-CAA plates. Gently spread the cells onto the plate using a sterilized spreader. Plate the rest of the transformed cells onto SD-CAA plates at a density of 200 μL/plate and once again spread gently. If necessary, allow fluid to be taken up by plates prior to incubation (see Note 8). Incubate the plates inverted at 30°C for 3–4 days until colonies grow.

9. Count the transformants on the titer plates. If larger library size is desired, more transformations can be done (see Note 9).
10. Add 3 mL SR-CAA to each library transformation plate and collect cells by scraping with a spreader. If there is any minor contaminating bacteria or fungus on the plates, it should be excised with a flame-sterilized scalpel prior to resuspending the cells. Pool transformants in a 50-mL conical flask using a pipetman. To prepare freezer stocks, add 1/2 volume of sterile 50% glycerol to the transformants (i.e., add 10 mL 50% glycerol to 20 mL cells), pipet 1-mL aliquots into cryotubes, and store at –80°C.
11. To test the freezer stocks, prewarm and dry six 150-mm SD-CAA plates in a 30°C incubator. Thaw a 1-mL aliquot of yeast surface-displayed cDNA library at RT. To titer the library aliquot, dilute 5 μL of the thawed library into 50 mL sterile H_2O, mix well, and then add 10 μL of the dilution to a 100-μL puddle of sterile H_2O on a 150-mm SD-CAA plate and gently spread. At the same time, plate the remaining library aliquot on the remaining five 150-mm SD-CAA plates (200 μL/plate). Incubate plates upside down at 30°C for 2–3 days. A solid lawn of cells should grow. Count colonies on the titer plate and multiply by 10^6 to obtain the total CFU for the aliquot. This number should be greater than 10× the estimated diversity of the original library (see Note 10). The recovered library can be kept for several weeks on the plates if necessary by sealing the plate edges with parafilm and storing at 4°C.

3.4. Test Library Induction

1. If there are any minor contaminating bacteria or fungus on the plates, it should be excised with a flame sterilized scalpel prior to resuspending the cells. Recover library from plates by adding 5 mL 2× SR-CAA to each plate, scraping with a flame-sterilized spreader, and then collect the resuspended cells into a 50-mL conical flask by pipetting. Determine cell number by taking an OD_{600} reading of a 1/50-diluted sample of the resuspended cells (1 $OD_{600} \approx 2 \times 10^7$ yeast cells/mL).
2. Using the resuspended library, start a 200-mL 0.5 OD_{600} culture (approximately 2×10^9 cells) in 2× SR-CAA. Grow at 30°C with shaking for 1 h. To induce surface expression of the library, add 22 mL sterile 20% galactose (final 2% galactose) and continue growing at 25°C with shaking for 16 h. The induced library can be stored for several weeks at 4°C.
3. To check induction of the library by FACS, spin down three tubes with 100 μL of the induced library (30 s maximum speed in a microcentrifuge) and wash twice with 1 mL PBS.

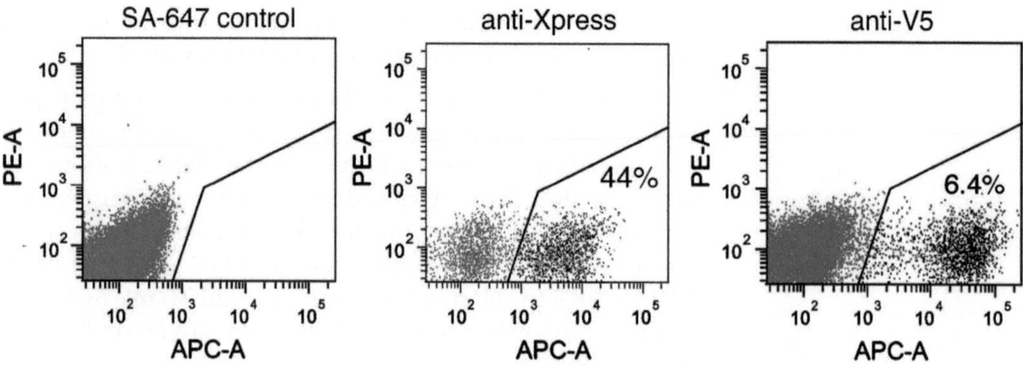

Fig. 2. Library induction is monitored using anti-Xpress and anti-V5 antibodies. The secondary detection reagent was SA-647 and expression is shown on the *x*-axis in the APC-A channel. In order to express the V5 epitope, the cDNA insert must have an open reading frame (ORF) along its entire length that is in frame with both the upstream *AGA2* coding region and the downstream V5 coding region. Typically, about 3–7% of cells from induced libraries are V5-positive, suggesting that about 10–20% of induced cells express an Aga2p fused ORF that runs the entire length of the insert.

4. Resuspend cells in 500 μL PBS and add 1 μL mouse anti-Xpress antibody to one tube, 1 μL mouse anti-V5 antibody to another tube, and nothing to the control tube. Incubate at RT with rotation for 1 h. Wash cells three times with 1 mL PBS.
5. Resuspend each tube in 500 μL 1:500 goat antimouse-CY5 in PBS and incubate at RT with rotation for 30 min. Wash three times with PBS, resuspend in 500 μL PBS, and place on ice until FACS analysis.
6. Analyze the control anti-Xpress and anti-V5 stained yeast cells by FACS. At least 40% of the population should be Xpress-positive and at least 3% of the population should be V5-positive (Fig. 2) (see Note 11). The induced library is now ready for selection experiments.

4. Notes

1. The pYD1 yeast display vector and the EBY100 yeast strain were commercially available from Invitrogen at the time of our library construction work. Please check with the vendor for current availability.
2. We used random-primed, size-selected (0.3–1.5 kB; average size, 0.75 kB) human cDNA libraries from Invitrogen as the source material for the libraries we have made so far. Other premade cDNA libraries can be used by modifying the cloning scheme.

3. This is a standard ligation condition with 1 μg DNA in 20 μL with a 3:1 insert to vector molar ratio, but could be optimized for each individual case.
4. These are the conditions suggested by Lucigen and they have worked well for us. We have generated similar results with Bio-Rad (Gene Pulser II) and Eppendorf (2510) electroporators.
5. If there are too many colonies on the vector control transformation plate, the vector digestion, dephosphorylation, and purification should be repeated.
6. Analyze the sequence to verify that the inserts are cDNA and diverse. Check to ensure that all three frameshift variants of pYD1 are present. Colony PCR on a larger number of colonies can be carried out to check the size distribution and percentage of empty vector (should be very low), but sequencing is the best method of quality control.
7. Transformation plates can be sealed with parafilm and stored at 4°C until enough colonies have been accumulated over the course of several days. Then they can be recovered simultaneously and pooled.
8. We found that a longer drying period for the plates allows cells to be more conveniently plated so that cells don't run down the side when plates are inverted.
9. We usually get greater than 5×10^7 transformants from ten transformations. Plates can be wrapped in parafilm and stored at 4°C until the desired number or transformants has been collected. Then the transformants can be simultaneously collected and pooled.
10. If the titer is close to the desired titer, more than one aliquot can be thawed for selection experiments. If the titer is too low for some reason, the library should be retransformed into yeast and aliquots refrozen.
11. There is always a negative population after induction when analyzed by FACS, and the maximum induction will vary from experiment to experiment. This is seen even when using pYD1-transformed EBY100 as a positive control. In order for the V5 epitope to be expressed, the cDNA insert must have an open reading frame (ORF) that spans its entire length and is in frame with both the upstream *AGA2* coding region and the downstream V5 coding region. Since only one-third of clones with a full-length ORF in frame with the *AGA2* coding region will also be in frame with the V5 epitope, the actual number of clones with a full-length *AGA2*-fused ORF will be approximately three times the number of V5-positive clones.

Acknowledgments

The work is supported by grants from the National Institute of Health (R01 CA118919, R01 CA129491, R21 CA137429, and R21 CA135586).

References

1. Fields, S., and Sternglanz, R. (1994) The two-hybrid system: an assay for protein–protein interactions. *Trends Genet.* **10**, 286–292.
2. Rhyner, C., Kodzius, R., and Crameri, R. (2002) Direct selection of cDNAs from filamentous phage surface display libraries: potential and limitations. *Curr. Pharm. Biotechnol.* **3**, 13–21.
3. Sidhu, S. S., Fairbrother, W. J., and Deshayes, K. (2003) Exploring protein–protein interactions with phage display. *Chembiochem.* **4**, 14–25.
4. Crameri, R., Achatz, G., Weichel, M., and Rhyner, C. (2002) Direct selection of cDNAs by phage display. *Methods Mol. Biol.* **185**, 461–469.
5. Danner, S., and Belasco, J. G. (2001) T7 phage display: a novel genetic selection system for cloning RNA-binding proteins from cDNA libraries. *Proc. Natl. Acad. Sci. U S A* **98**, 12954–12959.
6. Kurakin, A., Wu, S., and Bredesen, D. E. (2004) Target-assisted iterative screening of phage surface display cDNA libraries. *Methods Mol. Biol.* 264, 47–60.
7. Zucconi, A., Dente, L., Santonico, E., Castagnoli, L., and Cesareni, G. (2001) Selection of ligands by panning of domain libraries displayed on phage lambda reveals new potential partners of synaptojanin 1. *J. Mol. Biol.* **307**, 1329–1339.
8. Cicchini, C., Ansuini, H., Amicone, L., Alonzi, T., Nicosia, A., Cortese, R., Tripodi, M., and Luzzago, A. (2002) Searching for DNA–protein interactions by lambda phage display. *J. Mol. Biol.* **322**, 697–706.
9. Guo, D., Hazbun, T. R., Xu, X. J., Ng, S. L., Fields, S., and Kuo, M. H. (2004) A tethered catalysis, two-hybrid system to identify protein–protein interactions requiring post-translational modifications. *Nat. Biotechnol.* **22**, 888–892.
10. Boder, E. T., and Wittrup, K. D. (1997) Yeast surface display for screening combinatorial polypeptide libraries. *Nat. Biotechnol.* **5**, 553–557.
11. Cochran, J. R., Kim, Y. S., Olsen, M.J., Bhandari, R., and Wittrup, K. D. (2004) Domain-level antibody epitope mapping through yeast surface display of epidermal growth factor receptor fragments. *J. Immunol. Methods* **287**, 147–158.
12. Feldhaus, M. J., Siegel, R. W., Opresko, L. K., Coleman, J. R., Feldhaus, J. M., Yeung, Y. A., Cochran, J. R., Heinzelman, P., Colby, D., Swers, J., Graff, C., Wiley, H. S., and Wittrup, K. D. (2003) Flow-cytometric isolation of human antibodies from a nonimmune *Saccharomyces cerevisiae* surface display library. *Nat. Biotechnol.* **21**, 163–170.
13. Bidlingmaier, S., and Liu, B. (2006) Construction and application of a yeast surface-displayed human cDNA library to identify post-translational modification-dependent protein–protein interactions. *Mol. Cell Proteomics.* **5**, 533–540.
14. Bidlingmaier, S., and Liu, B. (2007) Interrogating yeast surface-displayed human proteome to identify small molecule-binding proteins. *Mol. Cell Proteomics.* **6**, 2012–2020.
15. Bidlingmaier, S., He, J., Wang, Y., An, F., Feng, J., Barbone, D., Gao, D., Franc, B., Broaddus, V. C., and Liu, B. (2009) Identification of MCAM/CD146 as the target antigen of a human monoclonal antibody that recognizes both epithelioid and sarcomatoid types of mesothelioma. *Cancer Res.* **69**, 1570–1577.
16. Wadle, A., Mischo, A., Imig, J., Wüllner, B., Hensel, D., Wätzig, K., Neumann, F., Kubuschok, B., Schmidt, W., Old, L. J., Pfreundschuh, M., and Renner, C. (2005) Serological identification of breast cancer-related antigens from a *Saccharomyces cerevisiae* surface display library. *Int. J. Cancer* **117**, 104–113.

Chapter 14

Identification of Protein/Target Molecule Interactions Using Yeast Surface-Displayed cDNA Libraries

Scott Bidlingmaier and Bin Liu

Abstract

We describe a novel expression cloning method based on screening yeast surface-displayed human cDNA libraries by direct affinity interaction to identify cellular proteins binding to a broad spectrum of target molecules. Being a eukaryote, yeast protein expression pathways are similar to those found in mammalian cells, and therefore, mammalian protein fragments displayed on the yeast cell wall are more likely to be properly folded and functional than proteins displayed using prokaryotic systems. Yeast surface-displayed human cDNA libraries have been successfully used to screen for proteins that bind to posttranslationally modified phosphorylated peptides, small signaling molecule phosphatidylinositides, and monoclonal antibodies. In this article, we describe protocols for using yeast surface-displayed cDNA libraries, coupled with fluorescence-activated cell sorting, to select protein fragments with affinity for various target molecules including posttranslationally modified peptides, small signaling molecules, monoclonal phage antibodies, and monoclonal IgG molecules.

Key words: Yeast surface cDNA display, Expression cloning, Phage antibody, Antigen identification, Small molecules, Posttranslationally modified ligands

1. Introduction

The external display of heterologous proteins or protein fragments incorporated into the *Saccharomyces cerevisiae* cell wall, termed yeast surface display, has been successfully utilized in various applications since the initial development of the technology by Boder and Wittrup (1). Yeast surface display technology has been most extensively applied to monoclonal antibody engineering, where it has been used to affinity mature human antibody fragments and map antibody-binding epitopes (2, 3).

Recently, yeast surface-displayed human cDNA libraries have been constructed and used to screen for protein fragments with

Chaofu Lu et al. (eds.), *cDNA Libraries: Methods and Applications*, Methods in Molecular Biology, vol. 729,
DOI 10.1007/978-1-61779-065-2_14, © Springer Science+Business Media, LLC 2011

affinity for various types of molecules. In this novel expression cloning system, ligands of any chemical and molecule compositions can be used as "baits" to identify binding cellular proteins, providing that the bait molecules can be labeled fluorescently or immobilized to a solid matrix. Yeast surface-displayed human cDNA libraries have been successfully used to screen for proteins that bind to posttranslational modifications (phosphorylated peptides) (4), small molecules (phosphatidylinositides) (5), monoclonal antibodies (6), and serum autoantibodies (7). In this article, we describe protocols for using yeast surface-displayed cDNA libraries, coupled with fluorescence-activated cell sorting (FACS), to select protein fragments with affinity for various soluble molecules.

2. Materials

2.1. Growth and Induction of Yeast Surface-Displayed Human cDNA Library

1. Yeast surface-displayed cDNA expression library (see Note 1).
2. 2× SR-CAA yeast growth media: 20 g raffinose, 14 g yeast nitrogen base w/o amino acids, 10 g bacto casamino acids, 5.4 g Na_2HPO_4, 7.4 g NaH_2PO_4; bring the volume to 1 L with ddH_2O and filter sterilize.
3. 10× SD-CAA for making plates: 70 g yeast nitrogen base w/o amino acids, 50 g bacto casamino acids, 100 g dextrose; bring the volume to 500 mL with ddH_2O and filter sterilize.
4. SD-CAA plates: 5.4 g Na_2HPO_2, 7.4 g NaH_2PO_4, 17 g agar. Bring the volume to 900 mL with ddH_2O, autoclave to sterilize, let the agar cool until it is not too hot to touch, add 100 mL 10× SD-CAA, pour into plates (100 and 150 mm) and allow to cool at RT, and store the plates at 4°C until ready to use.
5. 20% Galactose: 100 g galactose; bring the volume to 500 mL with ddH_2O and filter sterilize.
6. Mouse anti-Xpress (Invitrogen).
7. Goat anti-mouse-phycoerythrin (PE) (Jackson ImmunoResearch).

2.2. Yeast Surface-Displayed cDNA Library FACS and Analysis

1. PBS: 8 g of NaCl, 0.2 g of KCl, 1.44 g of Na_2HPO_4, 0.24 g of KH_2PO_4 ; bring the volume to 1 L with ddH_2O, adjust pH to 7.4, and sterilize by autoclaving.
2. Biotinylated phosphorylated target peptide, biotinylated nonphosphorylated target peptide, and nonbiotinylated, nonphosphorylated target peptide.
3. Streptavidin-PE (SA-PE) (Invitrogen).
4. Streptavidin-647 (SA-647) (Invitrogen).
5. Biotinylated target phosphatidylinositides.
6. Purified target phage antibody particles.

7. Purified control phage (M13 helper phage).
8. EZ-Link Sulfo-NHS-LC-Biotin (Pierce).
9. 20% PEG/2.5 M NaCl: 100 g PEG-8000, 73 g NaCl; bring the volume to 500 mL with ddH_2O and filter sterilize.
10. Biotinylated anti-fd bacteriophage (Sigma).
11. Target IgG.
12. Negative control IgG.
13. Goat antihuman IgG-PE (Jackson ImmunoResearch).

2.3. Plasmid Recovery from Yeast and Sequencing

1. Spin miniprep kit (Qiagen).
2. Acid-washed glass beads (Sigma).
3. Gap5 primer: 5′-TTAAGCTTCTGCAGGCTAGTG -3′.

3. Methods

The methods below are divided into six categories: Growth and induction of the yeast surface-displayed cDNA expression libraries (Subheading 3.1); FACS-based selection of phosphopeptide-binding protein fragments (Subheading 3.2); FACS-based selection of phosphatidylinositide-binding protein fragments (Subheading 3.3); FACS-based selection of scFv phage antibody-binding protein fragments (Subheading 3.4); FACS-based selection of mAb IgG-binding protein fragments (Subheading 3.5); plasmid recovery from individual binding clones (Subheading 3.6).

Although we provide specific examples using three different targets, these protocols can serve as a general template for designing methods to select protein fragments with affinity for any soluble molecule that can be fluorescently detected. However, each unique target molecule may require modifications or optimizations of the protocol. It is critical to prepare a full set of controls so that the selection progress can be monitored (see Note 2).

3.1. Induction of Yeast Surface-Displayed Human cDNA Library

1. Prewarm and dry four 150-mm SD-CAA plates in a 30°C incubator. Thaw 1 mL aliquot of yeast surface-displayed cDNA library at RT. To titer the library aliquot, dilute 2 μL of the thawed library into 10 mL sterile H_2O, mix well, and then evenly spread 5 μL of the dilution on a 150-mm SD-CAA plate. At the same time, plate the remaining library aliquot on the remaining three 150-mm SD-CAA plates (330 μL/plate). Incubate the plates upside down at 30°C for 3 days.
2. Count colonies on the titer plate and multiply by 10^6 to obtain the total colony forming units for the aliquot. This number should be >10× the size of the original library to

allow adequate representation of the diversity. The recovered library can be kept for several weeks on the plates if necessary by sealing the plate edges with parafilm and storing at 4°C.

3. Recover the library from plates by adding 5 mL of 2× SR-CAA to each plate and scraping with a flame-sterilized spreader and then collect the resuspended cells by pipetting. Determine cell number by taking an OD_{600} reading of a 1:50-diluted sample of the resuspended cells (1 $OD_{600} \approx 2 \times 10^7$ yeast cells/mL).
4. Using the resuspended library, start a 200 mL of 0.5 OD_{600} culture (approximately 2×10^9 cells) in 2× SR-CAA. Grow at 30°C with shaking for 1 h. To induce surface expression of the library, add 22 mL of sterile 20% galactose (final 2% galactose) and continue growing at 25°C with shaking for 16 h. The induced library can be stored for several weeks at 4°C.
5. Check induction of the library by FACS using the mouse anti-Xpress antibody. Spin down 100 μL of the induced library (10,000 rpm in microfuge) and wash twice with 1 mL PBS. Resuspend cells in 500 μL PBS and add 1 μL mouse anti-Xpress antibody. Incubate at RT with rotation for 1 h. Wash three times with 1 mL PBS. Resuspend in 500 μL PBS and add 1 μL goat anti-mouse PE. Incubate at RT with rotation for 30 min. Wash three times with PBS, resuspend in 500 μL PBS, and place on ice.
6. Analyze the cells by FACS. At least 40% of the population should be Xpress-positive (see Note 3). The induced library is now ready for FACS selection experiments.

3.2. FACS Selection of Phosphopeptide-Binding Protein Fragments

1. Check the OD_{600} of a 1:20 dilution of the induced library to determine cell density. For the first round of sorting, collect 2×10^8 yeast cells by centrifugation and wash them twice with PBS. For example, if the calculated OD_{600} of the undiluted induced library is 5, then 2 mL of culture will contain approximately 2×10^8 cells.
2. Resuspend the cells in 500 μL PBS and add 10 μM biotinylated phosphopeptide and 40 μM of the corresponding non-phosphorylated, nonbiotinylated peptide to compete away non-phospho-specific binding. As a negative control for the sorting, set up an incubation with the same number of cells but without the peptides. Incubate for 4 h at 4°C.
3. Wash the cells twice with ice-cold PBS and incubate with 500 μL of 1:500 diluted SA-PE for 30 min at 4°C (see Note 4). Wash the cells thrice with ice-cold PBS and resuspend in 3 mL PBS. Keep the cells on ice in the dark until sorting.
4. First, analyze the negative control incubation by FACS (see Note 5). This will give you a zero baseline for adjusting the FACS parameters. It is advisable to be less stringent in the first

round, and the sort gate should be placed directly above the point where the vast majority of the negative control cells cut off in the PE channel (Fig. 1). Use the "yield" or equivalent nonstringent sorting setting on your flow cytometer during the first round. The library should be comprehensively sorted for the first round of selection. Analyze at least 10^8 cells from the peptide selection incubation and sort PE-positive cells into an Eppendorf tube containing 100 μL PBS (see Note 6). Make a note of the total number of cells analyzed and the number sorted.

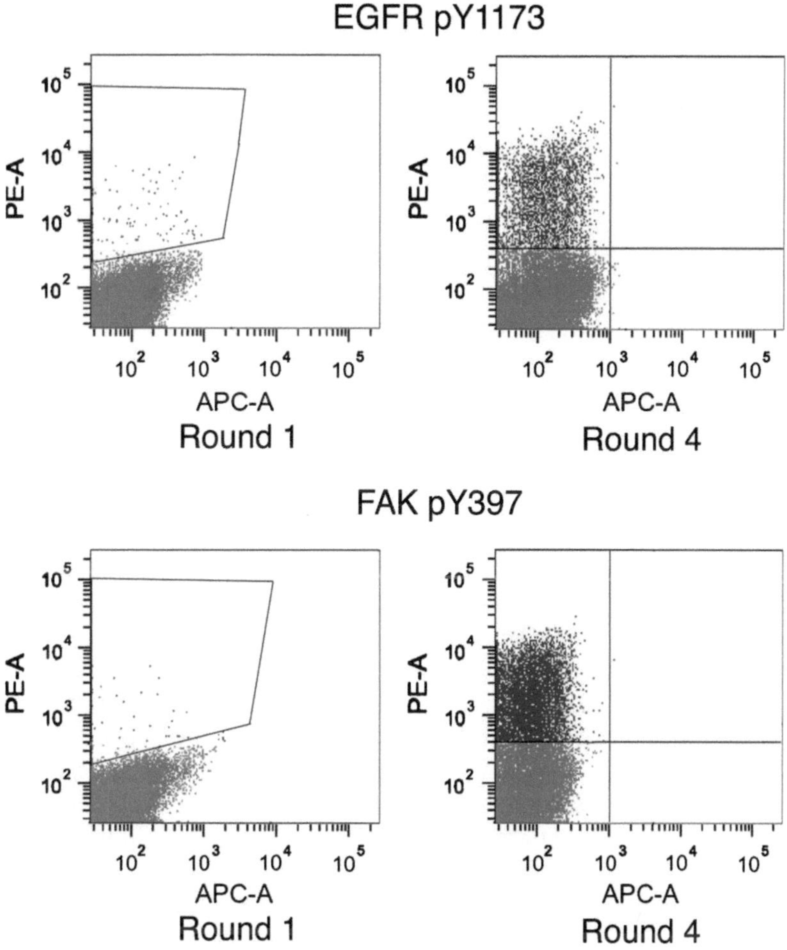

Fig. 1. Selection of phosphopeptide-binding clones from a yeast surface-displayed human cDNA library by FACS. Surface expression of the cDNA library was induced and the yeast cells were incubated with biotinylated, tyrosine-phosphorylated peptides derived from the major autophosphorylation sites of either EGFR (EGFRpY1173) or FAK (FAKpY397). Enrichment of EGFRpY1173- and FAKpY397-binding clones is achieved through four rounds of FACS. The sort window is shown for the first round. In the fourth round, the upper left quadrant was sorted and the output was used for individual clone analysis. (This research was originally published in Mol Cell Proteomics (4) © the American Society for Biochemistry and Molecular Biology).

5. Plate the sorted cells on one or several large pre-dried SD-CAA plates by gently spreading. Incubate the plates in an inverted position at 30°C until colonies are formed (3–4 days) (see Note 7).
6. Add 3 mL SR-CAA to the plate(s) and recover the cells by scraping with a sterile cell spreader. To prepare a freezer stock, add 250 μL of 50% glycerol to 500 μL cells and store at −80°C.
7. To induce the first-round output, inoculate a 10-mL culture at 0.5 OD_{600} in SR-CAA + 2% galactose using the remaining cells from the round-one output and grow at 25°C with shaking for 16 h.
8. Since the diversity is greatly reduced after the first round, it is not necessary to use as many cells during incubations in the subsequent rounds. Wash approximately 5×10^7 cells (see Note 8) from the induced first-round output and the starting library twice with PBS. Set up control and selection incubations in the same manner as that in the first round and incubate for 4 h at 4°C.
9. Wash the cells twice with ice-cold PBS and incubate with 500 μL of 1:500 diluted SA-647 for 20 min at 4°C. It is important to use a different secondary detection reagent than was used in the first round (see Note 9). Set up a secondary-only control with the first-round output using SA-647. After incubations with the secondary reagents, wash the cells three times with ice-cold PBS and resuspend in 1 mL PBS. Keep the cells on ice in the dark until sorting.
10. Analyze the selection incubation by FACS and compare it to the secondary-only control to decide where to place the sort gate. Since the goal is to recover the maximum diversity of specific binding clones, we usually place the sort gate directly above the point where the negative control cells cut off in the 647 (aka APC) channel, in the same manner as during the first-round sort. However, in the second and subsequent rounds, use the "purity" or equivalent more stringent sorting setting on your flow cytometer. If the selection is working, you are likely to observe increased binding in the selection incubation compared to that in the negative control. Analyze 10^7 cells from the second-round phosphorylated peptide selection incubation and sort 647-positive cells into an Eppendorf tube containing 100 μL PBS. Again, make a note of the total number of cells analyzed and the number sorted, and plate the sorted cells on one or several large SD-CAA plates. Incubate at 30°C until colonies form.
11. Add 3 mL SR-CAA to plate(s) and recover cells by scraping. Prepare a frozen stock as previously described and store at −80°C. Induce the second-round output by inoculating

a 10-mL culture at 0.5 OD_{600} in SR-CAA + 2% galactose using the remaining cells and grow at 25°C with shaking for 16 h.

12. The third round is carried out identically to second round, except for returning to SA-PE for the secondary detection in the phosphorylated peptide selection incubation. It is absolutely critical to set up a SA-PE secondary-only control and analyze the results carefully. It is also important to set up a control incubation with the biotinylated nonphosphorylated peptide control to determine if there is any non-phospho-specific binding. If the selection is working, there should be a significant population of binding clones (>5%) and very little background binding to the SA-PE negative control (<0.5%). If the specific binding population is >5% and the negative control background is comparably low (Fig. 1), the output of this round of sorting is ready to be screened using individual clones (see Note 10). Sort the binders in the induced second-round output and recover as previously described. If the intention is to screen individual clones from this sort output, some plates can be seeded at a lower density (500 cells/plate) based on the sorting data to facilitate the picking of individual colonies for screening. After the colonies have grown, make a glycerol freezer stock of the third-round sort output as described previously.
13. Protocols for the screening of individual clones are similar to those for the sorting, but are done on a smaller scale to save reagents and facilitate high throughput analysis (see Note 11). Inoculate 1 mL SR-CAA + 2% galactose cultures with a yeast colony from the SD-CAA sort output plates and grow for at least 16 h with shaking at 30°C.
14. Wash 200 μL of the cultures with PBS and pellet. Save the remaining culture at 4°C so that the binding clones can be identified and saved. Add 50 μL of 10 μM phosphorylated biotinylated peptide or 50 μL nonphosphorylated biotinylated peptide (control) in PBS to the cell pellets, resuspend by pipetting or agitation, and incubate for 2 h at 25°C or overnight at 4°C.
15. Wash twice with ice-cold PBS, add 50 μL 1:500 diluted SA-PE, and incubate at 25°C for 30 min. Wash twice with ice-cold PBS, resuspend in 300 μL PBS, and keep on ice until FACS analysis.
16. Analyze by FACS. Phospho-specific binding clones will exhibit binding in the phosphorylated peptide incubation, but not the nonphosphorylated control incubation (Fig. 2). Clones that exhibit phospho-specific binding can be patched onto SD-CAA plates for temporary storage and frozen in SR-CAA + 15% glycerol for permanent storage at −80°C. A protocol for plasmid recovery for sequencing of the clones is detailed in Subheading 3.6.

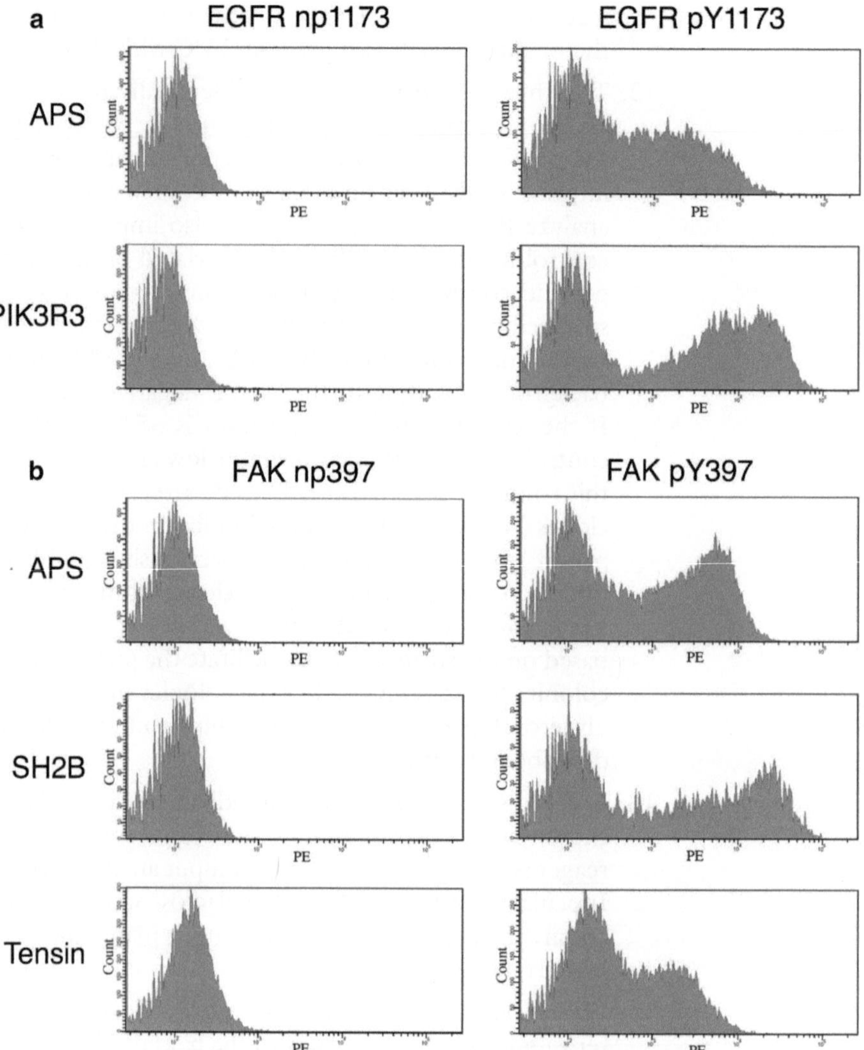

Fig. 2. Phosporylation-dependent binding of selected clones. Induced yeast clones were tested for binding with 10 μM of either phosphorylated (p) or nonphosphorylated (np) peptides. (**a**) Cellular proteins (APS and PIK3R3) binding specifically to EGFRpY1173. (**b**) Cellular proteins (APS, SH2B, and Tensin) binding specifically to FAKpY397. (This research was originally published in Mol Cell Proteomics (4) © the American Society for Biochemistry and Molecular Biology).

3.3. FACS Selection of Phosphatidylinositide-Binding Protein Fragments

1. The protocols for the selection of phosphatidylinositide-binding protein fragments are similar to the protocols described above using the phosphorylated peptides. For the first round of selection, collect 2×10^8 cells from the induced library, wash twice with PBS, and incubate in 500 μL of PBS with 2 μM biotinylated phosphatidylinositides for 4 h at 4°C.
2. Wash the cells twice with PBS and incubate with 500 μL of 1:500 diluted SA-PE for 20 min at 4°C. Wash the cells twice with PBS, analyze, sort, and recover the cells as described above.

3. Continue with subsequent rounds as described above, taking care to prepare appropriate negative controls and alternating secondary detection reagents between rounds. When the binding population is greater than 5% with very little background binding to the negative controls, individual colonies from the sort output from that round can be screened using 2 μM biotinylated phosphatidylinositides with a SA-PE secondary-only control. Binding clones should be patched onto SD-CAA plates for further analysis and stored in glycerol freezer stocks.

3.4. FACS Selection of Phage Antibody-Binding Protein Fragments

1. The protocols for selection of phage antibody-binding protein fragments are similar to the protocols described above. We present a protocol for the biotinylation of phage antibody particles, but protocols and reagents exist that are suitable for the labeling of a wide variety of target molecules. To biotinylate the phage antibodies, make a fresh 10 mM biotinylation solution by adding 180 μL of ddH_2O to 1 mg of EZ-Link sulfo-NHS-LC-biotin. Add 60 μL of this solution to 800 μL of fd phage antibody in PBS (see Note 12). Incubate at RT for 15 min and then stop the reaction by adding 140 μL 1 M Tris–HCl, pH 7.5.
2. Add 250 μL of 20% PEG/2.5 M NaCl solution and incubate on ice for at least 30 min to precipitate the labeled phage particles. Spin down precipitated phage at 16,000 × *g* at 4°C in a microcentrifuge for 20 min. Carefully remove as much of the PEG solution as possible from the phage pellet and resuspend in 1 mL PBS.
3. Precipitate the phage a second time using the same protocol, carefully remove the PEG solution, and resuspend in 1 mL PBS. Filter the labeled phage solution with a 0.45-μm syringe filter. The binding activity of the labeled phage should be confirmed using an appropriate assay.
4. The sorting process is similar to that previously described. In the first round of selection, collect 2×10^8 cells from the induced library, wash twice with PBS, and resuspend in 400 μL PBS. Add 50 μL labeled target phage and 250 μL unlabeled helper phage to compete away nonspecific phage-binding clones and incubate at RT for 1 h with rotation.
5. Wash the cells twice with PBS and incubate with 500 μL of 1:500 diluted SA-PE for 20 min. Wash the cells twice with PBS, analyze, sort, and recover binding clones as described above. In subsequent rounds, be sure to prepare appropriate controls (e.g., control phage, secondary only, and previous round outputs) and remember to alternate SA-647 with SA-PE or other detection reagents to minimize the selection of clones that bind the secondary detection reagent. When significant and specific

binding (>5%) is observed in the sorted population, individual clones can be tested for phage binding by FACS.

6. It is possible to use a scaled-down version of the above protocol with the biotinylated phage and appropriate secondary-only controls. However, for screening many samples, we routinely use a biotinylated anti-fd bacteriophage antibody. Grow overnight cultures for screening as described above and wash 200 μL of the cultures with PBS and pellet. Add 200 μL of a 1:10 dilution of target phage and negative control helper phage in PBS to the cell pellets and incubate for 2 h at 25°C or overnight at 4°C.
7. Wash the cells twice with PBS and incubate with 500 μL of 1:500 diluted anti-fd bacteriophage antibody for 1 h at RT.
8. Wash the cells twice with PBS and incubate with 500 μL of 1:500 diluted SA-PE for 20 min. Wash the cells twice with PBS and analyze by FACS.
9. Binding clones can be streaked on SD-CAA plates for temporary storage and should be stored permanently as glycerol stocks at –80°C as previously described.

3.5. FACS-Based Selection of mAb IgG-Binding Protein Fragments

1. To biotinylate the target and negative control IgGs, make a fresh 10 mM biotinylation solution by adding 180 μL of ddH_2O to 1 mg of EZ-Link sulfo-NHS-LC-biotin. Add 30 μL of this solution to 1 mL of 2 mg/mL IgG in PBS and incubate at RT for 30 min. Stop the reaction by adding 100 μL 1 M Tris–HCl, pH 7.5.
2. Remove excess biotin reagent by dialysis against PBS. The binding activity of the labeled IgG should be confirmed using an appropriate assay.
3. The sorting process is similar to that previously described. In the first round of selection, collect 2×10^8 cells from the induced library, wash twice with PBS, and resuspend in 400 μL PBS. Add 10 μL biotinylated target IgG and 100 μL 1 mg/mL unlabeled negative control IgG to compete away nonspecific IgG-binding clones and incubate at RT for 1 h with rotation.
4. Wash the cells twice with PBS and incubate with 500 μL of 1:500 diluted SA-PE for 20 min. Wash the cells twice with PBS, analyze, sort, and recover binding clones as described above. In subsequent rounds, be sure to prepare appropriate controls (e.g., negative control IgG, secondary only, and previous round outputs) and remember to alternate SA-647 with SA-PE or other detection reagents to minimize the selection of clones that bind the secondary detection reagent. When significant and specific binding (>5%) is observed in the sorted population, individual clones can be tested for target IgG binding by FACS.

5. To screen for binding clones in the sort output, grow overnight yeast cultures for screening as described previously and wash 200 μL of the cultures with PBS and pellet. Add 200 μL of a 1:100 dilution of target phage or negative control IgG in PBS to the cell pellets and incubate for 2 h at 25°C or overnight at 4°C.
6. Wash the cells twice with PBS and incubate with 500 μL of 1:500 diluted goat-antihuman PE for 20 min. Wash the cells twice with PBS and analyze by FACS.
7. Binding clones can be streaked on SD-CAA plates for temporary storage and should be stored permanently as glycerol stocks at −80°C as previously described.

3.6. Plasmid Recovery from Individual Binding Clones

1. The method we describe is based on the Qiagen spin miniprep kit. The buffers used (P1, P2, N3, and EB) are provided in the kit (see Note 13). Recover an approximately 50 μL yeast cell pellet and wash twice with ddH_2O. The cells can be recovered from liquid cultures or by scraping from plates.
2. Resuspend the pellet in 400 μL Qiagen buffer P1 and add approximately 200 μL glass beads. Vortex at high speed for 3 min and remove 250 μL of the cell slurry (leaving the glass beads behind) to a clean tube.
3. Add 250 μL buffer P2, gently mix by inverting, and incubate at RT for 5 min.
4. Add 350 μL buffer N3 (a cloudy precipitate will form) and spin at 16,000 × *g* in a microcentrifuge for 15 min.
5. Apply supernatant to a Qiagen spin miniprep column and spin at 16,000 × *g* for 1 min. Discard flow-through.
6. Add 750 μL buffer PE and spin at 16,000 × *g* for 1 min. Discard flow-through and spin at 16,000 × *g* for 2 min to remove residual buffer PE. Replace collection tube with a clean Eppendorf tube, add 50 μL EB, and spin at 16,000 × *g* for 1 min to elute.
7. Transform the recovered plasmids into bacteria using any standard transformation protocol and plate on LB-ampicillin plates. Prepare minipreps from the transformants using any standard method that is compatible with DNA sequencing. Use the Gap5 primer to sequence the plasmid's cDNA insert.

4. Notes

1. The construction and application of yeast surface-displayed cDNA libraries has been previously described (4–6) and a detailed protocol is presented elsewhere in this volume.

2. The controls should include incubations with secondary reagents only (previously used and used in the current round of selection) and incubations using all previous rounds so that the enrichment of target-specific-binding clones can be observed from round to round. It is also important to alternate the secondary detection agent between rounds to minimize the chances of selecting clones with affinity for them. The use of proper controls and careful monitoring of the selection process are critical to maximize the chances for success and to aid in troubleshooting.
3. There is always a negative population after induction when analyzed by FACS, and the maximum induction will vary from experiment to experiment.
4. We used a BD FACSAria for sorting and a BD LSRII for analysis. The protocols are written generally and other FACS equipment can be used.
5. We generally use SA-PE in the first round because it appears to give a cleaner background than SA-647 during sorting.
6. At a sort rate of 50,000 events/s, this should take about 35 min. The first round is sorted on "faith" – there should not be an obvious population of positive cells and there may not be any significant difference between the negative control and the selection incubation. It is critical to analyze enough cells to cover the full diversity of the library or else some binding clones may be lost immediately. Since the diversity is massively diminished after the first-round selection, subsequent rounds take significantly less time to sort. MACS bead selection (Miltenyi Biotec) could also be used in the first round, but we prefer to use FACS.
7. We recommend that the sorted cells be directly plated without spinning them down. We have observed a poor rate of recovery when we attempt to centrifuge the cells prior to plating, even when great care is taken. We sometimes dry the SD-CAA plates overnight at RT prior to plating the sort output so that a larger volume of cells can be absorbed more quickly. The efficiency of recovery of the sort output (colonies recovered vs. events sorted) should be closely monitored, especially in the first round. Any clones lost at this step are lost for good.
8. The cell number should be at least 20× more than the recovered output from the first round.
9. Alternation between two detection reagents is usually sufficient to prevent the enrichment of secondary reagent-binding clones. However, in the absence of real target-binding clones (e.g., failed sorts), you will almost always end up with secondary binders if enough rounds of sorting are carried out.

10. Sorting for too many rounds will reduce the output diversity and tend to favor high-affinity or high-expressing clones.
11. The most efficient screening protocol will depend on the equipment available to each individual researcher. We describe a protocol based on single-tube FACS analysis, which is the format likely to be available to the most researchers. It is very simple to adapt the protocols to a 96-well or other more high throughput format if this equipment is available. The appropriate screening method will depend on the experimental goals of each project (i.e., strong binders vs. diversity).
12. Phage particles must be in a solution that does not contain free primary amine groups that will interfere with the biotinylation reaction.
13. The protocol can also be scaled up to recover plasmids from a polyclonal population. These plasmids can then be sequenced as an alternative screening method. Plasmids of interest will have to be retransformed into EBY100 and tested for target binding.

Acknowledgments

The work is supported by grants from the National Institute of Health (R01 CA118919, R01 CA129491, R21 CA137429, and R21 CA135586).

References

1. Boder, E. T., and Wittrup, K. D. (1997) Yeast surface display for screening combinatorial polypeptide libraries. *Nat. Biotechnol.* **15**, 553–557.
2. Cochran, J. R., Kim, Y. S., Olsen, M. J., Bhandari, R., and Wittrup, K. D. (2004) Domain-level antibody epitope mapping through yeast surface display of epidermal growth factor receptor fragments. *J. Immunol. Methods* **287**, 147–158.
3. Feldhaus, M. J., Siegel, R. W., Opresko, L. K., Coleman, J. R., Feldhaus, J. M., Yeung, Y. A., Cochran, J. R., Heinzelman, P., Colby, D., Swers, J., Graff, C., Wiley, H. S., and Wittrup, K. D. (2003) Flow-cytometric isolation of human antibodies from a nonimmune *Saccharomyces cerevisiae* surface display library. *Nat. Biotechnol.* **21**, 163–170.
4. Bidlingmaier, S., and Liu, B. (2006) Construction and application of a yeast surface-displayed human cDNA library to identify post-translational modification-dependent protein–protein interactions. *Mol. Cell Proteomics.* **5**, 533–540.
5. Bidlingmaier, S, and Liu, B. (2007) Interrogating yeast surface-displayed human proteome to identify small molecule-binding proteins. *Mol. Cell Proteomics.* **6**, 2012–2020.
6. Bidlingmaier, S., He, J., Wang, Y., An, F., Feng, J., Barbone, D., Gao, D., Franc, B., Broaddus, V.C., and Liu, B. (2009) Identification of MCAM/CD146 as the target antigen of a human monoclonal antibody that recognizes both epithelioid and sarcomatoid types of mesothelioma. *Cancer Res.* **69**, 1570–1577.
7. Wadle, A., Mischo, A., Imig, J., Wüllner, B., Hensel, D., Wätzig, K., Neumann, F., Kubuschok, B., Schmidt, W., Old, L. J., Pfreundschuh, M., and Renner, C. (2005) Serological identification of breast cancer-related antigens from a *Saccharomyces cerevisiae* surface display library. *Int. J. Cancer* **117**, 104–113.

Chapter 15

SNP Discovery by Transcriptome Pyrosequencing

W. Brad Barbazuk and Patrick S. Schnable

Abstract

Single nucleotide polymorphisms (SNPs) are single base differences between haplotypes. SNPs are abundant in many species and valuable as markers for genetic map construction, modern molecular breeding programs, and quantitative genetic studies. SNPs are readily mined from genomic DNA or cDNA sequence obtained from individuals having two or more distinct genotypes. While automated Sanger sequencing has become less expensive over time, it is still costly to acquire deep Sanger sequence from several genotypes. "Next-generation" DNA sequencing technologies that utilize new chemistries and massively parallel approaches have enabled DNA sequences to be acquired at extremely high depths of coverage faster and for less cost than traditional sequencing. One such method is represented by the Roche/454 Life Sciences GS-FLX Titanium Series, which currently uses pyrosequencing to produce up to 400–600 million bases of DNA sequence/run (>1 million reads, ~400 bp/read). This chapter discusses the use of high-throughput pyrosequencing for SNP discovery by focusing on 454 sequencing of maize cDNA, the development of a computational pipeline for polymorphism detection, and the subsequent identification of over 7,000 putative SNPs between Mo17 and B73 maize. In addition, alternative alignment and polymorphism detection strategies that implement Illumina short reads, data processing and visualization tools, and reduced representation techniques that reduce the sequencing of repeat DNA, thus enabling efficient analysis of genome sequence, are discussed.

Key words: Next-generation sequencing, Pyrosequencing, SNP, Whole-genome SNP discovery, Computational biology, Maize

1. Introduction

Single nucleotide polymorphisms (SNPs) are single base differences between haplotypes. SNPs are abundant in many species and can be rapidly detected (1–3) making them ideal for large-scale genetic studies. Because they are abundant, it is possible to use SNPs to generate very dense genetic maps that are useful for marker assisted selection programs, to construct the specific genotypes required for quantitative genetic studies, and to enhance our

Chaofu Lu et al. (eds.), *cDNA Libraries: Methods and Applications*, Methods in Molecular Biology, vol. 729, DOI 10.1007/978-1-61779-065-2_15, © Springer Science+Business Media, LLC 2011

understanding of genome organization and function. In addition, SNPs are valuable for genome-wide linkage disequilibrium and associations studies. For example, the ultimate goal of the International HapMap consortium is to link human genetic variation to specific illnesses, which is expected to lead to new methods of preventing, diagnosing, and treating disease (4, 5). Furthermore, transcript-associated SNPs can be used to develop allele-specific assays for the examination of *cis*-regulatory variation within a species (6–10).

SNPs can be identified by sequencing candidate genes from a set of individuals representing genetic diversity in the species of interest; however, this is an expensive and slow process. Approaches adopted during construction of the human SNP map included identifying sequence polymorphisms within overlapping BAC clones derived from different individuals and shotgun sequencing of genomic fragments (11). This approach cannot be applied in many cases because most genome sequencing projects use DNA extracted from highly similar or inbred individuals. Instead, SNP-based markers are typically mined from whole-genome sequences or expressed sequence tags (ESTs) obtained from genetically diverse individuals. For example, SNPs have been identified by comparing genomic sequences from two or more genetically distinct inbred lines of mouse (12), the Indica and Japonica subspecies of rice (13), the Columbia and Landsberg ecotypes of *Arabidopsis* (14), and different lines of maize (15). EST collections from genetically dissimilar individuals have similarly been mined for SNPs in humans (16), pine (17), barley (18, 19), cassava (20), and maize (21, 22).

Mining SNPs requires genomic or transcriptome sequences from multiple genotypes. While automated Sanger sequencing has become less expensive over time, it is still costly to acquire deep Sanger sequence from several genotypes. Recent advances in high-throughput sequencing technology provide a rapid and cost-effective means to generate sequence data. This new paradigm, termed flow-cell sequencing (see review in ref. 23), consists of stepwise determination of DNA sequence by iterative cycles of nucleotide extensions done in parallel on huge numbers of clonally amplified template molecules. This massively parallel approach enables DNA sequence to be acquired at extremely high depths of coverage faster and for less cost than traditional sequencing. Currently, there are three next-generation flow-cell sequencing platforms available commercially: the Roche 454-FLX (23, 24), the Illumina Genome Analyzer (23, 25), and the Applied Biosystems SOLiD system (23). Each of these platforms clonally amplify template by PCR to produce enough substrate to provide sufficient signal for base determination, and each involves iterative nucleotide washes. The two most established commercially available next-generation platforms are the 454-FLX and the

Illumina Genome Analyzer systems. This chapter is primarily concerned with detecting SNP polymorphism within 454 sequences, although the Illumina Genome Analyzer has also proven effective in this capacity and will be discussed briefly. Because the ABI system became available commercially more recently and the community has less experience using it for SNP discovery, it will not be discussed further.

1.1. Next-Generation Sequence Platforms

The 454 system utilizes iterative pyrosequencing. Beads containing clonally amplified template and bead-tethered pyrosequencing enzymes (polymerase, luciferase sulfurylase) are deposited into the wells of a picotitre plate, and buffer containing one of four nucleotides is passed across the plate. Base incorporation results in the emission of a light signal (catalyzed by sulfurylase and luciferase) that is detected by the apparatus. A single run of the 454 GS-FLX titanium instrument takes 7 h and produces >1 million sequences, each of which is ~400–500 bp in length (500 Mb). While the original version of the 454 machine produced only 20 Gb of sequence with an average read length of 100 bp (24), the current and expected sequence lengths are long enough to permit accurate assembly with common sequence assembly tools such as CAP3 (26). However, alternate sequence assembly tools designed specifically for next-generation sequence are also available, e.g., MIRA http://chevreux.org/projects_mira.html, and SHRAP (27). In addition, the 454 sequencing platform ships with the Newbler assembler that assembles directly from the 454 signal output (24).

The per-base accuracy for 454 technology, as reported by the manufacturer, is ~99%, which is lower than the base call error for Sanger sequencing. However, accuracy increases with the greater sequence redundancy provided by the high output of these machines. While each machine cycle permits incorporation of only one nucleotide, all bases within a homopolymer (for example, AAA) will be incorporated during the same cycle. This results in a light signal emission that is proportional to the number of bases added. Unfortunately, the signal for a fixed number of incorporated bases varies substantially, and there is usually a nonzero signal even when no base is incorporated (28). Discrimination is reliable for up to four bases, while homopolymers approaching eight bases are typically irresolvable (24, 29).

The Illumina Genome Analyzer IIx platform differs from the Roche 454 GS-FLX by using a glass plate to which templates are anchored and clonally amplified in situ, and by employing nucleotides labeled with four-color fluorescent reversible terminators. This allows simultaneous addition of all four labeled nucleotides and permits only a single base extension each round. After extension and detection of the fluorescent signal, sequencing reagents are washed away, the fluorophore is cleaved, and the 3′ hydroxyl is unblocked in preparation for the next round of extension. A run of the Illumina Genome Analyzer requires ~3 days and can produce >100 million sequences, each of which is 30–36 bp in length (>4 Gb). Because all four fluorescently tagged bases are present, the risk of mis-incorporation is low and the base-by-base sequencing process provides information on the raw data collected at each base, enabling the derivation of quality values that are analogous to Sanger Phred (30) scores. In addition, unlike the 454-FLX system, a discrete signal is generated for every base, and accuracy is independent of sequence context. Per-base accuracy produced by the previous Illumina 1G system is reported to be >98.5% (23).

The main difficulties presented by short read technology are data handling (10^5–10^7 individual sequences per run; 10^7–10^9 bases per run) and assembly of short reads with less than perfect base call accuracy. De novo assembly has been quite successful with 454 reads (31–35), and longer read lengths will only improve assembly. De novo assembly for the short-read platforms remains challenging. Several algorithms have been published: SSAKE (36), VCAKE (37), SHRAP (27), SHARCGS (38), Edena (39), ALLPATHS (40), and VELVET (41).These assemblers can already build accurate contigs for bacterial genomes that are on the order of 10 kbp in length, but contig sizes produced by assembling complex eukaryotic genomes are significantly smaller (e.g., 2–3 kb produced by Velvet (41)).

Short read lengths can confound assembly programs, but the reduction in read length versus increased depth is an acceptable trade-off for many re-sequencing applications (25) such as mutation detection (42), transcript profiling (43), polymorphism discovery (21), and in vivo DNA binding site detection (44). The availability of a reference genome simply requires that each read is long enough, and accurately enough called, to align uniquely to the genome. In this instance, a high volume of short reads becomes very powerful in discriminating sequence variants, making these excellent systems for SNP discovery (21, 45, 46).

Recently Van Tassel (46) used the Illumina Genome Analyzer to identify over 62,000 SNPs from reduced representation genomic libraries (see below) representing 66 cattle. Due to its massive throughput, the Illumina system can be a powerful system for SNP discovery. However, the benefits of this throughput are best realized with a reference sequence in hand, since the very short lengths make it less than ideal for "de novo" sequencing. In contrast, assembling 454 sequence is less problematic, making it the high-throughput sequencing method of choice for species with few genomic resources (45). One method for SNP discovery in species with few genomic resources would utilize a hybrid sequencing approach. The Roche 454 sequencer could be used to generate transcriptome or genomic sequences that could be assembled and used as a reference sequence for subsequent alignment of Illumina or SOLiD short reads. This approach has been successfully applied to *Tragopogon* (47) and will be discussed later in this chapter.

Maize is highly polymorphic, and sequence comparison between genotypes provides a vast resource for SNPs. 454-based sequencing of the B73 and Mo17 maize shoot apical meristem (SAM) transcriptomes was expected to provide a large collection of diverse ESTs that could support high-throughput computational identification of gene-associated SNPs (48). This chapter concentrates on the development of an efficient computational SNP mining pipeline based on the POLYBAYES sequence polymorphism detection tool (16), and the subsequent identification of over 7,000 putative Mo17/B73 SNPs within over 280,000 Mo17 maize and 260,000 B73 maize ESTs (see Note 7) sequenced with 454 sequencing technology (21). The methods section details SNP discovery in 454 maize sequences, and discusses alternative alignment, polymorphism detection strategies, and tools. Since computational biology is highly experimental and context dependent, and there is often no single "one-size fits all" solution, the following discussions are meant to serve as guidelines, rather than explicit recipes.

2. Materials

2.1. Computational Tools

1. Polybayes 3.0 (16) polymorphism discovery tool: http://genome.wustl.edu/tools/software/polybayes.cgi.
2. Crossmatch (Phil Green, personal communication) genome sequence assembler: http://www.phrap.org/phredphrap-consed.html.
3. PYROBASE (28) 454 sequence call quality determination tool: http://bioinformatics.bc.edu/marthlab/PyroBayes.
4. Mosaik, next-generation sequence alignment tool: http://bioinformatics.bc.edu/marthlab/Mosaik.
5. GigaBayes (49) polymorphism discovery tool: http://bioinformatics.bc.edu/marthlab/GigaBayes.
6. CAP3 (26) DNA sequence assembler.
7. PaCE (50) DNA sequence assembler.
8. TIGR gene indices clustering tools (TGICL) EST (51): http://compbio.dfci.harvard.edu/tgi/software/.
9. EagleView (52) sequence alignment viewer: http://bioinformatics.bc.edu/marthlab/EagleView.
10. CONSED (53) sequence alignment viewer: http://bozeman.mbt.washington.edu/consed/consed.html.

3. Methods

3.1. Polymorphism Discovery in Maize Pyrosequences

Putative SNPs are identified as mismatches between multiply aligned allelic sequences. Several computational tools for SNP identification are available (16, 54–58). The computational methodology used to identify SNPs in 454 maize EST sequences (21) relied on POLYBAYES (16). POLYBAYES is a computational SNP discovery tool that uses a Bayesian statistical model that considers depth of coverage, sequence quality, and an a priori expected polymorphism rate to determine the probability that polymorphic sites within a multiple sequence alignment (MSA) are SNPs, rather than disagreements resulting from either sequencing errors or the alignment of paralogous (rather than allelic) sequences (16). In addition, POLYBAYES uses either genomic or EST sequence data as a template upon which all remaining sequences to be assessed are multiply aligned with CROSS_MATCH (P. Green, unpublished results) prior to scanning for polymorphisms. Template-based MSAs are often correct even in the presence of abundantly expressed or alternatively spliced transcripts (16), and therefore are more likely to overcome false polymorphism calls due to the higher sequence error rate associated with 454 sequence compared to Sanger sequenced reads.

454 ESTs were generated by sequencing SAM cDNA from the maize inbred lines B73 and Mo17 on the 454 Life Sciences GS-20 sequencing system. A flowchart illustrating POLYBAYES-mediated SNP mining in maize is presented in Fig. 1a.

1. The MSAs provided by CROSS_MATCH can be CPU intensive, so large 454 sequence sets are best preprocessed using BLAST to address each 454 sequence to a template; CROSS_MATCH is subsequently run on small sequence subsets consisting of the template and the aligning 454 ESTs (Fig. 1a). 454 ESTs can be assigned to maize genomic anchor sequences using BLAST by identifying the highest scoring alignment between each 454 EST and the collection of genomic sequence (1e-8 minimum E-value).

   High-quality assembled ESTs can be used as an anchor in place of genomic DNA (Fig. 1b) – the main requirement is that the anchor sequences should be of high quality since they drive the MSA (see Subheading 4 for a discussion of alternate approaches). Anchoring ESTs to reference sequences can be done using "best hit" criteria and poor alignments or alignments between paralogs will be caught either during formation of MSAs by CROSS_MATCH (see below) or by the internal paralog filter implemented within POLYBAYES. The maize genome currently has been sequenced (59) and would provide an excellent collection of anchor sequences. However, the maize genome sequencing project was less than 20% complete

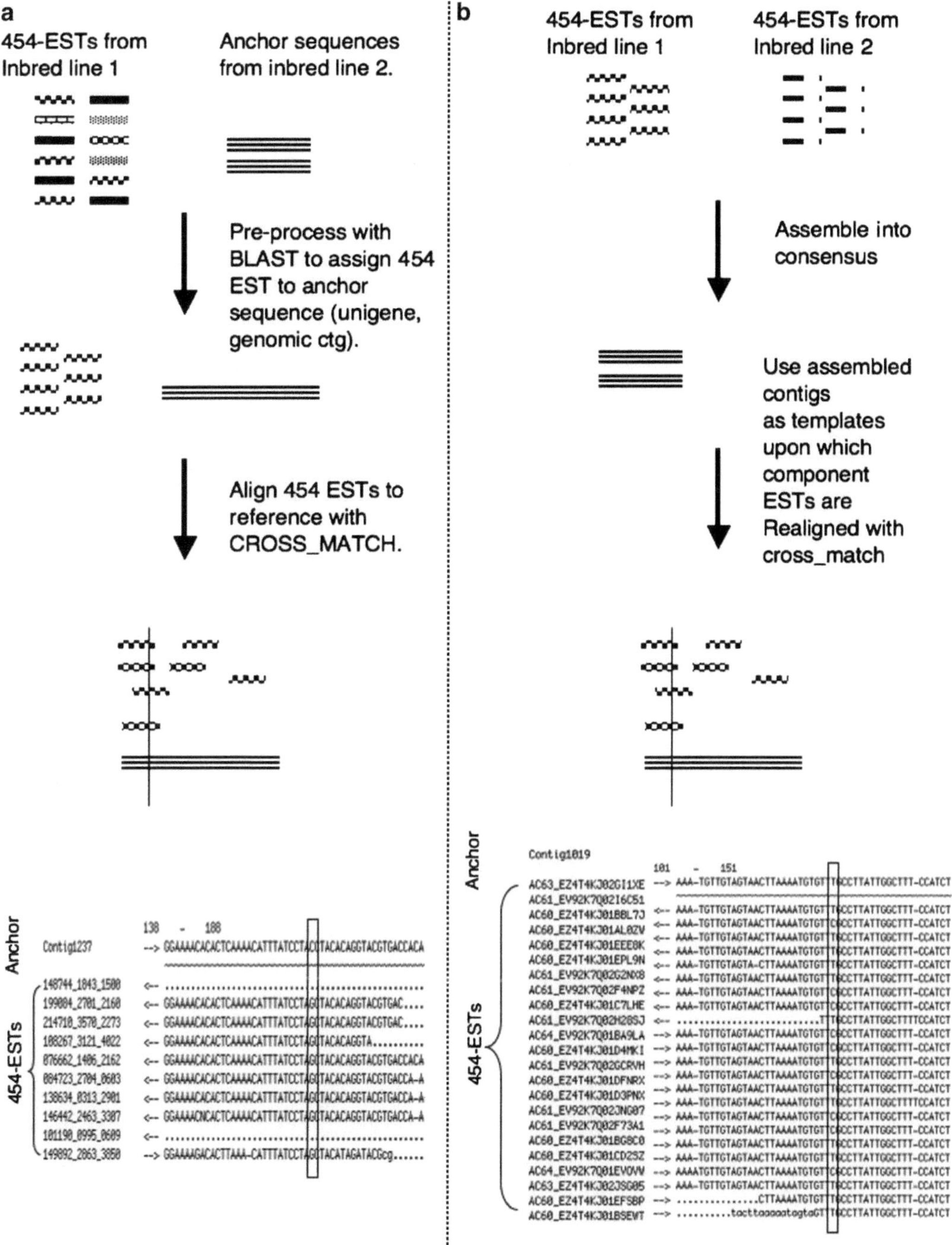

Fig. 1. A flowchart describing two methods for POLYBAYES-mediated SNP mining. When anchor sequences from one genotype are available either as high-quality genomic DNA or ESTs (**a**), 454 ESTs generated from a second genotype can be aligned directly to these anchors. The multiple sequence alignments (MSA) provided by CROSS_MATCH can be CPU intensive, so large 454 sequence sets are best preprocessed using BLAST to address each 454 sequence to a template; CROSS_MATCH is subsequently run on small sequence subsets consisting of the template and the aligning 454 ESTs. POLYBAYES evaluates the resulting MSA and calls SNPs. Sequence errors are not supported by a majority of the 454 sequences and are therefore easily identified in deep alignments. When anchor sequences are not available (**b**), 454 ESTs collected from two or more genotypes can be assembled to create consensus "anchors." The resulting consensus anchor sequences are then used as templates upon which cross match re-aligns the components of each consensus. POLYBAYES evaluates the resulting MSA and calls SNPs. This methodology is particularly well suited for SNP discovery within species lacking extensive genomic DNA resources.

when the maize SNP analysis presented in this chapter was conducted. Thus, a collection of high-quality (1 disagreement in 5,000 bp) maize genomic sequence assemblies (60) composed of B73 genomic survey sequences (GSSs) that are enriched for genes (61) was used. This collection is available from the MAGI website: http://magi.plantgenomics.iastate.edu/.

2. Run CROSS_MATCH on each anchor sequence and its associated 454 ESTs to create an anchored MSA. The following CROSS_MATCH parameters are recommended:

   --discrep_lists --tags --masklevel 5 --gap_init -1 --gap_ext -1.

   Low initiation (--gap_init) and gap extension (--gap_ext) are used to increase alignment tolerance between the short 454 ESTs and genomic anchors. Low values here become less critical as 454 read lengths increase. Substitute higher values for gap_init and gap_ext if the anchored MSA are unspliced (i.e., ESTs aligned to an EST anchor, or genomic sequence aligned to a genomic sequence anchor).

3. Run POLYBAYES on the MSA. Recommended POLYBAYES parameters for maize are:

   --maskAmbiguousMatches

   --nofilterParalogs

   --priorParalog 0.03

   --thresholdNative 0.75

   --screenSnps

   --considerAnchor

   --noconsiderTemplateConsensus

   --prescreenSnps

   --priorPoly 0.01

   --thresholdSnp 0.5

   The --considerAnchor argument considers the anchor base during SNP discovery. If the consensus anchor sequences have been created from several genotypes, or represent a single genotype that is not being considered during polymorphism mining, the --noconsiderAnchor argument should be used.

   It is necessary to include sequence quality files for the anchor sequence and the sequences aligned to it (member sequences). If these are unavailable or unreliable, set default quality values with:

   --anchorBaseQualityDefault (value)

   --memberBaseQualityDefault (value)

   Default quality values of 18 for each base within the 454 reads were used rather than relying on the quality scores assigned by the 454 sequencing machine. This corresponds to an error rate of ~1/65, which overcompensates for the error rate observed for current 454 sequencing (24, 48). As a result, sequence depth and relative allele proportions have the greatest influence on polymorphism detection and, based on this observation, potential SNPs were filtered by examining these statistics at each polymorphic site. (New tools that improve 454 quality score reliability are now available – see Subheading 4). POLYBAYES evaluates the resulting MSA and calls SNPs. Sequence errors such as insertions or substitutions that are not supported by a majority of sequences are easily identified in deep alignments. Likewise, true InDel polymorphisms are represented within a majority of the sequences and identified by POLYBAYES. However, this chapter is concerned only with SNP detection. The .ace file outputted by POLYBAYES allows for easy navigation and visualization of the alignments and called SNPs within the CONSED sequence assembly editor (53) (Fig. 2, see Note 2).

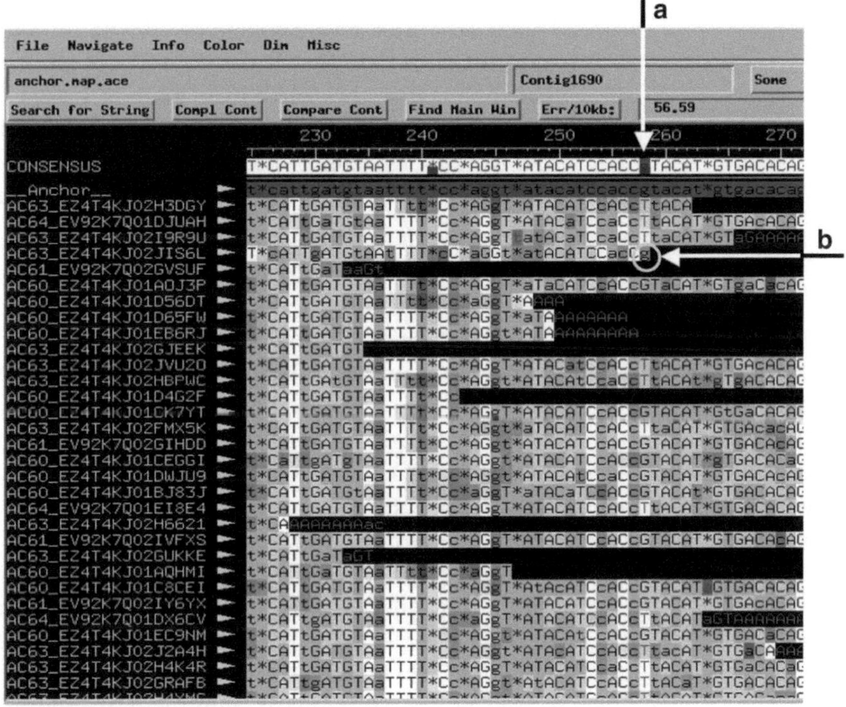

Fig. 2. The CONSED sequence assembly editor view of a *Brachypodium* 454 EST contig alignment and called SNPs. The .ace file outputted by POLYBAYES allows for easy manual navigation and visualization of the alignments, and aids in manually reviewing SNPs. PYROBASE corrected 454 base qualities were utilized by the Bayesian algorithm underlying POLYBAYES during SNP mining in *Brachypodium*. Utilization of individual base qualities enables "quality shading" in the CONSED display assisting manual review and confirmation of polymorphisms. EST bases within the multiple sequence alignment that are of high quality are displayed on *white* backgrounds, while progressively darker shades of *gray* indicate poorer quality. The .ace files provide SNP tags (dark blocks on the consensus sequence) at polymorphic positions (**a**) to aid in navigation. In this case, four different genotypes are represented in the alignment, distinguishable by the four-character prefix (AC60, AC61, AC63, or AC64) in each EST name. Genotypes AC60 and AC61 are represented by a G at this site, while AC63 and AC64 are represented by a T. One EST from AC63 has a G at this site (**b**); the dark shading of this base, however, indicates that it is of poor quality and is likely a miscall.

4. Postprocessing requires reading the POLYBAYES output files and deciding on appropriate rules to distinguish putative SNPs from false positives. In maize SNP mining experiments conducted by the authors, both Mo17 and B73 454 ESTs were available, and the B73 maize MAGI assemblies were used as alignment anchors (21). Because Mo17 and B73 are inbreds they should be monoallelic at every base position, with relatively rare exceptions caused by nearly identical paralogs (62). Hence, putative SNPs were filtered using the following rules designed to substantially decrease the rate of false positives within the context of this study:
   (a) Polymorphic sites require a minimum of 2× representation in the Mo17-454-ESTs.
   (b) All Mo17 base calls at sites that were polymorphic between Mo17-454-ESTs and the B73 MAGI anchors were expected to be identical. This ensures monoallelism within the Mo17-454-ESTs.
   (c) When B73-454-EST sequences also align across polymorphic sites that pass Rules 4a and 4b, all of the B73-454-ESTs and the MAGI3.1 anchor base calls must agree. This avoids polymorphisms resulting from incorrect MAGI base calls or the existence of duplicated genes (e.g., Nearly Identical Paralogs, NIPs (50)) within B73.

The above strategy identified over 7,000 potential SNPs between Mo17 and B73. Based on experimental validation, this collection contains at least 6,100 valid SNPs within 3,100 genes (21) (see Note 1). SNPs between B73 and Mo17 can be rapidly converted to accurate mapping markers using Sequenom technology (63).

3.2. Hybrid Sequencing Approaches for Use in Species with New Genomic Resources

Tragopogon (*Asteraceae*) is a natural allopolyploid evolutionary model system for which only limited genomic resources were available. A hybrid sequencing approach was designed to assay gene expression and identify potential gene losses within the allotetraploid *Tragopogon miscellus* versus its parental diploid species *Tragopogon dubius* and *Tragopogon pratensus* (47). This approach used the Roche 454 sequencer to generate transcriptome sequences from *T. dubius* which were assembled and used as reference sequences for subsequent alignment of Illumina short reads obtained from *T. miscellus*, *T. dubius*, and *T. pratensus* transcriptomes. This method gains maximum leverage from the longer read lengths of 454 sequencing for gene discovery and annotation, and the deeper coverage of Illumina reads for evaluation of gene expression and SNP discovery. 195 MB of 454 transcriptome sequences obtained from a normalized cDNA pool from *T. dubius* leaf tissue were assembled into 33,515 contigs (14.7 Mb) with the Roche 454 Newbler assembler (47). Non-normalized cDNA pools were obtained from *T. dubius*, *T. pratensis*, and

T. miscellus, and a single channel of Illumina GAII analyzer sequence was obtained for each, resulting in over 6.5 M 36 bp-reads from each sample. The next-generation sequence aligner Moasik (http://bioinformatics.bc.edu/marthlab/Mosaik) was used to align the pooled Illumina reads to the *T. dubius* 454 assembled EST reference sequences, and SNPs were discovered with the GigaBayes polymorphism detection system (49). Aligning with a mismatch tolerance of 2 bp followed by identification of polymorphic sites that were represented to a minimum of threefold redundancy in both *T. dubius* and *T. pratensis* revealed >45,000 potential SNPs within 10,428 contigs. Reducing the mismatch tolerance to only one within an Illumina read resulted in the identification of 24,078 potential SNPs between *T. dubius* and *T. pratensis* within 7,837 unique 454 EST contig reference sequences (47).

1. Use MOSAIK aligner to align Illumina reads to the 454 reference contigs: Illumina reads from the *T. dubius* and *T. pratensis* parents and the *T. miscellus* allotetraploid were labeled with species identifiers, pooled, and aligned to the *T. dubius* 454 FLX contigs with the MosaikAligner package (49) using the following MosaikAligner parameters: -a (alignment algorithm) all; -p (CPUs used) 8; -mm (maximum mismatch) 2 in a preliminary analysis and 1 in a final analysis; -m (alignment mode) unique; -hs (hash size) 15; and -mhp (maximum number of hash positions to use) 100. These alignment parameters ensured that each Illumina sequence aligned to a unique position within the 454 *T. dubius* EST assembly reference sequences and with no more than one base-pair mismatch in the final analysis. Illumina reads that did not align with the 454 contigs under these stringent conditions were discarded from the analysis (47).
2. SNP identification with the GigaBayes SNP detection package:
   SNPs were identified within the alignments with the GigaBayes package (49) (http://bioinformatics.bc.edu/marthlab/GigaBayes). GigaBayes is a reimplementation of the PolyBayes (16) SNP discovery tool that has been optimized for next-generation sequences. Arguments to GigaBayes were --D (pairwise nucleotide diversity) 0.001; --ploidy (sample ploidy) diploid; algorithm banded --sample (sequence source) multiple;--anchor; --CAL (minimum overall allele coverage) 3; --QRL (minimum base quality value) 20. GigaBayes output files are in GFF format and contain the site identification of each SNP, its representation within each of the three *Tragopogon* species (*T. dubius*, *T. pratensis*, and *T. miscellus*), and its allele usage (47).

3.3. Alternative Reference Assembly Strategies

The benefits of high-throughput pyrosequencing are most apparent for species with few genomic resources. Novaes et al. (45) demonstrated this in *Eucalyptus grandis* from which 148 Mb of 454 ESTs were acquired from a normalized cDNA pool of multiple tissues and genotypes. Their SNP discovery strategy was based on assembly of the 454 sequences to produce a contig consensus sequence. Each contig consensus sequence was potentially derived from sequences originating from two or more genotypes. The contig consensus sequence acts as an anchor sequence upon which those reads that went into reference sequence construction are re-aligned, in this particular case with the Roche-454s GS Reference Mapper software (Fig. 1b). SNPs are identified within the re-aligned collection. Based on simple rules that required at least 10% of the reads to cover the polymorphic locus and the consensus and variant alleles needed to be confirmed within at least two reads each, 23,742 SNPs were identified and validated at a rate of 83%.

Of particular interest is the assembly strategy. Adapters were added to the cDNA during the normalization protocol (45). Because a significant number of 454 sequence reads contained adapter sequences, and the Roche-454 Newbler assembler does not mask sequence repeats, a two-stage assembly approach was taken. The Newbler assembler considers the normalized base incorporation signal instead of individual base calls and is thus much better suited for assembling 454 pyrosequencing data (24). First, all reads without adapters were assembled with the Roche-454 Newbler assembler, and then the resulting consensus and 454 sequence reads containing adapter sequences were assembled with the Paracell transcript Assembler (45).

454 sequence collections that do not contain adapters can be assembled solely with the Roche-454 Newbler assembler. Outputs of the assembly process are consensus sequences, base probabilities of the consensus, and an .ace file detailing the position and orientation of the component sequences that make up the consensus. The .ace file can be parsed to identify each group of component sequences that assemble each consensus, and this relationship can be used to drive the multiple alignment for SNP detection. Chum Salmon 454 ESTs were assembled this way, and the resulting .ace file parsed to address each 454 sequence to a template (Fig. 1b; Barbazuk unpublished). These small bins consisting of a template and the aligning 454 ESTs easily feed into the POLYBAYES based pipeline used for SNP discovery in maize (21).

Similarly, SNP detection in cDNA sequences obtained from four inbred lines of *Brachypodium distachyon* was performed with POLYBAYES by detecting polymorphisms identified in multiple alignments of component 454 reads to consensus sequences (64) produced by CAP3 (26). The ~200 bp read lengths produced by the 454-FLX instrument are adequate for more traditional and well-established sequence assemblers (such as PaCE (65, 66)/ CAP3 (26)) to be used effectively. Related sequences were first pre-clustered using PaCE (65, 66). Individual clusters were then assembled using CAP3. Use of CAP3 to assemble short 454 sequences requires one to establish the percent identity of overlap (-*p*) stringently enough to ensure the fidelity of the overlap, but low enough that sequence errors do not compromise the assembly, or if assembling sequences from multiple genotypes, natural polymorphisms do not prevent construction of reference sequences (see Note 3). If assembling sequence from a single genotype, values of -*p* 98 will promote high fidelity assembly. If assembling sequences from multiple genotypes, -*p* values from 90 to 95 should be used depending on the degree of polymorphism expected. In addition, values for minimum allowed overlap (-0) should be chosen to maximize alignment of short reads. Values of 60 or less are recommended, but it is best to select several values and run multiple assemblies to assess performance. Values of 60 for minimum overlap and of 90 for alignment identity were chosen for the *Brachypodium* EST assembly.

As mentioned previously, the ability to bin sequence prior to consensus assembly will greatly reduce the complexity of the assembly. There are several packages designed to accomplish this. PaCE is a parallel assembler that will take full advantage of multiple compute nodes during the binning and assembly process (65, 66). This "divide and conquer" approach is also efficiently performed by the TIGR EST clustering tools (TGICL) (51) that are publicly available, can be adapted to 454 sequence collections, and may be better suited for investigators with access to only a single (or few) computer workstation.

New base callers, sequence alignment, and polymorphism identification tools that are designed specifically for next-generation sequences are being developed (see Notes 4–6). One such tool, PYROBAYES (28), is an improved 454 sequence base caller that has been demonstrated to reflect the true base quality more accurately. 454 pyrosequences are prone to over- and under-calls resulting in addition or deletion of bases, respectively. PYROBASE is a novel base caller that significantly improves the accuracy of the base calls (see Note 4). The re-calling procedure results in a 60% increase in the ability to distinguish true genetic polymorphisms from 454 substitution errors, thus allowing 454 sequences to be used reliably in applications such as polymorphism detection without requiring deep sequence coverage. These improved base qualities have enabled the identification of tens of thousands of high-quality SNPs in low-density 454 *Drosophila* genome sequence (28). PYROBASE outputs base quality scores as Phred files which are easily utilized by POLYBAYES. This strategy was used in the Brachypodium and chum salmon analyses discussed previously. Figure 2 illustrates the CONSED sequence assembly editor view of a Brachypodium 454 EST contig alignment and called SNPs. PYROBASE corrected 454 base qualities were utilized by the Bayesian algorithm underlying POLYBAYES during SNP detection, which increases reliability. Utilization of individual base qualities enables "quality shading" in the CONSED display assisting manual review and confirmation of polymorphism. The .ace file outputted by POLYBAYES allows for easy navigation and visualization of the alignments and called SNPs within the CONSED sequence assembly editor (53).

4. Notes

1. The maize 454 SNP discovery protocol described above required that both B73 and Mo17 inbreds be monoallelic at each nucleotide before accepting a given site as polymorphic. This is quite restrictive and has a high false negative rate. Consider a polymorphic site that is well sampled and monoallelic in Mo17 and represented by an additional ten B73 sequences, nine of which are concordant (but different than the Mo17 allele) and one which is not. This site would have been filtered out by these rules, although the one disconcordant B73 call is likely a sequence error. Therefore, further parsing of the SNP output files can be used to more accurately represent quality.
2. .ace file output can be activated/suppressed with the --(no) writeAce flag. The .ace file is most useful if individual quality values are used, thus enabling "quality shading" during assembly review in CONSED (Fig. 2). To enable quality shading, use the --writePhdFiles flag which will ensure that POLYBAYES creates quality files for viewing and specify the output path to these with the --phdFilePathOut <dir> flag.
3. When calling SNPs in alignments created by re-aligning sequences from several genotypes against a consensus reference sequence created by assembling these same sequences with Newbler, CAP3, PHRAP, or another assembler, it is best not to consider the consensus base during SNP discovery. This behavior is modified within POLYBAYES with the --(no) considerAnchor flag.
4. PYROBAYES takes the original SFF files outputted by the Roche 454 sequencer, and produces a .fasta file of sequences and a .quality file of associated quality scores.
5. MOSAIK, a new rapid sequence alignment tool for next-generation sequences is under development and is available for beta testing. This will align Illumina short reads or 454 reads to a reference, and produces .ace files. Similarly a short-read

optimized version of POLYBAYES is under development. Beta releases of these tools can be obtained at: http://bioinformatics.bc.edu/marthlab/Beta_Release. Descriptions of parameters for alignment and SNP detection are presented in steps 1 and 2 of Subheading 3.1 and Note 7.5 (below).

6. At the time of this writing, new releases of the Phrap/cross-match assembler (1.080721) and CONSED (v. 17) are available that provide support for next-generation sequencing reads.

7. It is sometimes advantageous to develop markers from collections of genomic sequence rather than transcriptome sequence. It is difficult to assemble and analyze repetitive DNA sequences that are often abundant in genomic sequence and they are poor reagents for developing uniquely mapping markers. For this reason, several approaches have been used to avoid their recovery while preferentially sampling unique or low-copy (genic) sequences.

   7.1 EST Sequencing
   Random cDNA clones can be sequenced to rapidly obtain expressed sequence information, regardless of the availability of genomic sequence information. EST data provides not only useful coding sequences, but also gene expression information. SNPs identified in ESTs make valuable map markers since each represents a transcribed sequence. However, EST data shows uneven representation of genes due to expression biases, resulting in an incomplete collection of gene sequences, regardless of the size of the data set. In addition, polymorphism-rich introns and untranscribed flanking regions are generally absent from ESTs.

   7.2 Methylation Filtration
   In general, plant repetitive DNA is heavily methylated at 5′-CG-3′ and 5′-CNG-3′ sequences while genes and other low-copy DNA elements are not (67–69). The differential methylation that exists in plant genomes can be exploited to enable preferential recovery of gene-rich DNA sequences.

   Methylation filtration uses the endogenous restriction-modification system of *Escherichia coli* McrBC to eliminate random genomic clones containing methylated (i.e., repetitive) DNA inserts (69), which results in libraries highly enriched for low-copy DNA fragments (69, 70). Methylation filtration has been successfully applied to maize (61) and sorghum (70), but the requirement of clone libraries makes it unattractive in combination with 454 sequencing. While cloned inserts could be removed prior to sequencing, this requires massive clone plating and pooling to ensure comprehensive representation of the unique fraction of the genome. An alternate approach could involve *in vitro* digestion of genomic DNA by McrBC followed by size selection. Lippman et al. (68)

used McrBC to isolate the methylation-depleted DNA fraction within *Arabidopsis* genomic DNA. While Lippman et al. (68) applied this to an *Arabidopsis* tilling array to map un-methylated regions of the *Arabidopsis* genome, this double-stranded DNA could be sequenced directly by 454 and should provide genome reduction and low-copy DNA representation similar to methylation filtration.

7.3 Methyl Sensitive Enzymes

Methylation-sensitive enzymes provide a simple, effective, and highly reproducible alternative to using methylation filtration to enrich for low-copy genomic DNA. This method has been documented in human (71) and has been demonstrated to facilitate the isolation of low-copy-number sequence (72). In order to increase the randomness of this method, partial restriction with multiple methylation-sensitive restriction enzymes (generally frequent cutters) may be used. This concept has been tested in maize and was called hypomethylated partial restriction (HMPR). Pilot HMPR studies in maize showed six- to sevenfold enrichment in genes relative to a library made with nonmethylation-sensitive restriction enzymes (73).

7.4 High Cot Selection

Another method, high Cot selection, relies on a DNA renaturation technology that was originally utilized to characterize the repetitive and unique DNA components of complex genomes (74).This technique uses DNA renaturation to examine the fraction of low-copy DNA within a genome. In a Cot experiment, the fully denatured genome will exhibit earliest renaturation by the most repetitive components and slowest renaturation by the single copy components. HC sequencing works by cloning the single copy components (74–76). Although the products of high Cot selection can be sequenced directly (no need for clone libraries), the multistep, high Cot selection process is difficult to control, and minor variations in annealing time and temperature can have profound effects on reproducibility (77).

7.5 Microarray-Based Sequence Capture

Microarray-based sequence capture has been effective for re-sequencing exons, large genomic loci, and candidate gene sets in mammalian genomes (78–81). The substantial enrichment of target sequences achieved via sequence capture makes it much less expensive than re-sequencing whole genomes or even the entire gene space. For SNP discovery, where it is possible to obtain large numbers of "random" SNPs via high-throughput sequencing of genomic fractions as it has been done for maize (21), typically large numbers of SNPs are not discovered within a set of specific genes or a defined genomic region. Microarray-based sequence capture enables targeted re-sequencing of specific intervals of genome. Essentially, probes are designed to a genomic region of interest based on available reference sequence, these are arrayed on a microarray and used to "capture" equivalent regions within closely related species or from individuals within a species. Captured sequence is "released" from the microarray and sequenced. Fu et al. (82) have recently demonstrated microarray-based sequence capture in maize. Probes were designed to a 2.2-Mb interval within the inbred B73 maize genome and used to capture sequences from inbred Mo17 maize, which were sequenced with 454. Over 1,600 SNPs were identified after aligning the captured Mo17 sequences to the maize B73 genome (82). Alignments were performed with Mosaik aligner (http://bioinformatics.bc.edu/marthlab/Mosaik) using parameters that accommodate the longer 454 reads, increased sequence error associated with 454 reads, and the high level of polymorphisms between maize inbreds; SNPs were detected with GigaBayes (49).

Parameters for MosaikAligner were: -a (alignment algorithm) all; -p (CPUs used) 8; -mmp (maximum percentage of read length to be mismatched) 0.05; –minp (specifies the minimum percentage of the read length aligned) 0.95; –mmal (uses the aligned read length rather than the original read length when counting errors); -m (alignment mode) unique; -hs (hash size) 15; -mhp (maximum number of hash positions to use) 100. These alignment parameters ensured that each 454 sequence read was uniquely aligned; sequences that failed to meet these criteria were discarded from the analysis (82).

SNPs were identified within the alignments with the GigaBayes package (49) (http://bioinformatics.bc.edu/marthlab/GigaBayes). Arguments to GigaBayes were: --D (pairwise nucleotide diversity) 0.003; --ploidy (sample ploidy) haploid; --algorithm recursive; --sample (sequence source) single; --anchor; --CAL (minimum overall allele coverage) 3; --QRL (minimum base quality value) 20. Potential SNP sites were required to be covered by a minimum of 3 Mo17 reads, and all Mo17 base calls at the polymorphic site were expected to be identical (82).

Acknowledgments

We thank Sanzhen Liu (Iowa State University), Yi Jia (Iowa State University), and Cheng-Ting Eddy Yeh (Iowa State University) for comments on the manuscript; Drs. Haiyan Wu (Iowa State University and China Agriculture University), Ananth Kalyanaraman (Washington State University), Wei Wu (Iowa State University), and An-Ping Hsia (Iowa State University) for sharing *Brachypodium* 454 EST data prior to publication; Drs. Richard Buggs (University of Florida), Doug Soltis (University of Florida), and Pam Soltis (University of Florida) for sharing *Tragopogon* data prior to publication; Dr. Nathan Springer (University of Minnesota) and Dr. Jeff Jeddeloh (Roche NimbleGen Inc.) for sharing maize sequence capture data prior to publication; Dr. Scott Emrich (University of Notre Dame) for stimulating discussions; and Marianne Smith (Iowa State University) and Lisa Coffey (Iowa State University) for technical assistance. This project was supported by competitive grants from the National Science Foundation Plant Genome Program to P.S.S. (DBI-0321711, DBI-0321595, and DBI-0919254) and W.B.B. (DBI-0501758 and DBI-0919254), by the National Research Initiative (NRI) Plant Genome Program of the USDA Cooperative State Research, Education and Extension Service (CSREES) to W.B.B., and by Hatch Act and State of Iowa funds to P.S.S.

References

1. Gut, I. G. (2001) Automation in genotyping of single nucleotide polymorphisms. *Hum. Mutat.* **17**, 475–492.
2. Kwok, P. Y. (2001) Methods for genotyping single nucleotide polymorphisms. *Annu. Rev. Genomics Hum. Genet.* **2**, 235–258.
3. Leushner, J. and Chiu, N. H. (2000) Automated mass spectrometry: a revolutionary technology for clinical diagnostics. *Mol. Diagn.* **5**, 341–348.
4. Consortium, T. I. H. (2003) The International HapMap Project. *Nature* **426**, 789–796.
5. Consortium, T. I. H. (2005) A haplotype map of the human genome. *Nature* **437**, 1299–1320.
6. Bray, N. J., Buckland, P. R., Owen, M. J., and O'Donovan, M. C. (2003) Cis-acting variation in the expression of a high proportion of genes in human brain. *Hum. Genet.* **113**, 149–153.
7. Cowles, C. R., Hirschhorn, J. N., Altshuler, D., and Lander, E. S. (2002) Detection of regulatory variation in mouse genes. *Nat. Genet.* **32**, 432–437.
8. Guo, M., Rupe, M. A., Zinselmeier, C., Habben, J., Bowen, B. A., and Smith, O. S. (2004) Allelic variation of gene expression in maize hybrids. *Plant Cell* **16**, 1707–1716.
9. Pastinen, T., Sladek, R., Gurd, S., Sammak, A., Ge, B., Lepage, P., Lavergne, K., Villeneuve, A., Gaudin, T., Brandstrom, H., Beck, A., Verner, A., Kingsley, J., Harmsen, E., Labuda, D., Morgan, K., Vohl, M. C., Naumova, A. K., Sinnett, D., and Hudson, T. J. (2004) A survey of genetic and epigenetic variation affecting human gene expression. *Physiol. Genomics* **16**, 184–193.
10. Stupar, R. M. and Springer, N. M. (2006) Cis-transcriptional variation in maize inbred lines B73 and Mo17 leads to additive expression patterns in the F1 hybrid. *Genetics* **173**, 2199–2210.
11. Sachidanandam, R., Weissman, D., Schmidt, S. C., et al. (2001) A map of human genome sequence variation containing 1.42 million single nucleotide polymorphisms. *Nature* **409**, 928–933.
12. Wiltshire, T., Pletcher, M. T., Batalov, S., Barnes, S. W., Tarantino, L. M., Cooke, M. P., Wu, H., Smylie, K., Santrosyan, A., Copeland, N. G., Jenkins, N. A., Kalush, F., Mural, R. J., Glynne, R. J., Kay, S. A., Adams, M. D., and Fletcher, C. F. (2003) Genome-wide single-nucleotide polymorphism analysis defines haplotype patterns in mouse. *Proc. Natl Acad. Sci. USA* **100**, 3380–3385.
13. Feltus, F. A., Wan, J., Schulze, S. R., Estill, J. C., Jiang, N., and Paterson, A. H. (2004) An SNP resource for rice genetics and breeding based on subspecies indica and japonica genome alignments. *Genome Res.* **14**, 1812–1819.
14. Jander, G., Norris, S. R., Rounsley, S. D., Bush, D. F., Levin, I. M., and Last, R. L.

(2002) *Arabidopsis* map-based cloning in the post-genome era. *Plant Physiol.* **129**, 440–450.

15. Yamasaki, M., Tenaillon, M. I., Bi, I. V., Schroeder, S. G., Sanchez-Villeda, H., Doebley, J. F., Gaut, B. S., and McMullen, M. D. (2005) A large-scale screen for artificial selection in maize identifies candidate agronomic loci for domestication and crop improvement. *Plant Cell* **17**, 2859–2872.
16. Marth, G. T., Korf, I., Yandell, M. D., Yeh, R. T., Gu, Z., Zakeri, H., Stitziel, N. O., Hillier, L., Kwok, P. Y., and Gish, W. R. (1999) A general approach to single-nucleotide polymorphism discovery. *Nat. Genet.* **23**, 452–456.
17. Dantec, L. L., Chagne, D., Pot, D., Cantin, O., Garnier-Gere, P., Bedon, F., Frigerio, J. M., Chaumeil, P., Leger, P., Garcia, V., Laigret, F., De Daruvar, A., and Plomion, C. (2004) Automated SNP detection in expressed sequence tags: statistical considerations and application to maritime pine sequences. *Plant Mol. Biol.* **54**, 461–470.
18. Kota, R., Rudd, S., Facius, A., Kolesov, G., Thiel, T., Zhang, H., Stein, N., Mayer, K., and Graner, A. (2003) Snipping polymorphisms from large EST collections in barley (Hordeum vulgare L.). *Mol. Genet. Genomics* **270**, 24–33.
19. Kota, R., Varshney, R. K., Thiel, T., Dehmer, K. J., and Graner, A. (2001) Generation and comparison of EST-derived SSRs and SNPs in barley (Hordeum vulgare L.). *Hereditas* **135**, 145–151.
20. Lopez, C., Piegu, B., Cooke, R., Delseny, M., Tohme, J., and Verdier, V. (2005) Using cDNA and genomic sequences as tools to develop SNP strategies in cassava (Manihot esculenta Crantz). *Theor. Appl. Genet.* **110**, 425–431.
21. Barbazuk, W. B., Emrich, S. J., Chen, H. D., Li, L., and Schnable, P. S. (2007) SNP discovery via 454 transcriptome sequencing. *Plant J.* **51**, 910–918.
22. Batley, J., Barker, G., O'Sullivan, H., Edwards, K. J., and Edwards, D. (2003) Mining for single nucleotide polymorphisms and insertions/deletions in maize expressed sequence tag data. *Plant Physiol.* **132**, 84–91.
23. Holt, R. A. and Jones, S. J. (2008) The new paradigm of flow cell sequencing. *Genome Res.* **18**, 839–846.
24. Margulies, M., Egholm, M., Altman, W. E., et al. (2005) Genome sequencing in microfabricated high-density picolitre reactors. *Nature* **437**, 376–380.
25. Bentley, D. R. (2006) Whole-genome re-sequencing, *Curr. Opin. Genet. Dev.* **16**, 545–552.
26. Huang, X. and Madan, A. (1999) CAP3: a DNA sequence assembly program. *Genome Res.* **9**, 868–877.
27. Sundquist, A., Ronaghi, M., Tang, H., Pevzner, P., and Batzoglou, S. (2007) Whole-genome sequencing and assembly with high-throughput, short-read technologies. *PLoS One* **2**, e484.
28. Quinlan, A. R., Stewart, D. A., Stromberg, M. P., and Marth, G. T. (2008) Pyrobayes: an improved base caller for SNP discovery in pyrosequences. *Nat. Methods* **5**, 179–181.
29. Huse, S. M., Huber, J. A., Morrison, H. G., Sogin, M. L., and Welch, D. M. (2007) Accuracy and quality of massively parallel DNA pyrosequencing. *Genome Biol.* **8**, R143.
30. Ewing, B., Hillier, L., Wendl, M. C., and Green, P. (1998) Base-calling of automated sequencer traces using phred. I. Accuracy assessment. *Genome Res.* **8**, 175–185.
31. Goldberg, S. M., Johnson, J., Busam, D., et al. (2006) A Sanger/pyrosequencing hybrid approach for the generation of high-quality draft assemblies of marine microbial genomes. *Proc. Natl Acad. Sci. USA* **103**, 11240–11245.
32. Hiller, N. L., Janto, B., Hogg, J. S., Boissy, R., Yu, S., Powell, E., Keefe, R., Ehrlich, N. E., Shen, K., Hayes, J., Barbadora, K., Klimke, W., Dernovoy, D., Tatusova, T., Parkhill, J., Bentley, S. D., Post, J. C., Ehrlich, G. D., and Hu, F. Z. (2007) Comparative genomic analyses of seventeen *Streptococcus pneumoniae* strains: insights into the pneumococcal supragenome. *J. Bacteriol.* **189**, 8186–8195.
33. Hofreuter, D., Tsai, J., Watson, R. O., Novik, V., Altman, B., Benitez, M., Clark, C., Perbost, C., Jarvie, T., Du, L., and Galan, J. E. (2006) Unique features of a highly pathogenic *Campylobacter jejuni* strain. *Infect. Immun.* **74**, 4694–4707.
34. Pearson, B. M., Gaskin, D. J., Segers, R. P., Wells, J. M., Nuijten, P. J., and van Vliet, A. H. (2007) The complete genome sequence of *Campylobacter jejuni* strain 81116 (NCTC11828). *J. Bacteriol.* **189**, 8402–8403.
35. Smith, M. G., Gianoulis, T. A., Pukatzki, S., Mekalanos, J. J., Ornston, L. N., Gerstein, M., and Snyder, M. (2007) New insights into *Acinetobacter baumannii* pathogenesis revealed by high-density pyrosequencing and transposon mutagenesis. *Genes Dev.* **21**, 601–614.

36. Warren, R. L., Sutton, G. G., Jones, S. J., and Holt, R. A. (2007) Assembling millions of short DNA sequences using SSAKE. *Bioinformatics* **23**, 500–501.
37. Jeck, W. R., Reinhardt, J. A., Baltrus, D. A., Hickenbotham, M. T., Magrini, V., Mardis, E. R., Dangl, J. L., and Jones, C. D. (2007) Extending assembly of short DNA sequences to handle error. *Bioinformatics* **23**, 2942–2944.
38. Dohm, J. C., Lottaz, C., Borodina, T., and Himmelbauer, H. (2007) SHARCGS, a fast and highly accurate short-read assembly algorithm for de novo genomic sequencing. *Genome Res.* **17**, 1697–1706.
39. Henderson, I. R., Zhang, X., Lu, C., Johnson, L., Meyers, B. C., Green, P. J., and Jacobsen, S. E. (2006) Dissecting *Arabidopsis thaliana* DICER function in small RNA processing, gene silencing and DNA methylation patterning. *Nat. Genet.* **38**, 721–725.
40. Butler, J., MacCallum, I., Kleber, M., Shlyakhter, I. A., Belmonte, M. K., Lander, E. S., Nusbaum, C., and Jaffe, D. B. (2008) ALLPATHS: de novo assembly of whole-genome shotgun microreads. *Genome Res.* **18**, 810–820.
41. Zerbino, D. R. and Birney, E. (2008) Velvet: algorithms for de novo short read assembly using de Bruijn graphs. *Genome Res.* **18**, 821–829.
42. Andries, K., Verhasselt, P., Guillemont, J., Gohlmann, H. W., Neefs, J. M., Winkler, H., Van Gestel, J., Timmerman, P., Zhu, M., Lee, E., Williams, P., de Chaffoy, D., Huitric, E., Hoffner, S., Cambau, E., Truffot-Pernot, C., Lounis, N., and Jarlier, V. (2005) A diarylquinoline drug active on the ATP synthase of *Mycobacterium tuberculosis. Science* **307**, 223–227.
43. Eveland, A. L., McCarty, D. R., and Koch, K. E. (2008) Transcript profiling by 3 -untranslated region sequencing resolves expression of gene families. *Plant Physiol.* **146**, 32–44.
44. Johnson, D. S., Mortazavi, A., Myers, R. M., and Wold, B. (2007) Genome-wide mapping of in vivo protein-DNA interactions. *Science* **316**, 1497–1502.
45. Novaes, E., Drost, D. R., Farmerie, W. G., Pappas, G. J., Jr., Grattapaglia, D., Sederoff, R. R., and Kirst, M. (2008) High-throughput gene and SNP discovery in *Eucalyptus grandis*, an uncharacterized genome. *BMC Genomics* **9**, 312.
46. Van Tassell, C. P., Smith, T. P., Matukumalli, L. K., Taylor, J. F., Schnabel, R. D., Lawley, C. T., Haudenschild, C. D., Moore, S. S., Warren, W. C., and Sonstegard, T. S. (2008) SNP discovery and allele frequency estimation by deep sequencing of reduced representation libraries. *Nat. Methods* **5**, 247–252.
47. Buggs, R. J. A., Chamala, S., Wu, W., Gao, L., May, G. D., Schnable, P. S., Soltis, D. E., Soltis, P. S., and Barbazuk, W. B. (2010) Characterization of duplicate gene evolution in the recent natural allopolyploid *Tragopogon miscellus* by next-generation sequencing and Sequenom iPLEX MassARRAY genotyping. *Mol. Ecol.* **19**(S1), 132–146.
48. Emrich, S. J., Barbazuk, W. B., Li, L., and Schnable, P. S. (2007) Gene discovery and annotation using LCM-454 transcriptome sequencing. *Genome Res.* **17**, 69–73.
49. Hillier, L. W., Marth, G. T., Quinlan, A. R., Dooling, D., Fewell, G., Barnett, D., Fox, P., Glasscock, J. I., Hickenbotham, M., Huang, W., Magrini, V. J., Richt, R. J., Sander, S. N., Stewart, D. A., Stromberg, M., Tsung, E. F., Wylie, T., Schedl, T., Wilson, R. K., and Mardis, E. R. (2008) Whole-genome sequencing and variant discovery in *C. elegans. Nat. Methods* **5**, 183–188.
50. Emrich, S. J., Aluru, S., Fu, Y., Wen, T. J., Narayanan, M., Guo, L., Ashlock, D. A., and Schnable, P. S. (2004) A strategy for assembling the maize (Zea mays L.) genome. *Bioinformatics* **20**, 140–147.
51. Pertea, G., Huang, X., Liang, F., Antonescu, V., Sultana, R., Karamycheva, S., Lee, Y., White, J., Cheung, F., Parvizi, B., Tsai, J., and Quackenbush, J. (2003) TIGR Gene Indices clustering tools (TGICL): a software system for fast clustering of large EST datasets. *Bioinformatics* **19**, 651–652.
52. Huang, W. and Marth, G. (2008) EagleView: a genome assembly viewer for next-generation sequencing technologies. *Genome Res.* **18**, 1538–1543
53. Gordon, D., Abajian, C., and Green, P. (1998) Consed: a graphical tool for sequence finishing. *Genome Res.* **8**, 195–202.
54. Manaster, C., Zheng, W., Teuber, M., Wachter, S., Doring, F., Schreiber, S., and Hampe, J. (2005) InSNP: a tool for automated detection and visualization of SNPs and InDels. *Hum. Mutat.* **26**, 11–19.
55. Nickerson, D. A., Tobe, V. O., and Taylor, S. L. (1997) PolyPhred: automating the detection and genotyping of single nucleotide substitutions using fluorescence-based resequencing. *Nucleic Acids Res.* **25**, 2745–2751.
56. Wang, J. and Huang, X. (2005) A method for finding single-nucleotide polymorphisms with allele frequencies in sequences of deep coverage. *BMC Bioinform.* **6**, 220.

57. Weckx, S., Del-Favero, J., Rademakers, R., Claes, L., Cruts, M., De Jonghe, P., Van Broeckhoven, C., and De Rijk, P. (2005) novoSNP, a novel computational tool for sequence variation discovery. *Genome Res.* **15**, 436–442.

58. Zhang, J., Wheeler, D. A., Yakub, I., Wei, S., Sood, R., Rowe, W., Liu, P. P., Gibbs, R. A., and Buetow, K. H. (2005) SNP detector: A software tool for sensitive and accurate SNP detection. *PLoS Comput. Biol.* **1**, e53.

59. Schnable, P. S., Ware, D., Fulton, R. S., et al. (2009) The B73 maize genome: complexity, diversity, and dynamics. *Science* **326**, 1112–1115.

60. Fu, Y., Emrich, S. J., Guo, L., Wen, T. J., Ashlock, D. A., Aluru, S., and Schnable, P. S. (2005) Quality assessment of maize assembled genomic islands (MAGIs) and large-scale experimental verification of predicted genes. *Proc. Natl Acad. Sci. USA* **102**, 12282–12287.

61. Whitelaw, C. A., Barbazuk, W. B., Pertea, G., Chan, A. P., Cheung, F., Lee, Y., Zheng, L., van Heeringen, S., Karamycheva, S., Bennetzen, J. L., SanMiguel, P., Lakey, N., Bedell, J., Yuan, Y., Budiman, M. A., Resnick, A., Van Aken, S., Utterback, T., Riedmuller, S., Williams, M., Feldblyum, T., Schubert, K., Beachy, R., Fraser, C. M., and Quackenbush, J. (2003) Enrichment of gene-coding sequences in maize by genome filtration. *Science* **302**, 2118–2120.

62. Emrich, S. J., Li, L., Wen, T. J., Yandeau-Nelson, M. D., Fu, Y., Guo, L., Chou, H. H., Aluru, S., Ashlock, D. A., and Schnable, P. S. (2007) Nearly identical paralogs: implications for maize (Zea mays L.) genome evolution. *Genetics* **175**, 429–439.

63. Liu, S., Chen, H. D., Makarevitch, I., Shirmer, R., Emrich, S. J., Dietrich, C. R., Barbazuk, W. B., Springer, N. M., and Schnable, P. S. (2009) High-throughput genetic mapping of mutants via quantitative SNP-typing. *Genetics* **184**, 19–26.

64. The International *Brachypodium* Initiative. (2010) Genome sequence and analysis of the model grass *Brachypodium distachyon. Nature* **463**, 763–768.

65. Kalyanaraman, A., Aluru, S., Kothari, S., and Brendel, V. (2003) Efficient clustering of large EST data sets on parallel computers. *Nucleic Acids Res.* **31**, 2963–2974.

66. Kalyanaraman, A., Aluru, S., Kothari, S., and Brendel, V. (2003) Space and time efficient parallel algorithms and software for EST clustering. *IEEE Transactions on Parallel and Distributed Systems (TPDS)* **14**, 1209–1221.

67. Bennetzen, J. L., Schrick, K., Springer, P. S., Brown, W. E., and SanMiguel, P. (1994) Active maize genes are unmodified and flanked by diverse classes of modified, highly repetitive DNA. *Genome* **37**, 565–576.

68. Lippman, Z., Gendrel, A. V., Black, M., Vaughn, M. W., Dedhia, N., McCombie, W. R., Lavine, K., Mittal, V., May, B., Kasschau, K. D., Carrington, J. C., Doerge, R. W., Colot, V., and Martienssen, R. (2004) Role of transposable elements in heterochromatin and epigenetic control. *Nature* **430**, 471–476.

69. Rabinowicz, P. D., Schutz, K., Dedhia, N., Yordan, C., Parnell, L. D., Stein, L., McCombie, W. R., and Martienssen, R. A. (1999) Differential methylation of genes and retrotransposons facilitates shotgun sequencing of the maize genome. *Nat. Genet.* **23**, 305–308.

70. Bedell, J., Budiman, M., Nunberg, A., Citek, R., Robbins, D., et al., (2005) Sorghum genome sequencing by methylation filtration. *PloS Biol.* **3**, e13.

71. Altshuler, D., Pollara, V. J., Cowles, C. R., Van Etten, W. J., Baldwin, J., Linton, L., and Lander, E. S. (2000) An SNP map of the human genome generated by reduced representation shotgun sequencing. *Nature* **407**, 513–516.

72. Burr, B., Burr, F. A., Thompson, K. H., Albertson, M. C., and Stuber, C. W. (1988) Gene mapping with recombinant inbreds in maize. *Genetics* **118**, 519–526.

73. Emberton, J., Ma, J., Yuan, Y., SanMiguel, P., and Bennetzen, J. L. (2005) Gene enrichment in maize with hypomethylated partial restriction (HMPR) libraries. *Genome Res.* **10**, 1441–1446.

74. Britten, R. J., Graham, D. E., and Neufeld, B. R. (1974) Analysis of repeating DNA sequences by reassociation. *Methods Enzymol.* **29**, 363–418.

75. Peterson, D. G., Wessler, S. R., and Paterson, A. H. (2002) Efficient capture of unique sequences from eukaryotic genomes. *Trends Genet.* **18**, 547–550.

76. Yuan, Y., SanMiguel, P. J., and Bennetzen, J. L. (2003) High-Cot sequence analysis of the maize genome. *Plant J.* **34**, 249–255.

77. Barbazuk, W. B., Bedell, J. A., and Rabinowicz, P. D. (2005) Reduced representation sequencing: a success in maize and a promise for other plant genomes. *Bioessays* **27**, 839–848.

78. Albert, T. J., Molla, M. N., Muzny, D. M., Nazareth, L., Wheeler, D., Song, X., Richmond, T. A., Middle, C. M., Rodesch, M.

J., Packard, C. J., Weinstock, G. M., and Gibbs, R. A. (2007) Direct selection of human genomic loci by microarray hybridization. *Nat. Methods* **4**, 903–905.

79. D'Ascenzo, M., Meacham, C., Kitzman, J., Middle, C., Knight, J., Winer, R., Kukricar, M., Richmond, T., Albert, T. J., Czechanski, A., Donahue, L. R., Affourtit, J., Jeddeloh, J. A., and Reinholdt, L. (2009) Mutation discovery in the mouse using genetically guided array capture and resequencing. *Mamm. Genome* **20**, 424–436.

80. Hodges, E., Xuan, Z., Balija, V., Kramer, M., Molla, M. N., Smith, S. W., Middle, C. M., Rodesch, M. J., Albert, T. J., Hannon, G. J., and McCombie, W. R. (2007) Genome-wide in situ exon capture for selective resequencing. *Nat. Genet.* **39**, 1522–1527.

81. Okou, D. T., Steinberg, K. M., Middle, C., Cutler, D. J., Albert, T. J., and Zwick, M. E. (2007) Microarray-based genomic selection for high-throughput resequencing. *Nat. Methods* **4**, 907–909.

82. Fu, Y., Springer, N. M., Gerhardt, D. J., Ying, K., Yeh, C-T., Wu, W., Swanson-Wagner, R., D'Ascenzo, D., Millard, T., Freeberg, L., Aoyama, N., Kitzman, J., Burgess, D., Richmond, T., Albert, T. J., Barbazuk, W. B., Jeddeloh, J. A., and Schnable, P. S. (2010) Repeat subtraction-mediated sequence capture from a complex genome. *Plant J.* **62**, 898–909.

Chapter 16

RNA-Seq Analysis of Gene Expression and Alternative Splicing by Double-Random Priming Strategy

Michael T. Lovci, Hai-Ri Li, Xiang-Dong Fu, and Gene W. Yeo

Abstract

Transcriptome analysis by deep sequencing, more commonly known as RNA-seq is, becoming the method of choice for gene discovery and quantitative splicing detection. We published a double-random priming RNA-seq approach capable of generating strand-specific information [Li et al., Proc Natl Acad Sci USA 105:20179–20184, 2008]. Poly(A)$^+$ RNA from a treated and an untreated sample were utilized to generate RNA-seq libraries that were sequenced on the Illumina GA1 analyzer. Statistical analysis of approximately ten million sequence reads generated from both control and treated cells suggests that this tag density is sufficient for quantitative analysis of gene expression. We were also able to detect a large fraction of reads corresponding to annotated alternative exons, with a subset of the reads matching known and detecting new splice junctions. In this chapter, we provide a detailed, bench-ready protocol for the double-random priming method and provide user-friendly templates for the curve-fitting model described in the paper to estimate the tag density needed for optimal detection of regulated gene expression and alternative splicing.

Key words: Gene expression, RNA-seq, Alternative splicing

1. Introduction

We have devised a procedure based on double-random priming and solid phase selection to produce libraries for high-throughput sequencing on the Illumina Genome Analyzer (1). In order to sequence these libraries, P1 and P2 adapter sequences must be added to the ends of the DNA of interest. In this protocol, double poly(A)-selected RNA is first primed with an oligonucleotide that contains a random octamer and the P1 adapter sequence. This first primer also carries a biotin moiety at the 5′ end, which allows for the capture of extended cDNA product on streptavidin beads. A second random primer linked to the other sequencing

Chaofu Lu et al. (eds.), *cDNA Libraries: Methods and Applications*, Methods in Molecular Biology, vol. 729,
DOI 10.1007/978-1-61779-065-2_16, © Springer Science+Business Media, LLC 2011

primer (P2) adapter sequence is next added to the cDNA bound to the streptavidin-coated magnetic beads. After extensive washes, potential P2 dimers are eliminated and the second random primed products are released from the beads by heat, leaving behind unused P1 primer, P1-extended cDNA, and potential P1 dimers. The released products are PCR amplified, gel purified to enrich for amplicons in the size range of 100–300nt, quantified, and subjected to sequencing (from the P1 primer side) on the Illumina/Solexa flow cell.

This procedure has several advantages compared to previous published protocols. First, it provides strand-specific information, as opposed to other methods that convert RNA to cDNA before primer addition. Second, sequencing a short region right after the first random priming reaction avoids cDNA artifacts resulting from extension of the hairpins formed after the first strand synthesis (2), which may account for artifactual "antisense transcripts" seen in previous large-scale mRNA sequencing and tiling analysis (3,4). Third, the built-in random primer region retains the molecular memory for originally primed products, allowing computational elimination of sequenced reads amplified by PCR, because all PCR products from the same initial amplicon will have identical sequences in the randomized region. This strategy permits the use of PCR amplification without distorting the representation of the transcriptome, a feature critical for quantitative analysis on a limited population of cells.

2. Materials

2.1. Total RNA Extraction Reagents

1. RNAbee (amsbio).

2.2. Double-Random Priming Reagents

1. RT buffer (Invitrogen): First-strand buffer (5×), DTT (0.1 M), RNase inhibitor, Superscript III reverse transcriptase, 10 mM dNTPs, and RNAase-free water (Invitrogen Superscript III kit).
2. QIAquick PCR purification kit (Qiagen):
   (a) Qiagen PCR purification buffer.
   (b) Qiagen purification columns.
   (c) Qiagen binding buffer.
   (d) Qiagen wash buffer.
   (e) Qiagen elution buffer: 10 mM Tris–HCl, pH 8.5.
3. NaOH (0.1 M).
4. 10 mM dNTPs.
5. 130 mM ddNTPs.

6. Adaptor 1: Biotinylated random oligo with Solexa Adaptor P1: (Bio-P1-N(8)) *OR* biotinylated oligo-dT with Solexa Adaptor P1 (Bio-P1-poly(T)$^{+}$), 50 μM (see Note 3).
7. Terminal transferase (NEB).
8. 10× Terminal transferase buffer (NEB).
9. EDTA.
10. Beads: Streptavidin-coated magnetic beads (SeraMag beads of Seradyne or Dynal beads).
11. Magnetic stand (Dynal).
12. Adaptor 2: Random oligo-linked Solexa Adaptor P2 (P2-N(8)) (100 μM).
13. PCR buffer: 10× standard Taq DNA polymerase buffer (NEB).
14. Wash buffer: 10 mM Tris-HCl, pH 7.5, 1 mM EDTA, 0.1% Tween-80 or Triton-X 100.
15. Taq DNA polymerase (NEB).
16. Agarose (NuSieve).
17. PicoGreen (Invitrogen).

3. Methods

3.1. Double-Random Priming Method

3.1.1. Reverse Transcription

The bench-ready protocol is described as follows:

1. Add 1 μl of Adaptor 1 (reagent 6) to 10 pg–5 μg of total RNA, 1 μl of dNTP mix, and RNase-free water to 13 μl per reaction.
2. Heat the mixture to 65°C for 5 min and incubate on ice for at least 1 min.
3. Add 4 μl of RT buffer.
4. Incubate at 50°C for 30–60 min.
5. Add deionized water to a total volume of 100 μl and inactivate the reaction by heating at 70°C for 15 min.
6. To remove the free biotin-labeled oligos, add 500 μl of Qiagen PCR purification buffer before transferring the mixture to a Qiagen purification column. Wash the Qiagen column once with the binding buffer and twice with the wash buffer. Elute with 50 μl of Qiagen elution buffer to a clean tube.

3.1.2. First Primer Blocking and Random Primer Extension Reaction

1. Transfer the eluate to PCR tubes. Add 15 μl of terminal transferase buffer, 3 μl of ddNTP mix and DI water to make up the volume to 150 μl. Add 2 μl of terminal transferase enzyme. Incubate at 37°C for 1 h (see Note 1).

2. Add 20 mM of EDTA.
3. Add 5 μl of beads and incubate the mixture at room temperature for 20 min.

   Collect the beads with a magnetic stand and discard the supernatant (see Note 4).

   Remove the tubes containing beads from the magnetic stand. Wash the beads with 100 μl of NaOH solution by drawing beads to oneside of the tube then the other with the magnetic stand (see Note 4). Incubate for 5 min at room temperature.
4. Collect the beads with the magnetic stand and wash with DI water twice, removing the tubes from the magnetic stand between washes to wash completely.
5. Off the magnetic stand, add 1 μl of Adaptor 2 to the beads, 5 μl of PCR buffer, 1 μl of dNTPs, add DI water to make up the volume to 49 μl. Add 1 μl of Taq DNA polymerase (5 U).
6. Incubate the tubes at 25°C for 1 h. Heat to 72°C for 30 s and then raise the temperature to 75°C for 5 min. Add 10 mM of EDTA to to stop the polymerization reaction.
7. Collect the beads and wash twice with 150 μl of wash buffer, removing the tubes from magnetic stand during washes.
8. On the stand, add 20 μl of water and heat for 5 min at 95°C. Collect the extended DNA in the supernatant.
9. Amplify the extended DNA with PCR using Solexa Adaptors 1 and 2 as primers (without poly(T)$^+$ or N(8)).
10. Run the library on an agarose gel and excise the band corresponding to 75–125 nt. Gel extract the band to elute DNA library.
11. Quantify DNA using PicoGreen or quantitative PCR prior to sequencing. A typical sequencing run uses 10–20 ng of DNA.

3.2. Transcript Databases for Gene Expression and Alternative Splicing Detection

In order to utilize RNA-seq reads to measure gene expression quantitatively, it is imperative to first define our concept of genes. To that end, we have developed detailed annotations of gene structures based on publicly available annotations downloaded from the University of California, Santa Cruz (UCSC) (5). We have also generated alignable sequence databases that can be used with data generated from high-throughput sequencing and for the purpose of aligning sequencing reads to spliced mRNA transcripts. Basic notes on the acquisition and processing of data such as these are outlined here. Please review our previously published work for more detailed information (6).

3.2.1. Building an Aggregate Gene Model (Fig. 1)

Genome sequences of human (hg17) and annotation for protein-coding genes were obtained from the UCSC. The lists of known human genes (knownGene containing 43,401 entries) and knownisoforms (knownIsoforms containing 43,286 entries in 21,397 unique isoform clusters) with annotated exon alignments to human hg17 genomic sequence were processed as follows. Knowngenes that were mapped to >1 isoform clusters were discarded. All mRNAs aligned to the human genome that were >300 bases long were clustered together with the knownisoforms. For the purposes of measuring differential gene expression, all genes were considered. For the purposes of inferring alternative splicing, genes containing <3 exons were not considered. Exons with canonical splice signals (GT-AG, AT-AC, and GC-AG) were retained, resulting in a total of 213,736 exons. Of these, 92% of all exons were constitutive exons, 7% had evidence of exon skipping, 1% of exons were mutually exclusive alternative events, 3% of exons had alternative 3′ splice sites, and 2% exons had alternative 5′ splice sites (Fig. 1). A total of 2.7 million spliced ESTs were mapped onto the 17,478 high-quality gene clusters to identify alternative splicing. To eliminate redundancies in this analysis, final annotated gene regions were clustered together so that any overlapping portion of these databases was defined by a single genomic position.

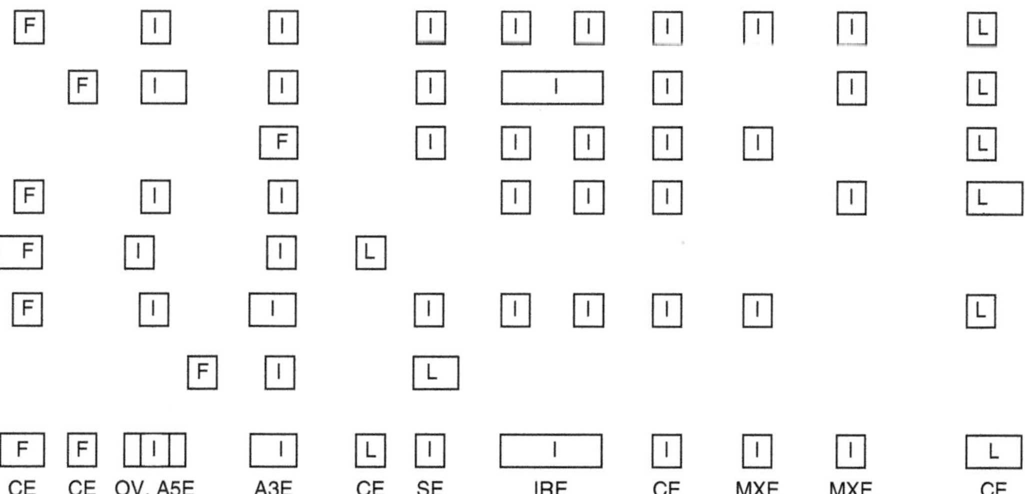

Fig. 1. Cartoon depicting construction of an aggregate gene model. Exons are depicted as boxes labeled as internal (I), first (F), or last (L). Region classifications are listed at the bottom of the schematic. Classifications of splicing were defined as follows: overlap (OV), skipped exons (SE), alternative 5′/3′ exons (A5E/3E), constitutive exons (CE), mutually skipped exons (MXE), and intron retentions (IRE).

3.2.2. Building an Exon-Junction Database

Exons with canonical splice signals (GT-AG, AT-AC, and GC-AG) were used to create an exon-junction database (EJDB). For each protein-coding gene, the 35 bases at the 3′ end of each exon were concatenated with the 35 bases at the 5′ end of the downstream exon. This was repeated, joining every exon of a gene to every exon downstream. This approach produced 1,929,065 theoretical splicing junctions. An equal number of "impossible" junctions were generated by joining the 35-base exon-junction sequences in reverse order.

3.3. Metrics for Differential Gene Expression

3.3.1. Alignment

MosaikAligner (7), using a maximum of 2 mismatches over 95% alignment of the tag (34nt) and a hash size of 15, was used to align reads to the human genome (hg17). However, since the publication of this work, several new alignment algorithms have been made available that offer other options for this step (such as QPalma (8) Bowtie (9) or RazerS (10)). To determine the number of reads contained within protein-coding genes, promoter, and intergenic regions, we arbitrarily defined promoter regions as regions 3-kb upstream of the transcriptional start site of the gene, and intergenic regions as unannotated regions in the genome.

Alignments to our EJDB were also done using the same alignment algorithm and mapping requirements, with the added requirement that reads map at least 4 nt across the exon–exon junction.

3.3.2. Evaluation of Differential Gene Expression

Differentially expressed transcripts were identified by enumerating the number of reads that mapped within the spliced mRNA transcript in untreated and hormone-treated cells, using the total number of reads mapped to exons in each condition as a basis for determining significance by the χ^2 statistic.

The χ^2 statistic was calculated for genes with ≥ 5 reads in each experimental condition, and the value of the χ^2 statistic was computed using a 2×2 square with the reads within a particular gene in both conditions on the top row and the reads not within that gene in both conditions on the bottom row.

After the number of reads mapped in each condition and the statistical significance are determined, each gene can be plotted as a scatter plot as in Fig. 2 for visualization purposes.

3.3.3. Detection of Alternative Splicing

Alternative splicing was detected by using reads mapped across exon junctions. We were able to detect both annotated and novel splice junctions. The type of exon–exon junction (i.e., constitutive or alternative) was determined based on our aggregate gene model (see above). False-discovery rate (FDR) was assessed by mapping reads to a set of "impossible" junctions that were created by reversing the order of exons in the EJDB (e.g., if exons 1 and 2 of a particular gene that are in the EJDB are joined $1 \rightarrow 2$,

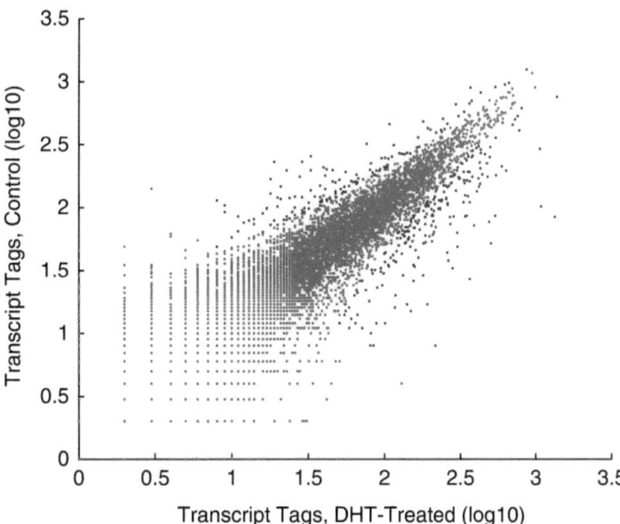

Fig. 2. Digital analysis of androgen-regulated gene expression in LNCaP cells. Scatter plot of gene expression in mock-treated and DHT-induced cells. Differential expressed genes were (in *light gray*) based on χ^2 analysis ($P<0.01$).

the impossible version of this would be the same exons joined in the reverse order, $2 \rightarrow 1$).

3.4. Power Curve Analysis

To establish the depth of sequencing required to examine several transcriptome features, we devised a method to predict not only the number of reads required to analyze a particular feature, but also the number of features observable at that sequencing depth. Reads were randomly sampled into subsets representing 10, 20%, etc., of the total number of sequence reads available using custom Perl scripts. These were aligned as described above and the number of features detected was assessed. To determine the number of sequence reads required to reach a user-defined threshold for saturation, the percentage change in discovering additional features was determined as follows:

$$T(n) = sn,$$

$$C(n) = \left(\frac{F(n) - F(n-1)}{F(n-1)} \right),$$

where $T(n)$ is the number of reads, s is the sampling size (in our case, two million reads), n is a constant multiplier, $C(n)$ is the empirical change in number of features detected, and $F(n)$ is the number of empirical features detected at n. A scatter plot of $C(n)$ to $T(n)$ was fitted with a power curve of the form $c(n) = a \times T(n)^b$ and an exponential curve of the form $c(n) = ae^{bT(n)}$, where $c(n)$ is the change estimated by the curve fitting.

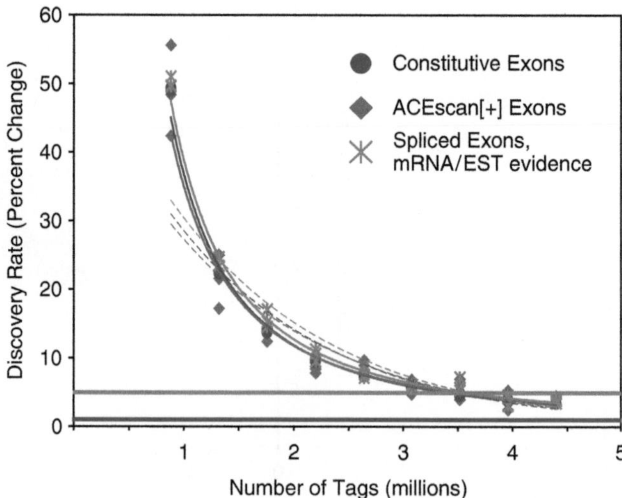

Fig. 3. Curve fitting the change in the number of exons and splice junctions detected against increasing tag densities. *Dashed line* indicates exponential curve; *solid line* indicates power curve. Decline in the rate of identifying additional exons as a function of increasing tag density.

The equation that had the best fit, indicated by r^2, was used to extrapolate the tag density required to achieve a defined change in the number of features detected. The number of estimated features was calculated by

$$f(n) = \sum_{i=m}^{n} f(i-1) + f(i-1) \times c(i),$$

where m is user defined (in our case, $m=6$). This will compute the predicted number of features observable based on observed change in feature detection, extrapolated from an area in the middle of the curve. Fig. 3 depicts one such fitted curve.

These calculations can be done easily using the "Data Analysis" ToolPak for Microsoft Excel. An example worksheet that calculates features using data from three independent samplings (labeled X, Y, and Z) can be downloaded from http://yeolab.ucsd.edu/yeolab/Papers_files/EXAMPLE.xls

4. Notes

1. Ensure that the beads do not dry out throughout the protocol.
2. Ensure that the areas used to perform experiments with RNA are free of RNAase contaminants.

3. Check the quality of adaptors by running them on an agarose gel (there should be one band) and be sure that they are PAGE purified.
4. When washing beads on the magnetic stand, it is useful to spin the tubes in the stand to get them to transfer from one side of the tube to the other; the beads tend to stick to the wall of the tube and this makes washes faster and more thorough.

Acknowledgments

The authors would like to thank the members of the Yeo and Fu laboratories for critical reading of this manuscript. This research was supported by grants to G.W.Y. and X.D.F. from the US National Institutes of Health (HG004659 and GM084317 and GM052872) for funding this research and the development of this protocol.

References

1. Li, H., Lovci, M. T., Kwon, Y. S., Rosenfeld, M. G., Fu, X. D., and Yeo, G. W. (2008) Determination of tag density required for digital transcriptome analysis: application to an androgen-sensitive prostate cancer model. *Proc Natl Acad Sci USA.* **105**, 20179–20184.
2. Perocchi, F., Xu, Z., Clauder-Munster, S., and Steinmetz, L. M. (2007) Antisense artifacts in transcriptome microarray experiments are resolved by actinomycin D. *Nucleic Acids Res.* **35**, e128.
3. Carninci, P., Kasukawa, T., Katayama, S., et al. (2005) The transcriptional landscape of the mammalian genome. *Science* **309**, 1559–1563.
4. Cheng, J., Kapranov, P., Drenkow, J., Dike, S., Brubaker, S., Patel, S., Long, J., Stern, D., Tammana, H., Helt, G., Sementchenko, V., Piccolboni, A., Bekiranov, S., Bailey, D. K., Ganesh, M., Ghosh, S., Bell, I., Gerhard, D. S., and Gingeras, T. R. (2005) Transcriptional maps of 10 human chromosomes at 5-nucleotide resolution. *Science* **308**, 1149–1154.
5. Karolchik, D., Baertsch, R., Diekhans, M., Furey, T. S., Hinrichs, A., Lu, Y. T., Roskin, K. M., Schwartz, M., Sugnet, C. W., Thomas, D. J., Weber, R. J., Haussler, D., and Kent, W. J. (2003) The UCSC genome browser database. *Nucleic Acids Res.* **31**, 51–54.
6. Yeo, G. W., Van Nostrand, E. L., and Liang, T. Y. (2007) Discovery and analysis of evolutionarily conserved intronic splicing regulatory elements. *PLoS Genet.* **3**, e85.
7. Hillier, L. W., Marth, G. T., Quinlan, A. R., et al. (2008) Whole-genome sequencing and variant discovery in *C. elegans. Nat. Meth.* **5**, 183–188.
8. De Bona, F., Ossowski, S., Schneeberger, K., and Ratsch, G. (2008) Optimal spliced alignments of short sequence reads. *Bioinformatics* **24**, i174–i180.
9. Langmead, B., Trapnell, C., Pop, M., and Salzberg, S. L. (2009) Ultrafast and memory-efficient alignment of short DNA sequences to the human genome. *Genome Biol.* **10**, R25.
10. Weese, D., Emde, A. K., Rausch, T., Doring, A., and Reinert, K. (2009) RazerS–fast read mapping with sensitivity control. *Genome Res.* **19**, 1646–1654.

Chapter 17

Generation of a Large Catalog of Unique Transcripts for Whole-Genome Expression Analysis in Nonmodel Species

Diana Bellin, Alberto Ferrarini, and Massimo Delledonne

Abstract

Next-generation sequencing technologies have allowed new tools to be developed for transcriptome characterization.

We describe here the methods for the preparation of a library from pooled RNAs for 454 sequencing of 3′-cDNA fragments. We also describe how the read sequences obtained can be used to generate, through *de novo* reads assembly, a large catalog of unique transcripts in organisms for which a comprehensive collection of transcripts or the complete genome sequence is not available. Finally, this catalog can be used efficiently to design oligonucleotide probes for setting up a comprehensive microarray for global analysis of gene expression.

Key words: cDNA library, 454 Sequencing, Microarray, Expression analysis, Nonmodel species, Transcriptome characterization

1. Introduction

Global analysis of gene expression is one of the most widely used tools in functional genomics. Hybridization to DNA microarrays is still the standard method, but its application is limited to organisms for which the complete genome sequence or a large cDNA collection is available (1, 2).

Ultra-high-throughput sequencing of the transcriptome using the different deep sequencing technologies available to date, such as the Illumina Genome Analyzer, Applied Biosystem SOLiD and Roche 454 Life Science, is emerging as a powerful and attractive alternative approach for expression profiling (3–9). Nevertheless, these strategies are still very expensive and require a reference genome for mapping the short reads produced.

Chaofu Lu et al. (eds.), *cDNA Libraries: Methods and Applications*, Methods in Molecular Biology, vol. 729, DOI 10.1007/978-1-61779-065-2_17, © Springer Science+Business Media, LLC 2011

For nonmodel species lacking the necessary sequence information, alternative technologies based either on cDNA fragment analysis (like cDNA-AFLP) or cDNA Sanger sequencing (ESTs programs) have been applied to transcriptome analysis but both imply a tremendous effort and present some drawbacks (10–12). Here, we provide a method based on next-generation sequencing of cDNA libraries generated on 3′-ends, the most polymorphic region of the transcripts, as well as the region used for designing gene-specific probes (13). This approach produces an extensive catalog of unique transcripts that can be used for the setting up of a comprehensive microarray, making large-scale expression analysis affordable in any species, without the need for prior sequence knowledge.

We have confirmed, in a previous study (14), the potential of this method applying it to monitor expression profiling during berry maturation in grape as if no other sequence information was available for this species. Taking advantage of the availability of the grape genome sequence (15) and the extensive EST collection, we could assess the quality of the catalog of transcripts produced using this approach and have shown that a microarray designed on grape berries' transcriptome, derived using the method described here, was more informative than one of the most comprehensive grape microarrays available (14).

2. Materials

2.1. RNA Extraction and Pooling

1. Trizol Reagent (Invitrogen, Carlsbad, CA). This reagent is toxic and can cause burns. Use gloves and eye protection. Avoid contact with skin or clothing and use in a chemical fume hood.
2. Chloroform.
3. Isopropyl alcohol.
4. Ethanol. Prepare a 75% solution in water.
5. RNA 6000 Nano Chip kit (Agilent Technologies, Waldbronn, Germany).

2.2. Generation of 3′-cDNA Library and 454 Sequencing

1. Primers: oligo$(dT)_{11}$ conjugated to the 3′-adapter sequence (5′-ACTAATCAGGCAGAGGACGAGAAGTTTTTTTTTTT-3′) and 5′-adapter sequence conjugated to a forward random $(N)_6$ primer (5′-ACTACTGGAACCGACAGTGAGTAN_6-3′).
2. Superscript II Reverse Transcriptase (200 U/μL) (Invitrogen, Carlsbad, CA).
3. dNTPs. Prepare a dNTPs solution 10 mM each. Store at –20°C.

4. RNaseH (5 U/μL).
5. QIAquick PCR purification kit (Qiagen, Chatsworth, CA).
6. QIAEX II Gel Extraction kit (Qiagen, Chatsworth, CA).
7. DNA Polymerase I Large (Klenow) fragment (50 U/μL) (NEB, Beverly, MA).
8. Pfu DNA Polymerase.
9. Agarose for molecular biology.

2.3. Analysis of 454 Reads and Generation of a Comprehensive Transcript Catalog

1. GS De Novo Assembler software (454 Life Sciences, Roche Branford, CT).
2. NCBI BLAST (16).

2.4. Design of Microarray Oligos on the 454-Derived Transcript Catalog

1. OligoArray 2.1 (17).

3. Methods

In order to develop a microarray for large-scale expression analysis of the majority of genes expressed in particular tissues or conditions, also in nonmodel organisms, it is of crucial importance to generate a comprehensive and adequate catalog of transcripts to be used for microarray generation (18). Due to the large number of reads afforded, 454 DNA sequencing technology is effective in revealing the expression of a large number of genes and has a great potential for discovering rare transcripts. This, combined with adequate pooling of samples, allows coverage of the obtained catalog to be maximized.

454 sequencing of cDNA is expected to produce transcript overestimation associated with shotgun sequencing of multiple nonoverlapping 454-ESTs per transcript (13, 19). In order to limit the number of contigs for the same transcript and, consequently, the redundancy of the probe sets, we developed a strategy which allows enrichment for 3′-cDNA ends prior sequencing. This enables resolution of a catalog of unique transcripts when these sequences are assembled. In addition, the specificity of the 3′-UTR-sequence facilitates the development of highly specific oligonucleotide probes (14). The efficiency of the developed enrichment strategy was confirmed by mapping reads obtained using this approach to cDNAs or transcript sequences in grape, a species for which the genome sequence and an extensive cDNA collection are available (Fig. 1).

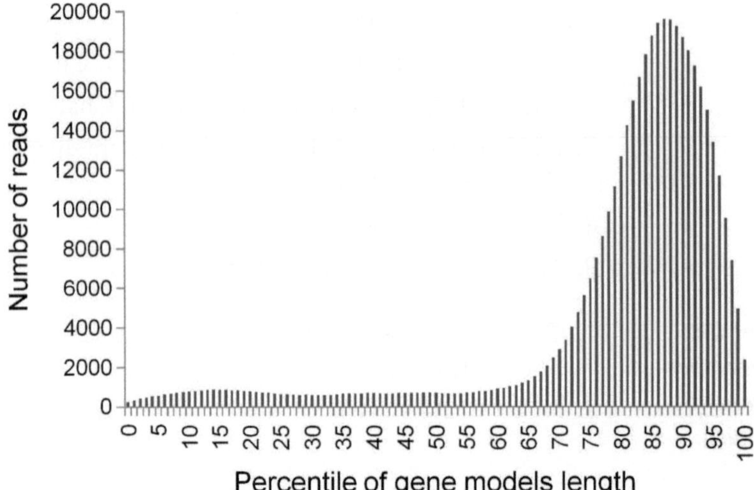

Fig. 1. Position of 21,512 reads from a grape 454 library matching 3,749 gene models of at least 1,000 nt in length, expressed as a percentile of the length of the gene model to which the read mapped. The library was generated from pooled RNAs from berries of *Vitis vinifera* cv Corvina (clone 48) harvested at six different time-points (14). Half of a 454 sequencing run produced 266,575 reads for a total of 127 Mbp of grape expressed sequences. The mapping confirmed that the two libraries were representing the 3′-ends of transcripts.

3.1. RNA Extraction and Pooling

1. Samples to be used for the generation of the transcript catalog should be carefully selected in order to maximize the coverage of the catalog. In order to represent a nearly complete transcriptome, total RNA is extracted from the largest possible pool of different tissues and samples. Samples are collected and either frozen in liquid nitrogen and stored at –80°C, or frozen and directly used for RNA extraction.
2. These instructions assume the use of the Trizol Reagent-based extraction system as protocol for RNA extraction (see Note 1). RNA can be extracted using this procedure from either tissues or cells grown in suspension. Grind frozen tissue in liquid nitrogen with pre-chilled mortar and pestle and then homogenize the tissue sample in Trizol Reagent. The sample should not exceed 10% of the volume of Trizol Reagent used for homogenization. In order to maximize the quantity and quality of extracted RNA, the material must be kept frozen at all times and the powder obtained must be as fine as possible. If working with cells in suspension, pellet the cells before extraction by centrifugation. Pellets can be frozen and stored at –70°C. Add the Trizol Reagent buffer directly to the pellet. Use 1 mL of the reagent per $5–10 \times 10^6$ animal, plant, or yeast cells or per 1×10^7 bacterial cells. Incubate the homogenate for 5 min at room temperature to allow the complete dissociation of nucleoprotein complexes.
3. Add 0.2 mL of chloroform for each milliliter of Trizol Reagent used for homogenization. Mix and incubate the samples at

room temperature for 3 min. Centrifuge the samples at 12,000 × *g* for 15 min at 4°C. The mixture separates into three phases and RNA remains exclusively in the upper aqueous phase. Transfer this into a new tube.

4. Precipitate the RNA by adding an equal volume of isopropyl alcohol and mixing. After an incubation of 10 min at room temperature, recover the RNA by centrifugation at 12,000 × *g* for 10 min at 4°C.
5. Discard the supernatant and wash the pellet once with 1 mL of 75% ethanol for each milliliter of Trizol Reagent used for the initial homogenization. Centrifuge again at 7,500 × *g* for 5 min at 4°C.
6. Discard the supernatant. Dry the pellet for 5 min at room temperature and re-suspend it in 33 μL RNAase-free water (see Note 2).
7. We recommend careful evaluation of RNA concentration (see Note 3) and integrity by agarose gel electrophoresis on an Agilent Bioanalyzer using RNA Nano 6000 kit according to the manufacturer's instructions. RNA integrity number (RIN) must be >7.5 for optimal results. The 25S and 18S ribosomal RNAs ratio should be approximately 2:1 (see Note 4).
8. Equal quantities of total RNA are pooled together from different samples for a total of 50 μg of RNA. To increase the amount of mRNA, the pool can be enriched in poly-A+ RNA before being used for preparation of the cDNA library. Any enrichment procedure can be used for this purpose.

3.2. Generation of 3′-cDNA Library and 454 Sequencing

1. RNA is subjected to double-strand cDNA synthesis. First-strand synthesis is performed from pooled RNA using as primer an oligo(dT)$_{11}$ conjugated to an adapter sequence and the reverse transcriptase SuperscriptII. The oligo(dT)$_{11}$-adapter primer (100 pmol) and pooled RNA (10 μg) are assembled together in a reaction tube. The sample is incubated at 65°C for 5 min and then immediately chilled in ice.
2. 5× First-strand buffer of SuperscriptII, 0.1 M DTT, and dNTPs 10 mM are added according to the manufacturer's indication, and the sample is incubated at 42°C for 2 min to equilibrate. SuperscriptII (400 U) is then added to the sample and this is incubated for 1 h at 42°C. Reaction inactivation is achieved by incubation at 70°C for 15 min.
3. To remove complementary RNA, the sample is treated with 5 U of RNaseH at 37°C for 20 min in its 10× Buffer. Alternatively, RNA hydrolysis using NaOH followed by equilibration with HCl can be used for the removal of complementary RNA.

4. Purification of the first strand is performed using the QIAquick PCR purification kit and protocol therein. Elution is performed in 30 μL of water.
5. Second-strand cDNA synthesis is achieved using an adapter sequence conjugated to the forward random $(N)_6$ primer and the Klenow polymerase. Primers (1OD) and the purified first-strand cDNA are mixed together in a tube. Then 5 μL of 10 mM dNTPs and 100 U of Klenow polymerase are added to a final volume of 100 μL. The reaction is incubated at 37°C for 2 h.
6. Double-strand cDNA is purified using the QIAquick PCR purification kit and protocol therein.
7. Amplification of double-strand cDNA is performed using primers complementary to the forward and reverse adaptor sequences introduced during first- and second-strand synthesis. Fifteen PCR cycles are recommended using a Pfu DNA polymerase. PCR is performed in 100 μL of final volume.

   Normalization has been shown to improve the described method allowing the production of a larger and more comprehensive catalog of transcripts (14). Details about one possible normalization procedure are given in Note 5. If normalization is performed, 8 PCR cycles are recommended instead of 15 cycles.
8. Amplified fragments in the 450–550 bp size range are purified from preparative 1.5% agarose gels. For this purpose, the entire PCR product is loaded in an agarose gel together with a marker for molecular weight. A gel slice in the proper area is removed using a clean gel excision tip.

   Size-selected cDNA fragments are then eluted from the gel using the QIAEX II Gel Extraction kit (QIAGEN) according to the manufacturer's instructions. After elution, an aliquot of the collected cDNA is run on a 1.5% agarose gel to confirm the size of eluted fragments. This procedure allows the selective enrichment in 3′-cDNA ends, as shown in Fig. 1.

   The obtained cDNA library will present the following structure *ACTACTGGAACCGACAGTGAGTA* + $(\text{NNNNNNNNNNNNNNNNNNNNNNNNNNN})_{(400\sim500\text{nt})}$+AAAAAAAAAAA + *CTTCTCGTCCTCTGCCTGATTAGT* (library specific 5′-adapter and 3′-adapter in italic) and a size of 450–550 bp (see Note 6).
9. The library is then subjected to 454 sequencing according to the standard manufacturer's instructions (see Note 7). Sequencing is performed from the 5′-end to retain the directional orientation of the reads and minimize the degrading performances of pyrosequencing technology when sequence

extension reaches the poly-A tail. We recommend using the 454 GS FLX Titanium series reagents for sequencing to allow the generation of longer reads (400–500 nt) which improves the sequence assembly process.

3.3. Analysis of 454 Reads and Generation of a Comprehensive Transcript Catalog

1. Assemble the reads produced by the basecalling process using the GS De Novo Assembler software (see Note 8) provided with the Genome Sequencer FLX hardware. Use the SFF files provided by the sequencing service. Alternatively, the FASTA (*.fna) files can be used as input. Consensus basecalls are then generated and quality scores are computed (see Note 9). After creating a new project using the New Project button, add your sequencing data to the project (see Note 10).
2. Switch to the parameters tab in order to customize assembly parameters. Set the parameters *Large or complex genome*, *Heterozygotic mode*, and *Expected depth* in order to fit your input dataset. Other parameters which must be set are a minimum overlap length of 40 bp, a minimum identity of overlapping reads of 90%, a seed step of 12 bp, and a seed length of 16 bp (see Note 11). Set the *All contig threshold* parameter to 100 in order to discard contigs with a length <100 nt (see Note 12).
3. Singleton sequences are not automatically saved by the GS De Novo Assembler and must be extracted from the sequence files in order to be added to the contig sequences to build the complete catalog of transcripts represented by the library analyzed. Identify IDs of singleton sequences which were not assembled in the assembly process by filtering the file 454ReadStatus.txt generated by the assembler for lines on which the column ReadStatus is set to Singleton. This can be easily performed by using the command awk '$2 ~ /^Singleton$/'<filename> |cut -f 1 >singletons.txt.
4. Format the FASTA files supplied by the sequencing service provider as BLAST databases using the NCBI BLAST command formatdb -i<filename.fna> -o T -p F. This step is needed in order to allow the extraction of singleton sequences whose ids have been saved in the file singletons.txt.
5. Extract the singleton reads sequences in FASTA format with the command fastacmd -i singletons.txt -d filename.fna >singletons.fas.
6. Concatenate the singletons and contigs FASTA files with the command cat 454AllContigs.fna singletons.fas >unigenes.fas. The file generated contains the complete catalog of transcripts represented in the library analyzed and

will be used to design the microarray oligo set. As an example, half of a 454 sequencing run of a cDNA library generated from pooled RNA of grape berries harvested at six different time-points generated a total of 266,575 reads (14). The assembly of these reads according to the strategy described here generated a catalog of unique transcripts constituted of 17,595 contigs and 12,032 singletons.

3.4. Design of Microarray Oligos on the 454-Derived Transcript Catalog

The virtual transcriptome constituted by the 454 unigene set (contigs and singletons) obtained by 454 sequencing and reads assembly of the normalized 3′-cDNA library is used to design microarray oligonucleotide probes using the OligoArray 2.1 software (17).

1. Prepare foreground and background FASTA files. The foreground file contains the sequences on which oligonucleotides must be designed. The background file is a nonredundant dataset of sequences against which to check the specificity of the probes. Both files should include all the sequences of unigenes in FASTA format. If you need to design spike sequences for the purpose of normalization of array performance assessment, you need to add the sequences in FASTA format to both the background and foreground files.
2. OligoArray will not work properly with long FASTA description lines. In order to strip away descriptions from the unigenes.fas file, you can use the command sed "s/\(>.*\)\s.*/\1/"<filename> >foreground.fas.
3. Format the background file as a BLAST database using the formatdb command: formatdb -i background.fas -o T -p F. This is required in order to allow OligoArray to automatically search for putative cross-hybridizations.
4. Run the OligoArray software with parameters optimized according to the microarray platform which will be used for the microarray analysis. The design process must be started with the command java -jar OligoArray2.jar -i foreground.fas -d background.fas -o oligo.txt -r rejected.fas -R OligoArray.log -n 1 -l 35 -L 40 -t 80 -T 86 -s 65 -x 65 -p 40 -P 60 -m "GGGGG;CCCCC;AAAAA;TTTTT" -N 10 (see Note 13 for an explanation of parameters used).
5. Select a set of sequences from species which are nonphylogenetically related to the one being analyzed. Construct a new foreground file with the negative sequences and run the OligoArray program on it using the same background file and parameters used for the design of the microarray probes for the 454-derived unigenes. The parameter -n of OligoArray can be increased to design more probes on negative sequences.

Generally, a set of 1,000–1,500 negative control spots are added to each array in random positions.

6. Process the OligoArray oligo.txt output file with the custom python script `filter_oligos.py n oligo.txt >oligo.filter.txt` (available at URL: http://ddlab.sci.univr.it/ddl/software.php) where n is the maximum number of cross-hybridizations allowed. Set a reasonable value of 15 or less in order to exclude oligos matching highly redundant sequences which will be scarcely informative and which could create biases in the normalization process of microarray data. The output file contains four columns, the first of which is a unique ID which is generated by concatenating the target ID, the distance of the 5′-end of the probe from the 5′-end of the target sequence, the probe length and a suffix equal to S in case the probe has only one target or Xn, where n is the number of targets, in case the probe has multiple targets. The second column indicates the target sequence on which

Table 1
Mapping statistics of 29,393 oligonucleotide probes designed on 29,627 454-derived unigenes sequenced from a normalized library of pooled RNAs from berries at six different developmental stages and of 32,771 oligonucleotide probes designed on 33,638 Tentative Consensus sequences (TCs) assembled from grape ESTs and included in the *Vitis vinifera* Gene Index 6.0 database (http://compbio.dfci.harvard.edu/cgi-bin/tgi/gimain.pl?gudb=grape)

	454 Contigs + singletons	TCs
Unigenes	29,627	33,638
Designed oligonucleotide probes	*29,393*	*32,771*
Probes specific for 1 sequence	28,187	25,658
Probes recognizing 2–3 sequences	1,177	6,326
Probes recognizing 4–5 sequences	29	787
Oligonucleotide probes mapping to genome	*25,879*	*26,578*
Probes mapping to a unique position	21,869	21,831
Probes mapping to 2–3 positions	2,654	2,848
Probes mapping to 4–5 positions	506	566
Probes mapping to more than 5 positions	850	1,333
Probes not mapping to genome	3,514	6,193

The percentage of oligonucleotides mapping to a unique location on the grape genome was 79% in the case of the 454-derived oligonucleotides. By comparison, the percentage of oligonucleotides designed on the TCs from the VvGI mapping to a unique location was 67%. The specificity of 454-derived oligonucleotides was thus higher than for the oligonucleotides designed on the extensive collection of ESTs available for this plant the probe has been designed. The third column indicates the putative targets of the oligo. Finally, the fourth column contains the probe sequences. An example of the results of the design process is shown in Table 1.

7. In order to produce a microarray with the designed oligonucleotides contact a microarray manufacturer providing a custom microarray production service such as CombiMatrix (http://www.combimatrix.com) or Nimblegen (http://www.nimblegen.com/).

4. Notes

1. Different RNA isolation procedures might be required for particular samples obtained from organisms or tissues with high contents of proteins, fat, polysaccharides, or extracellular material.
2. An incubation step at 55–60°C for 10 min can be performed to facilitate re-suspension of the RNA.
3. Typically, the RNA extraction procedure described here yields 1–15 μg of RNA per milligram of tissue or per 1×10^6 of cultured cells, depending on the starting material type.
4. If this instrument is not available, RNA quality and concentration can be normally verified by spectrophotometric analysis and denaturing agarose gel electrophoresis.
5. Normalization can be performed by one cycle of denaturation and reassociation of the cDNA (cot curve). The reassociated double-strand cDNA is separated from the remaining single-strand cDNA (i.e., the normalized cDNA) by passing the mixture over a hydroxyl apatite column (20).
6. MWG Biotech (http://www.mwg-biotech.com/), or similar sequencing companies, offers 3′-cDNA normalized library generation service.
7. MWG Biotech (http://www.mwg-biotech.com/), or similar sequencing companies, offers 454 sequencing service.
8. As an alternative to the proprietary software GS De Novo Assembler, open source sequence assembly software suitable for the *de novo* assembly of short and long reads such as MIRA is available (21). Interestingly, it has been shown that in some cases MIRA can also outperform the GS De Novo Assembler (22).
9. The assembler functions by identifying pairwise overlaps between reads, constructing multiple alignments of overlapping reads, and dividing them into multiple alignments in regions where consistent differences are found between different sets of reads.

10. The software can be run both by a command line interface (CLI) and by a graphical user interface (GUI). The same parameters can be applied to both software versions. GUI is better suited for first-time users, while the CLI interface is better when the process is run on a remote server. The commands shown here apply to the GUI version of the assembler.

11. Reducing the minimum reads overlap length down to 25 bp does not significantly increase the sensitivity of the assembly process. Be sure to check the Include consensus option to allow the generation of FASTA and quality score files.

12. A detailed explanation of all the files generated is shown in the Genome Sequencer FLX System Software Manual provided with the software. Files 454AllContigs.fna 454AllContigs.qual and 454LargeContigs.fna and 454LargeContigs.qual contain the FASTA and base quality score data of all the contigs generated and of contigs larger than 500 bp, respectively. The generated FASTA files contain the sequences of contigs assembled with a comment line indicating the automatically generated contig id, the length of the contig, and the number of reads used to assemble the contig. Quality files provide quality scores for the consensus sequence generated for each contig in numeric phred format.

13. The parameters are shown just as an example and are optimized for CombiMatrix 90 K microarrays with an oligonucleotide length range at 35–40 nt (–l and –L parameters), a melting temperature (Tm) range of between 80 (–t parameter) and 86°C (–T parameter), a secondary structure temperature threshold set at 65°C (–s parameter), and cross-hybridization threshold temperature at 65°C (–x parameter), with a GC content range of 40–60% (-p is the minimum GC threshold and –P is the maximum GC threshold). Sequences with stretches of five identical nucleotides are rejected (–m). Oligo length can be adjusted to match the one used by the platform of choice. GC range and Tm temperature ranges must be adjusted in order to match the GC content of the organism analyzed. GC content of the transcriptome of the organism analyzed can be easily calculated from the unigenes FASTA file (unigenes.fas) using the command geecee from the open source EMBOSS software suite (http://emboss.sourceforge.net/). The textual output file of geecee can be imported into a spreadsheet software to calculate the GC range of the transcript fragments sequenced. When setting the OligoArray temperature range keep in mind that the Tm is calculated using Nearest Neighbor model for DNA and Na+concentrations of 1 μM and 1 M, respectively. If you use different DNA and salts concentrations you should correct the Tm appropriately. For convenience, a Tm range calculator based on GC

content range and desired oligo length is provided at URL: http://berry.engin.umich.edu/oligoarray/. Cross-hybridization and secondary structure temperature thresholds must generally be set at a temperature 15°C lower than the minimum of the melting temperature range set to ensure a good specificity and performance of the probes.

Acknowledgments

This work was supported by the Project "Functional Genomics in Plants" granted by CARIVERONA Bank Foundation. In addition, funding was provided by the Project: "Structural and functional characterization of the grapevine genome (Vigna)" granted by the Italian Ministry of Agricultural and Forestry Policies (MIPAF).

References

1. Chen, J. J., Wu, R., Yang, P. C., Huang, J. Y., Sher, Y. P., Han, M. H., Kao, W. C., Lee, P. J., Chiu, T. F., Chang, F., Chu, Y. W., Wu, C. W., and Peck, K. (1998) Profiling expression patterns and isolating differentially expressed genes by cDNA microarray system with colorimetry detection. *Genomics* **51**, 313–324.
2. Lockhart, D. J., Brown, E. L., Wong, G. G., Chee, M., and Gingeras, T. R. (2004) Expression monitoring by hybridization to high density oligonucleotide arrays. *Nat. Biotechnol.* **14**, 1675–1680.
3. Cloonan, N., Forrest, A. R., Kolle, G., Gardiner, B. B., Faulkner, G. J., Brown, M. K., Taylor, D. F., Steptoe, A. L., Wani, S., Bethel, G., Robertson, A. J., Perkins, A. C., Bruce, S. J., Lee, C. C., Ranade, S. S., Peckham, H. E., Manning, J. M., McKernan, K. J., and Grimmond, S. M. (2008) Stem cell transcriptome profiling via massive-scale mRNA sequencing. *Nat. Meth.* **5**, 613–619.
4. Lister, R., O'Malley, R. C., Tonti-Filippini, J., Gregory, B. D., Berry, C. C., Millar, A. H., and Ecker, J. R. (2008) Highly integrated single-base resolution maps of the epigenome in *Arabidopsis*. *Cell* **133**, 523–536.
5. Marioni, J. C., Mason, C. E., Mane, S. M., Stephens, M., and Gilad, Y. (2008) RNA-seq: an assessment of technical reproducibility and comparison with gene expression arrays. *Genome Res.* **18**, 1509–1517.
6. Mortazavi, A., Williams, B. A., McCue, K., Schaeffer, L., and Wold, B. (2008) Mapping and quantifying mammalian transcriptomes by RNA-Seq. *Nat. Meth.* **5**, 621–628.
7. Nagalakshmi, U., Wang, Z., Waern, K., Shou, C., Raha, D., Gerstein, M., and Snyder, M. (2008) The transcriptional landscape of the yeast genome defined by RNA sequencing. *Science* **320**, 1344–1349.
8. Sultan, M., Schulz, M. H., Richard, H., Magen, A., Klingenhoff, A., Scherf, M., Seifert, M., Borodina, T., Soldatov, A., Parkhomchuk, D., Schmidt, D., O'Keeffe, S., Haas, S., Vingron, M., Lehrach, H., and Yaspo, M. L. (2008) A global view of gene activity and alternative splicing by deep sequencing of the human transcriptome. *Science* **321**, 956–960.
9. Wilhelm, B. T., Marguerat, S., Watt, S., Schubert, F., Wood, V., Goodhead, I., Penkett, C. J., Rogers, J., and Bahler, J. (2008) Dynamic repertoire of a eukaryotic transcriptome surveyed at single-nucleotide resolution. *Nature* **453**, 1239–1243.
10. Breyne, P., Dreesen, R., Cannoot, B., Rombaut, D., Vandepoele, K., Rombauts, S., Vanderhaeghen, R., Inze, D., and Zabeau, M. (2003) Quantitative cDNA-AFLP analysis for genome-wide expression studies. *Mol. Genet. Genomics* **269**, 173–179.
11. Vuylsteke, M., Peleman, J. D., and van Eijk, M. J. (2007) AFLP-based transcript profiling (cDNA-AFLP) for genome-wide expression analysis. *Nat. Protoc.* **2**, 1399–1413.
12. Weber, A. P., Weber, K. L., Carr, K., Wilkerson, C., and Ohlrogge, J. B. (2007) Sampling the

Arabidopsis transcriptome with massively parallel pyrosequencing. *Plant Physiol.* **144**, 32–42.

13. Eveland, A. L., McCarty, D. R., and Koch, K. E. (2008) Transcript profiling by 3′-untranslated region sequencing resolves expression of gene families. *Plant Physiol.* **146**, 32–44.
14. Bellin, D., Ferrarini, A., Chimento, A., Kaiser, O., Levenkova, N., Bouffard, P., and Delledonne, M. (2009) Combining next-generation pyrosequencing with microarray for large scale expression analysis in non-model species. *BMC Genomics* **10**, 555.
15. Jaillon, C. O., Aury, J., Noel, B., Policriti, A., Clepet, C., Casagrande, A., Choisne, N., Aubourg, S., Vitulo, N., and Jubin, C. (2007) The grapevine genome sequence suggests ancestral hexaploidization in major angiosperm phyla. *Nature* **449**, 463–467.
16. Altschul, S. F., Gish, W., Miller, W., Myers, E. W., and Lipman, D. J. (1990) Basic local alignment search tool. *J. Mol. Biol.* **215**, 403–410.
17. Rouillard, J. M., Zuker, M., and Gulari, E. (2003) OligoArray 2.0: design of oligonucleotide probes for DNA microarrays using a thermodynamic approach. *Nucleic Acids Res.* **31**, 3057–3062.
18. Vera, J. C., Wheat, C. W., Fescemyer, H. W., Frilander, M. J., Crawford, D. L., Hanski, I., and Marden, J. H. (2008) Rapid transcriptome characterization for a nonmodel organism using 454 pyrosequencing. *Mol. Ecol.* **17**, 1636–1647.
19. Torres, T. T., Metta, M., Ottenwalder, B., and Schlotterer, C. (2008) Gene expression profiling by massively parallel sequencing. *Genome Res.* **18**, 172–177.
20. Ko, M. S. H. (1990) An "equalized cDNA library" by the reassociation of short double-stranded cDNAs. *Nucleic Acids Res.* **18**, 5705.
21. Chevreux, B., Pfisterer, T., Drescher, B., Driesel, A. J., Müller, W. E. G., Wetter, T., and Suhai, S. (2004) Using the miraEST assembler for reliable and automated mRNA transcript assembly and SNP detection in sequenced ESTs. *Genome Res.* **14**, 1147–1159.
22. Papanicolaou, A., Stierli, R., Ffrench-Constant, R. H., and Heckel, D. G. (2009) Next generation transcriptomes for next generation genomes using est2assembly. *BMC Bioinform.* **10**, 447.

Index

Chaofu Lu et al. (eds.), *cDNA Libraries: Methods and Applications*, Methods in Molecular Biology, vol. 729, DOI 10.1007/978-1-61779-065-2, © Springer Science+Business Media, LLC 2011

MIX
Papier aus verantwortungsvollen Quellen
Paper from responsible sources
FSC® C105338

If you have any concerns about our products,
you can contact us on
ProductSafety@springernature.com

In case Publisher is established outside the EU,
the EU authorized representative is:
Springer Nature Customer Service Center GmbH
Europaplatz 3, 69115 Heidelberg, Germany

Printed by Libri Plureos GmbH
in Hamburg, Germany